Recent Advances in Natural Products Chemistry Related to Metabolites and Microbiomes

Recent Advances in Natural Products Chemistry Related to Metabolites and Microbiomes

Editors

Francesco Vinale
Maria Luisa Balestrieri

MDPI • Basel • Beijing • Wuhan • Barcelona • Belgrade • Manchester • Tokyo • Cluj • Tianjin

Editors
Francesco Vinale
Università degli Studi di Napoli Federico II
Italy

Maria Luisa Balestrieri
University of Campania "Luigi Vanvitelli"
Italy

Editorial Office
MDPI
St. Alban-Anlage 66
4052 Basel, Switzerland

This is a reprint of articles from the Special Issue published online in the open access journal *Molecules* (ISSN 1420-3049) (available at: https://www.mdpi.com/journal/molecules/special_issues/Metabolites_Microbiomes).

For citation purposes, cite each article independently as indicated on the article page online and as indicated below:

LastName, A.A.; LastName, B.B.; LastName, C.C. Article Title. *Journal Name* **Year**, *Article Number*, Page Range.

ISBN 978-3-03936-623-1 (Hbk)
ISBN 978-3-03936-624-8 (PDF)

Contents

About the Editors

Francesco Vinale, Ph.D, is Assistant Professor in Plant Pathology and Plant Protection at the Department of Veterinary Medicine and Animal Production, University of Naples Federico II, Naples, Italy and Associate Researcher at Institute for Sustainable Plant Protection, National Research Council of Italy (CNR). He is author of 4 patents and over 100 articles published on scientific journals, reviews, book chapters, abstracts, and proceedings of national and international congresses. Vinale serves as reviewer for several international journals and collaborates with various companies involved in the development and commercialization of biopesticides and that develop methods to remediate contaminated soil and water through the use of microorganisms. He is responsible for or involved in numerous research projects financed by the Italian Ministry of University and Research and the European Union. Research interests: i) purification and characterization of bioactive microbial metabolites; ii) study of the role of microbial secondary metabolites in complex plant/antagonist/pathogen interactions; iii) study of the three-way interaction between microbial antagonists, plant and pathogens by using proteomic or metabolomic approaches; iv) effect of bioactive microbial metabolites on plant growth promotion and induction of disease resistance; v) application of beneficial microorganisms and / or their metabolites in agriculture and industry; vi) study of the role of microbial metabolites involved in pathogenic events; vii) study of the role of microbial enzymes or metabolites in decontamination of polluted soil and water (bioremediation); viii) biochemical characterization of fungal antagonists or pathogens; ix) metabolomics.

Maria Luisa Balestrieri, Ph.D, is a Full Professor in Biochemistry at the Department of Precision Medicine, School of Medicine, University of Campania L. Vanvitelli, Naples, Italy where she has been a faculty member since 2004. She completed her Ph.D. in Biochemistry in 1997 at the University of Naples Federico II, Italy. She was Research Assistant in Biochemistry at the Basic and Applied Research Unit, Oak Ridge Associated Universities, Oak Ridge, Tennessee (USA) in 1998. From 1999 to 2003, she was Research Assistant in Biochemistry at the Department of Biochemistry and Biophysics, University of Campania L. Vanvitelli, Italy. Her major research interests lie in the area of molecular mechanism of cellular redox homeostasis in the control of human chronic degenerative diseases, with particular regard to cancer and type 2 diabetes, as well as in the biochemistry of bioactive naturally occurring food metabolites with antioxidant and anticancer properties. She serves as an expert researcher evaluator of REPRISE and is the research unit principal investigator of current research grants in the fields of life science and agrifood. She is a member of the Italian Society of Biochemistry and Molecular Biology.

Preface to "Recent Advances in Natural Products Chemistry Related to Metabolites and Microbiomes"

The characterization of molecules, metabolomes, and related microbiomes in recent years has paved the way for the identification of novel secondary metabolites with potential for agricultural applications both directly (increased yield and nutritional value of crops) and indirectly (effects on the environment, agricultural practices, etc.). The abundance and variety of metabolites in plant and animal extracts makes challenging the isolation of single secondary bioactive metabolites whose synergism with other biomolecules is particularly attractive in the field of natural products with health-promoting properties or active molecules for plant protection. In this context, it is now recognized that future agricultural production will be characterized by sustainable systems that produce food (and/or feed) with nutritional and functional properties suitable for sustaining human health and wellbeing.

In this Special Issue, aspects related to the production of bioactive microbial metabolites, characterization of plant metabolomes, and studies of the metabolome profiles of animal products have been explored.

We are confident that this book will highlight the need for natural products chemistry research related to metabolites and microbiomes, serving the scientific community and also being useful to teachers, researchers, and students.

We are grateful to the authors, to the colleagues that have reviewed the submitted articles, to the publisher MDPI, and to all the editorial staff of this journal.

Francesco Vinale, Maria Luisa Balestrieri
Editors

molecules

Review

Structures and Bioactive Properties of Myrtucommulones and Related Acylphloroglucinols from Myrtaceae

Rosario Nicoletti [1,2], Maria Michela Salvatore [3], Pasquale Ferranti [2] and Anna Andolfi [3,*]

[1] Council for Agricultural Research and Economics, Research Centre for Olive, Citrus and Tree Fruit, 81100 Caserta, Italy; rosario.nicoletti@crea.gov.it
[2] Department of Agriculture, University of Naples 'Federico II', 80055 Portici, Italy; ferranti@unina.it
[3] Department of Chemical Sciences, University of Naples 'Federico II', 80126 Naples, Italy; mariamichela.salvatore@unina.it
* Correspondence: andolfi@unina.it; Tel.: +39-081-2539179

Academic Editors: Francesco Vinale and Maria Luisa Balestrieri
Received: 2 December 2018; Accepted: 17 December 2018; Published: 19 December 2018

Abstract: Myrtaceae are a group of plants that include a number of renowned species used in ethnomedicine in many areas worldwide. Their valuable therapeutic properties have stimulated a fruitful research activity addressed to the identification of the bioactive components of their extracts yielding a great diversity of terpenes; polyphenols; and other exclusive products. Among the latter, starting with the discovery of myrtucommulone A from myrtle (*Myrtus communis*), a series of structurally-related acylphloroglucinol compounds have been characterized from several species that represent the basic active principles to be considered in view of possible drug development. Aspects concerning chemical and biological properties of these products are reviewed in the present paper.

Keywords: myrtucommulone; acylphloroglucinols; Myrtaceae; plant extracts; biological activities

1. Introduction

Myrtle (*Myrtus communis*) is a typical shrub of maquis and coastal bushes native of the Mediterranean area and Western Asia. It is well-known in traditional medicine, and for centuries its leaves and berries have found ethnomedical application in the treatment of several disorders of the digestive apparatus, as well as pulmonary and skin diseases [1,2]. More recently, experimental studies have provided an indication for a broader spectrum of pharmacological and therapeutic effects based on antioxidant, antiviral, antibiotic, antitumor, antidiabetic, hepatoprotective, and neuroprotective properties of extracts from this plant [2–4]. Many assorted compounds are considered as the possible bioactive components within the myrtle metabolome, such as terpenes occurring in the essential oils, α-tocopherol, anthocyanes, flavanols, and a series of acylphloroglucinols related to the myrtucommulones [4–10]. Particularly, the number of the latter compounds is continuously increasing as a result of a recent fruitful research activity carried out on other plant species belonging to the Myrtaceae in several independent laboratories worldwide. Structural aspects and the valuable bioactive properties that are pointed out for these compounds are reviewed in the present paper.

2. Biological Sources

The family Myrtaceae includes approximately 6000 species in 132 genera, with a wide distribution in tropical and warm-temperate regions of the world [11]. Until recently, these species were grouped in the two subfamilies of the Myrtoideae, including species with fleshy fruits and the Leptospermoideae, whose members produce dry capsules; however, this traditional arrangement has been disrupted in the current taxonomic schemes based on phylogenetic analysis resulting from DNA sequencing [12].

Historically relevant and well-known for its already-introduced multiple usage, *M. communis* represents the type species, and probably the best studied with reference to its biochemical properties. However, investigations in the field are developing more and more, enlarging the panorama of secondary metabolites produced by the Myrtaceae [2,8,13–15]. Particularly considered are aspects concerning the antibiotic properties, which have even led to proposing the use of some species in soil sanitization [16]. In this respect, a leading position pertains to the myrtucommulones, originally extracted from myrtle leaves [5,17]. More refined analytical studies have later shown their presence in fruits [7], which preludes the possible dietary intake of these products following the use of berries in gastronomy and in the preparation a digestive liquor typical of Sardinia [18,19].

After the pioneering reports concerning *M. communis*, compounds belonging to this class have been reported in species from other genera of Myrtaceae spread in the Australasian supercontinent. More particularly, species of *Callistemon*, which are now incorporated in the Linnaean genus *Melaleuca* [20], *Angophora, Baeckea, Corymbia, Eucalyptus, Kunzea, Lophomyrtus, Rhodomyrtus*, and *Syncarpia*, while *Myrciaria dubia* is endemic to the neotropical region (Table 1).

3. Structures and Chemical Properties

Within the large family of phloroglucinols [21], a series of alike natural products have been reported from plant species belonging to the Myrtaceae which are characterized by a molecular structure that is built on a phloroglucinol nucleus coupled with one or more syncarpic acid residues (Table 1). Myrtucommulone A (**1**), 4,4′-[(2,4,6-trihydroxy-5-isobutyryl-1,3-phenylene)bis (2-methylpropylidene)]bis(5-hydroxy-2,2,6,6-tetramethylcyclohex-4-ene-1,3-dione), represents the founder product of this series [17].

Table 1. Myrtucommulones and related compounds reported from plant species belonging to the Myrtaceae. Compounds are listed according to the chronological order of discovery.

Code	Compound Name	Formula, Nominal Mass (U)	Source	Ref.
1	Myrtucommulone A	$C_{38}H_{52}O_{10}$, 668	*Myrtus communis* *Melaleuca citrina* [1] *Corymbia scabrida*	[17] [22] [23]
2	Myrtucommulone B	$C_{25}H_{32}O_5$, 412	*Myrtus communis* *Melaleuca citrina* [1] *Melaleuca salicina* [1]	[17] [24] [25]
3	4-Cyclohexene-1,3-dioxo-5-hydroxy-2,2,6,6-tetramethy1-4-{1-[2,6-dihydroxy-4-methoxy-3-(3-methyl-1-oxo-butyl)phenyl]-3-methylbutyl}	$C_{26}H_{36}O_7$, 460	*Kunzea ericoides* *Kunzea sinclairii*	[26]
4	4-Cyclohexene-l,3-dioxo-5-hydroxy-2,2,6,6-tetramethyl-4-{l-[2,6-dihydroxy-4-methoxy-3-(2-methyl-1-oxopropyl)pheny1]-3-methylbutyl}	$C_{25}H_{34}O_7$, 446	*Kunzea ericoides* *Kunzea sinclairii*	[26]
5	Isomyrtucommulone B	$C_{24}H_{30}O_6$, 414	*Myrtus communis* *Melaleuca salicina* [1] *Myrciaria dubia*	[5] [25] [27]
6	Semimyrtucommulone	$C_{25}H_{34}O_7$, 446	*Myrtus communis*	[5]
7	Rhodomyrtone A	$C_{26}H_{34}O_6$, 442	*Rhodomyrtus tomentosa* *Eucalyptus globulus* *Angophora woodsiana* *Myrciaria dubia*	[28] [29] [30] [27]
8	Bullataketal A	$C_{37}H_{48}O_7$, 604	*Lophomyrtus bullata*	[31]
9	Bullataketal B	$C_{37}H_{48}O_7$, 604	*Lophomyrtus obcordata*	[32]
10	Myrtucommulone C	$C_{38}H_{50}O_9$, 650	*Myrtus communis*	[33]
11	Myrtucommulone D	$C_{38}H_{50}O_9$, 650	*Myrtus communis* *Corymbia scabrida* *Melaleuca salicina* [1]	[33] [23] [25]
12	Myrtucommulone E	$C_{38}H_{48}O_8$, 632	*Myrtus communis*	[33]
13	Eucalyptone G	$C_{40}H_{52}O_9$, 676	*Eucalyptus globulus*	[29]

Table 1. *Cont.*

Code	Compound Name	Formula, Nominal Mass (U)	Source	Ref.
14	Myrtucommulone F	$C_{40}H_{56}O_{10}$, 696	*Corymbia scabrida*	[23]
15	Myrtucommulone G	$C_{40}H_{54}O_{9}$, 678		
16	Myrtucommulone H	$C_{41}H_{58}O_{10}$, 710		
17	Myrtucommulone I	$C_{41}H_{56}O_{9}$, 692		
18	Rhodomyrtosone A	$C_{26}H_{32}O_{7}$, 456	*Rhodomyrtus tomentosa* *Angophora woodsiana*	[34] [30]
19	Rhodomyrtosone B	$C_{26}H_{34}O_{6}$, 442	*Rhodomyrtus tomentosa*	[34]
20	Rhodomyrtosone C	$C_{41}H_{54}O_{8}$, 674		
21	Myrtucommulone J	$C_{38}H_{52}O_{8}$, 636	*Myrtus communis*	[35]
22	Rhodomyrtosone I	$C_{28}H_{30}O_{6}$, 462	*Rhodomyrtus tomentosa*	[36]
23	Tomentosone A	$C_{41}H_{52}O_{9}$, 688	*Rhodomyrtus tomentosa*	[37]
24	Tomentosone B	$C_{41}H_{52}O_{9}$, 688		
25	Myrtucommulone M	$C_{49}H_{60}O_{12}$, 840	*Myrtus communis*	[38]
26	Myrtucommuacetalone	$C_{38}H_{52}O_{9}$, 652		
27	Callistenone A	$C_{25}H_{32}O_{6}$, 428	*Melaleuca citrina* [1]	[39]
28	Callistenone B	$C_{25}H_{32}O_{6}$, 428	*Melaleuca citrina* [1] *Melaleuca salicina* [1]	[39] [25]
29	Callistenone C	$C_{29}H_{40}O_{8}$, 516	*Melaleuca citrina* [1]	[39]
30	Callistenone D	$C_{39}H_{52}O_{9}$, 664		
31	Callistenone E	$C_{40}H_{54}O_{8}$, 662		
32	Watsonianone B	$C_{31}H_{34}O_{7}$, 518	*Corymbia watsoniana*	[40]

Table 1. *Cont.*

Code	Compound Name	Formula, Nominal Mass (U)	Source	Ref.
33	Rhodomyrtosone E	$C_{30}H_{34}O_6$, 490	*Eucalyptus citriodora*	[41]
34	Nor-semimyrtucommulone	$C_{24}H_{32}O_7$, 432	*Myrtus communis*	[42]
35	Rhodomyrtosone F	$C_{27}H_{36}O_6$, 456	*Syncarpia glomulifera*	[43]
36	Callistenone F	$C_{26}H_{34}O_6$, 442	*Melaleuca viminalis* [1]	[44]
37	Callistenone G	$C_{26}H_{34}O_6$, 442		
38	Callistenone H	$C_{26}H_{34}O_6$, 442	*Melaleuca viminalis* [1] *Melaleuca salicina* [1]	[44] [25]
39	Callistenone I	$C_{27}H_{36}O_6$, 456	*Melaleuca viminalis* [1]	[44]
40	Callistrilone A	$C_{33}H_{42}O_7$, 550	*Melaleuca linearis* [1]	[45]
41	Callistrilone B	$C_{35}H_{46}O_6$, 562		
42	Tomentosone C	$C_{38}H_{48}O_8$, 632	*Rhodomyrtus tomentosa*	[46]
43	Rhodomyrtosone G	$C_{26}H_{32}O_7$, 456	*Rhodomyrtus tomentosa*	[47]
44	Rhodomyrtosone H	$C_{26}H_{34}O_6$, 442		
45	Callistenone L	$C_{27}H_{34}O_7$, 470	*Melaleuca viminalis* [1]	[48]
46	Callistenone M	$C_{27}H_{34}O_7$, 470		
47	Callistenone N	$C_{26}H_{34}O_6$, 442		
48	Callistenone O	$C_{25}H_{32}O_6$, 428		
49	Callistenone P	$C_{27}H_{36}O_6$, 456		
50	Callisalignone B	$C_{25}H_{32}O_6$, 428	*Melaleuca salicina* [1]	[25]
51	Callisalignone C	$C_{26}H_{34}O_6$, 442		
52	Myrciarone A	$C_{25}H_{32}O_6$, 428	*Myrciaria dubia*	[27]
53	Myrciarone B	$C_{25}H_{32}O_6$, 428		

Table 1. *Cont.*

Code	Compound Name	Formula, Nominal Mass (U)	Source	Ref.
54	Callistrilone F	$C_{35}H_{48}O_8$, 596	*Melaleuca linearis* [1]	[49]
55	Callistrilone G	$C_{35}H_{48}O_7$, 580		
56	Callistrilone H	$C_{35}H_{52}O_7$, 608		
57	Callistrilone I	$C_{36}H_{50}O_7$, 594		
58	Callistrilone J	$C_{37}H_{52}O_7$, 608		
59	Callistrilone K	$C_{36}H_{48}O_6$, 576		
60	Tomentodione S	$C_{28}H_{32}O_5$, 448	*Rhodomyrtus tomentosa*	[50]
61	Tomentodione T	$C_{28}H_{32}O_5$, 448		
62	6-Methylisomyrtucommulone B	$C_{25}H_{32}O_6$, 428	*Myrtus communis*	[51]
63	4-Methylmyrtucommulone B	$C_{25}H_{32}O_6$, 428		
64	Baefrutone A	$C_{35}H_{48}O_7$, 580	*Baeckea frutescens*	[52]
65	Baefrutone B	$C_{35}H_{48}O_7$, 580		
66	Baefrutone C	$C_{35}H_{48}O_7$, 580		
67	Baefrutone D	$C_{35}H_{48}O_7$, 580		
68	Baefrutone E	$C_{40}H_{56}O_7$, 648		
69	Baefrutone F	$C_{40}H_{56}O_7$, 648		

[1] Species names have been updated according to their current taxonomic status [20].

Despite the use of an inconsistent nomenclature based on the plant species from which the different compounds were originally extracted, all of these structures are clearly related to the myrtucommulone skeleton, and they most likely are assembled through common biosynthetic pathways. A plausible scheme considers the residues of isobutyrylphloroglucinol and isobutylidenesyncarpic acid deriving from the intramolecular Claisen reaction of a polyketidic intermediate obtained from three residues of malonyl CoA and one of isobutyryl CoA [45,53]. In particular, the cyclic intermediate 2-isobutyrylcyclohexane-1,3,5-trione is thought to form isobutyrylphloroglucinol through a double geminal dimethylation, passing through the formation of flavesone [45]. In this hypothesis, the syncarpic acid residue itself is derived from acylphloroglucinol. On the other hand, it has been previously reported that tetramethylation of the polyketidic chain is followed by intramolecular cyclization to form isobutylidenesyncarpic acid [54]. The latter standpoint is supported by what has been assessed for the biosynthesis of several well-known compounds, such as 5-methylorsellic acid [53].

Acylphloroglucinol and substituted syncarpic acid (one or two residues) components are probably coupled via the Michael reaction to form dimeric and trimeric structures [45]. This reaction might be non-enzymatic because the carbon between these residues may lead to the formation of a couple of enantiomers (without further chiral centers) or epimers (with further chiral centers). In fact, in a few cases the enantiomeric composition of myrtucommulones and related compounds is not fully defined. More thorough assessments in this respect are also desirable in view of ascertaining whether or not the bioactivity of myrtucommulones depends on chirality of the molecules.

By means of derivatization, CD spectroscopy and enantiomeric analysis, it has been shown that natural myrtucommulone A is a mixture of three stereoisomers, a racemate, and a meso form in a 1:1 ratio c.a., while myrtucommulone B (**2**) and nor-semimyrtucommulone (**34**) are both racemates. Furthermore, many isomeric forms may occur due to a tautomeric equilibrium of the enolized syncarpic moiety [42].

Further intramolecular reactions could involve the phenolic or ketonic group to form, respectively, pyranic or furanic rings. Prenylation has also been reported for compounds from this class [25,27,45,49,52].

On account of such a wide extent of structural variation, a convenient discussion on the properties of oligomeric acylphloroglucinols should be based on their grouping in subclasses, depending on the number of structural units and the eventual occurrence of additional cycles, such as furan and pyrane. In this respect we distinguished eight homogeneous groups whose members further differ upon the variation in the acyl functionalities occurring in the different moieties (e.g., isobutyryl, isovaleryl, methylbutyryl, and hexanoyl substituents).

Compounds in the first subclass (Figure 1) present a dimeric structure, modeled on semimyrtucommulone (**6**) and two closely related compounds (**3–4**) reported from *Kunzea* species. Other compounds in this group present additional cyclic structures. In particular, the bullataketals (**8–9**) have a phenyl-oxabycyclooctane system, myrtucommulone J (**21**) is characterized by a dipyrancyclopentanone moiety, while myrtucommuacetalone (**26**) contains an unprecedented bridged furochromene moiety.

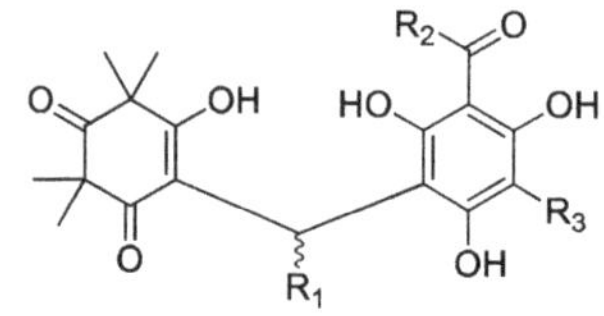

	∿∿	R1	R2	R3
3	—	iBu	iBu	H
4	—	iBu	iPr	H
6	—	iPr	iPr	Me
29	◀	iPr	iBu	COiPr
34	—	iPr	iPr	H

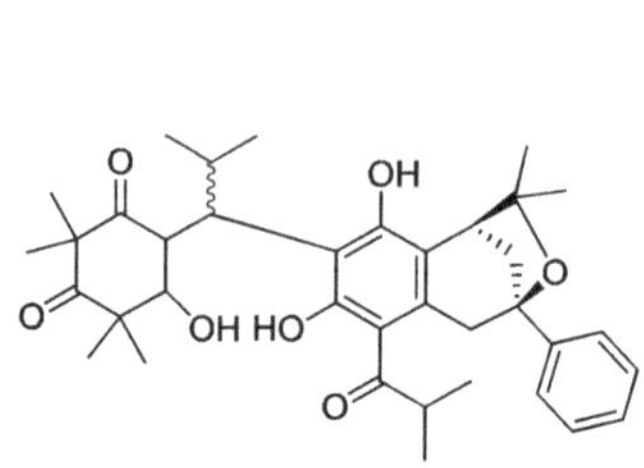

	∿∿
8	⋯⋯
9	◀

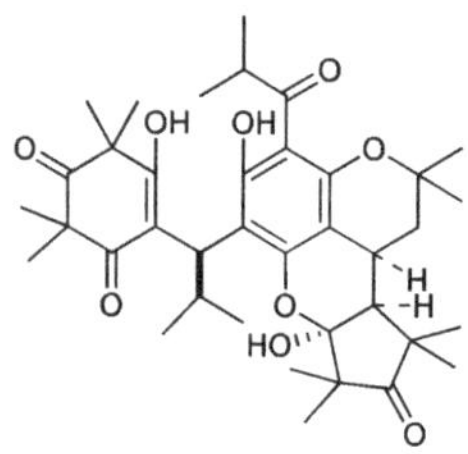

21

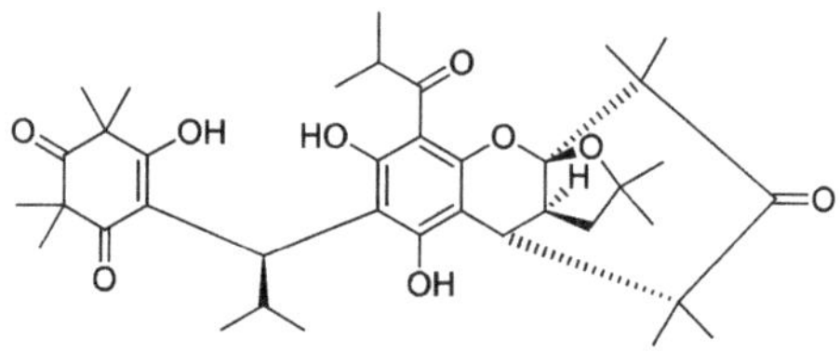

26

Figure 1. Structures of compounds of the dimeric type.

The dimeric-monopyrane skeleton is shared by over one-third of the compounds examined in this review, which mainly differ in the assortment of the acyl functionalities (Figure 2). This group includes compounds that exhibit a methylated phenolic group on the phloroglucinol residue. Interestingly, myrtucommulone M (**25**) consists in two myrtucommulone B (**2**) moieties that are linked together through a methylene bridge to form a symmetrical structure.

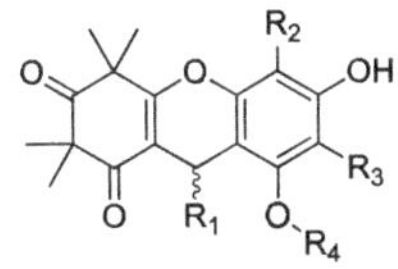

	〰	R1	R2	R3	R4
2	wedge	iPr	COiPr	H	H
5	—	iPr	H	COiPr	H
7	—	iBu	H	COiBu	H
19	wedge	iBu	COiBu	H	H
22	—	Ph	H	COiBu	H
27	wedge	iPr	H	COiBu	H
28	wedge	iPr	COiBu	H	H
33	—	iPr	H	COEtPh	H
35	—	iBu	COnPent	H	H
36	wedge	iPr	COsecBu	H	Me
37	wedge	iPr	H	COsecBu	Me
38	wedge	iPr	COiBu	Me	H
39	wedge	iBu	COiBu	H	Me
44	wedge	iBu	H	COsecBu	H
47	wedge	iPr	COsecBu	H	Me
48	wedge	iPr	COiPr	H	Me
49	—	secBu	COiBu	H	Me
(+) 50	wedge	iPr	COiPr	H	Me
(-) 50	……	iPr	COiPr	H	Me
51	……	iPr	(*R*)-COsecBu	H	Me
52	wedge	iBu	H	COiPr	H
53	—	iPr	H	COsecBu	H
60	—	Ph	H	COsecBu	H
61	wedge	iBu	COsecBu	H	H
62	—	iPr	COMe	COiPr	H
63	—	iPr	COiPr	MeCH3	H

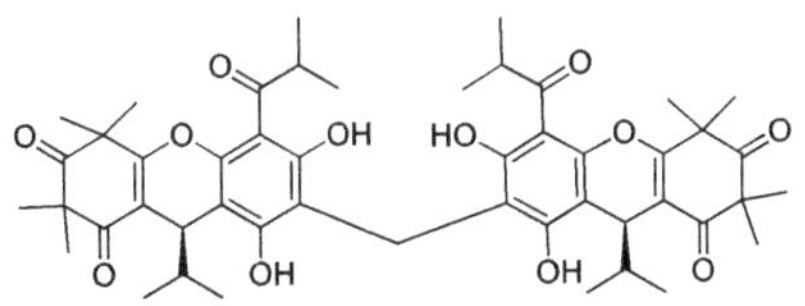

25

Figure 2. Structures of compounds of the dimeric-monopyrane type.

The founder compound, myrtucommulone A (**1**), and the related myrtucommulones F (**14**) and H (**16**), presenting a hexanoyl residue on the phloroglucinol ring, are characterized by a trimeric structure (Figure 3). This kind of skeleton can be modified by additional cyclization, with the formation of mono and dipyranic analogues that are separated in the following subclasses.

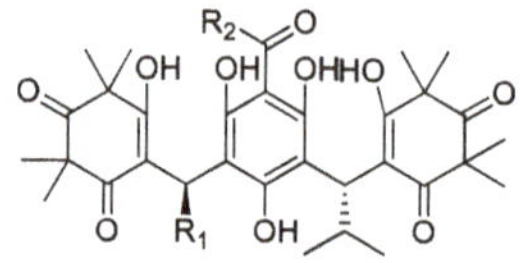

	R1	R2
1	iPr	iPr
14	iPr	*n*-Pent
16	iBu	*n*-Pent

Figure 3. Structures of trimeric compounds. Compound **1** is reported as (*R*,*R*)-stereoisomer.

Myrtucommulone C (**10**), with a trimeric-monopyrane structure (Figure 4), was isolated as single stereoisomer presenting an isobutanoylic residue. Eucalyptone G (**13**) is also a member of this subclass.

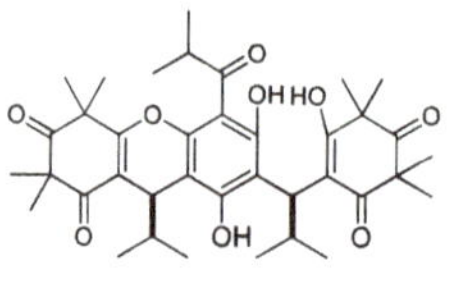

10

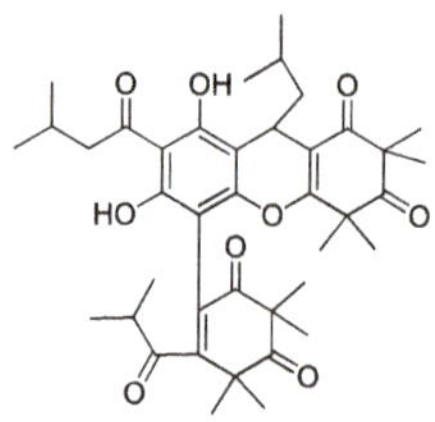

13

Figure 4. Structures of compounds of the trimeric-monopyrane-type.

In products of the trimeric-dipyrane-type (Figure 5), the presence of a pentacyclic structure can be observed where cycles may have an orientation that is similar to pentacene in the case of tomentosone C (**42**), while it is similar to pentaphene for other compounds (**11**, **12**, **15**, **17**, **20**, **30**, **31**). The different orientation is due to diverse phenolic and enolic/ketonic groups that are involved in the formation of pyranic rings of the trimer intermediate.

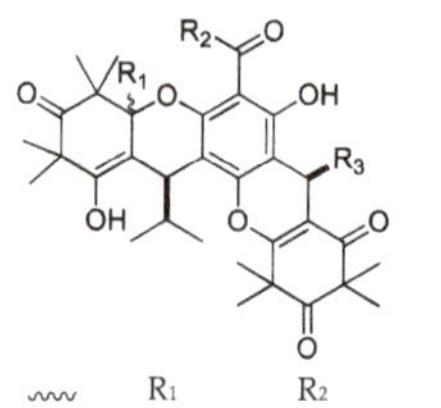

	〰	R1	R2	R3
11	▬	OH	iPr	iPr
12	—	H	*n*-Pent	iPr
15	⋯⋯	OH	*n*-Pent	iPr
17	⋯⋯	OH	*n*-Pent	iBu

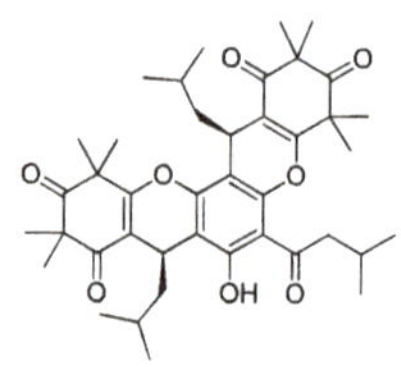

20

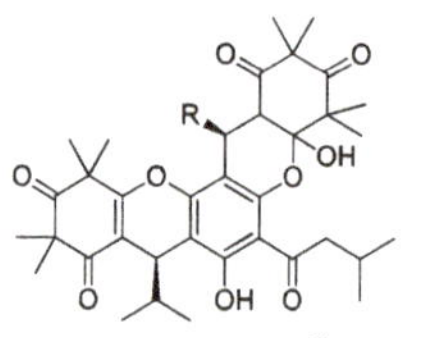

	R
30	iPr
31	iBu

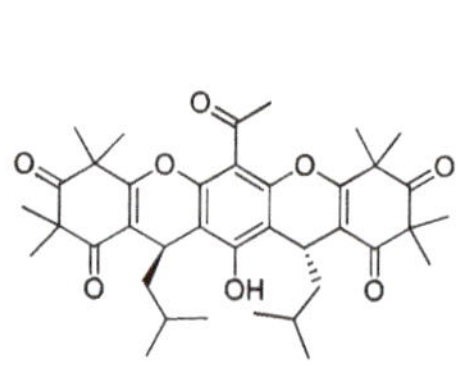

42

Figure 5. Structures of compounds of the trimeric-dipyrane type.

Rhodomyrtosone A (**18**) from *R. tomentosa* is the first natural product possessing a bisfurane fused ring (Figure 6). The dimeric compound **3** may represent its possible biosynthetic precursor, based on oxidation of the isobutyl side chain followed by the formation of benzofuran via cyclization and dehydration. Afterwards, several related compounds (**32, 43, 45, 46**) have been characterized from other species in the Myrtaceae, indicating a possible wider occurrence of this peculiar structure.

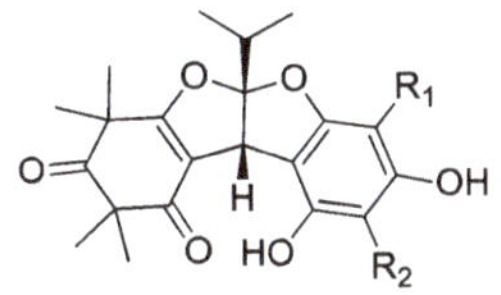

	R_1	R_2
18	COiBu	H
32	COEtPh	Me
43	H	COiBu
45	CoiBu	Me
46	Me	COiBu

Figure 6. Structures of compounds of the dimeric-bisfurane type.

Tomentosones A and B (**23, 24**) are two epimers possessing a novel hexacyclic ring system (Figure 7) whose structures present a bisfuranic group and a hexacyclic ring.

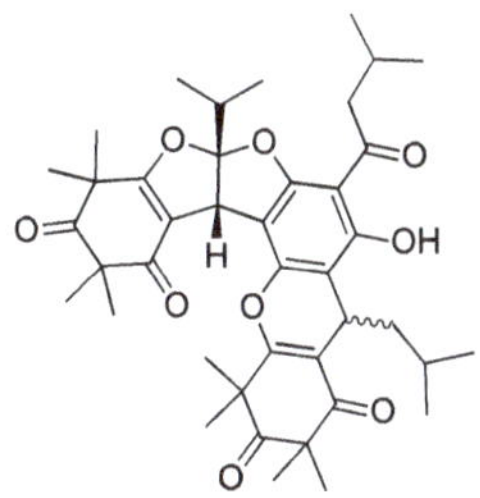

23
24

Figure 7. Structures of compounds of the trimeric-bisfurane-pyrane-type.

Callistrilones (**40, 41,** and **54–59**) represent the first syncarpic-phloroglucinol-monoterpene compounds that were isolated from a natural source. These compounds are characterized by the presence of a residue similar to phellandrene, which is fused through a furan ring to the phloroglucinol unit (Figure 8). Other compounds belonging to this class are the baefrutones, four of which (**64–67**) show the presence of an iridane skeleton, while **68** and **69** are sesquiterpene adducts.

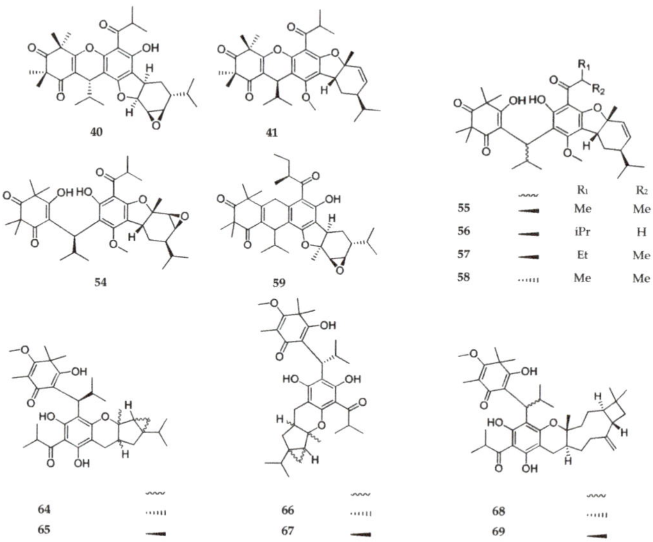

Figure 8. Structures of compounds of the terpene-adduct type.

The available literature concerning Myrtaceae also reports the existence of products that are not classifiable as acylphloroglucinol oligomers, and are hence not included in this review. Particularly, monomers of either acylphloroglucinol (e.g., callisalignene A–C [25], xanchryone A–D [55], operculatol A–B [56]) or syncarpic acid (e.g., myrtucommulone K [35,57], callistiviminene A–O [58]), and flavonoids conjugated to a syncarpic acid residue (e.g., kunzeanones A–C [59], myrtocummunines A–D [51]).

A huge laboratory activity has been carried out on the synthesis of phloroglucinol compounds [60]. As an answer to the rising interest for pharmaceutical applications of myrtucommulones and related compounds, in the last decade several independent approaches have been developed in order to synthesize compounds belonging to this class. In particular, myrtucommulone A was first obtained from commercially available precursors [61], and later through stereoselective synthesis [62,63]. Other analogs of the series have been synthetically obtained (Table 2), particularly in the last couple of years, which possibly preludes further achievements in this respect in the short term.

Table 2. Myrtucommulone-related compounds obtained synthetically.

Code	Compound Name	Subclass	Ref.
1	Myrtucommulone A	Trimeric type	[61–63]
7	Rhodomyrtone A	Dimeric-monopyrane	[64,65]
18	Rhodomyrtosone A	Dimeric-bisfurane type	[66]
19	Rhodomyrtosone B	Dimeric-monopyrane type	[64–66]
21	Myrtucommulone J	Dimeric type	[67]
26	Myrtucommuacetalone	Dimeric type	[67,68]
40	Callistrilone A	Terpene-adduct type	[68–70]
55, 58	Callistrilone G, J	Terpene-adduct type	[71]

4. Other Biological Sources

Following recent discoveries concerning a number of valuable plant-derived drugs that have been also detected as fermentation products of endophytic fungal strains [72], myrtucommulones A and D (**1, 11**) were extracted from the culture filtrates of a strain of *Neofusicoccum australe* endophytically associated with myrtle [73]. This finding represents the start point for new search terms addressed to a comparative elucidation of the genetic base of myrtucommulone biosynthesis, and possible applicative opportunities for more economic production to be exploited in view of drug development. Actually, while our laboratory investigations are in progress, we can anticipate that this extraordinary aptitude is shared with more endophytic strains isolated from myrtle in several locations (Figure 9). A few of these strains have been taxonomically identified and were found to belong to infrequent species, such as *Neosetophoma italica*, *Neocucurbitaria cava*, *Colletotrichum karsti*, and *Helminthosporium asterinum*. Following this preliminary assessment concerning myrtle, more evidences in this respect may be expected if metabolomic investigations are extended to endophytic microorganisms from other species in the Myrtaceae.

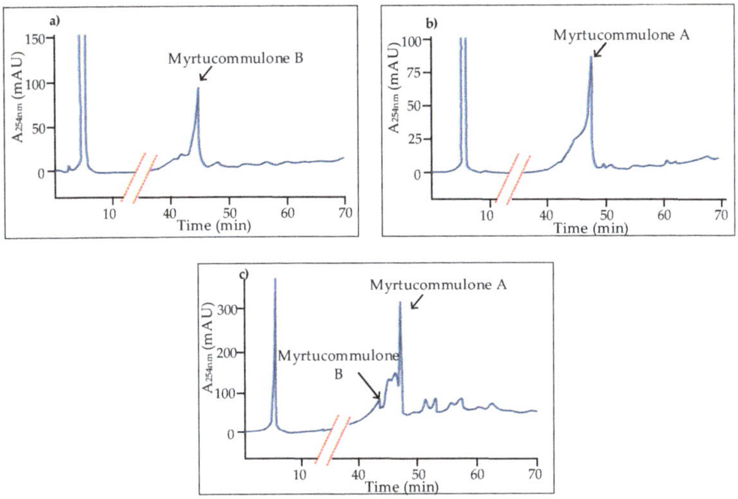

Figure 9. Detection of myrtucommulones A and B through HPLC-DAD analysis in culture extracts of endophytic fungi isolated from *M. communis*. (**a**) A1304B (*N. australe*); (**b**) A1306A (*H. asterinum*); and, (**c**) M15M2B (*N. italica*). Methods for culturing, extraction and chromatography have been previously reported [73].

Finally, quite interesting is the finding of myrtucommulone I (**17**), together with syncarpic acid and some more identified and unidentified alkylated phloroglucinols in propolis of the Australian stingless bee *Tetragonula carbonaria*, which has been taken into account to explain the antibacterial properties of this bee product [74].

5. Biological Activities

5.1. Antibacterial Activity

Myrtucommulones A and B (1–2), the founder compounds in this review, were preliminarily characterized for their antibiotic activity in agar plate diffusion assays against Gram-positive bacteria, namely *Staphylococcus aureus*, *Staphylococcus epidermidis*, *Bacillus subtilis*, *Bacillus pumilus*, *Enterococcus faecalis*, *Corynebacterium diphtheriae*, and *Corynebacterium xerosis* [75]. Similar assays provided concurrent indications in this respect for a few more products, such as the bullataketals (8–9) [31], myrtucommulones C–E (10–12) [33], and eucalyptone G (13) [29]. Afterwards, effectiveness against Gram-positive bacteria has been repeatedly reported for other related compounds, and circumstantiated with details concerning their minimum inhibitory (MICs) (Table 3). Conversely, assays against Gram-negative bacteria have been generally unfruitful, with a few questionable exceptions reporting inhibitory effects against *Escherichia coli* for rhodomyrtone A (7) [28,46], eucalyptone G [29], isomyrtucommulone B (5) [25], and callistenone H (38) [48]. Effects against Gram-negative species other than *E. coli*, such as *Pseudomonas aeruginosa*, *Salmonella typhi*, and *Shigella flexneri*, were reported for myrtucommulones C–E (10–12) [33].

Table 3. Bioactivity of myrtucommulones and related compounds resulting from assays carried out on Gram-positive bacteria.

	Bc	*Bs*	*Cd*	*Ef*	*Ml*	*Pa*	*Sa*	**MRS***a*	*Se*	*Sg*	*Sm*	*Spn*	*Spy*	*Ss*	**Ref.**
1							1–2	0.5							[5]
2							7.813								[25]
	8–16						16–32	16–32							[51]
5	1.56	0.78					1.56		6.25		3.13				[27]
						0.8									[10]
							0.488								[25]
	1–2						1–2	0.5–1							[51]
6							32–64	32							[5]
7	0.39	0.39		1.56			0.39	0.39–0.78	0.39	0.19	0.19	0.39	0.39	0.39	[76]
							0.5–1		0.25–1						[77,78]
							2	0.5							[79]
							0.5	0.5							[80]
							0.78				0.39				[81]
	0.5														[82]
			0.62–2.5												[83]
													0.78		[84]
						0.12–0.5									[85,86]
							1.83								[46]
				1–32			0.5–1	0.5–1							[87]
							0.5	0.5							[80]
	0.78	0.78			0.78		0.78		0.78		1.56				[27]
							0.5–1		0.5–1						[88]
												0.5–1			[89]
11							1.953	0.975							[25]
19				2.5			0.62–1.25	0.62–1.25							[90]
21							0.38								[35]
27							0.5	1							[39]

Table 3. *Cont.*

	Bc	*Bs*	*Cd*	*Ef*	*Ml*	*Pa*	*Sa*	**MRS*a***	*Se*	*Sg*	*Sm*	*Spn*	*Spy*	*Ss*	**Ref.**
28							8	8							[39]
29							8	8							[39]
38							20.3								[48]
40				32			16	16							[45]
41				64											[45]
42							3.66								[46]
52	1.56	1.56			1.56		1.56		3.13		3.13				[27]
53	1.56	1.56					1.56		3.13		1.56				[27]
62	8						8	8–16							[51]

MIC_{50} is reported in μg/mL. Bc = Bacillus cereus; Bs = Bacillus subtilis; Cd = Clostridium difficile; Ef = Enterococcus faecalis or E. faecium; Ml = Micrococcus luteus; Pa = Propionibacterium acnes; Sa = Staphylococcus aureus; MRSa = methicillin-resistant S. aureus; Se = Staphylococcus epidermidis; Sg = Streptococcus gordonii; Sm = Streptococcus mutans; Spn = Streptococcus pneumoniae; Spy = Streptococcus pyogenes; Ss = Streptococcus salivarius.

Particularly interesting is the activity against multi-resistant bacterial strains, especially methicillin-resistant *S. aureus* (MRS*a*) and vancomycin-resistant *E. faecalis*, exhibited by products, such as callistrilone A (**40**) [45], and rhodomyrtone A (**7**), which did not induce resistance even after 45 passages in vitro [87]. Moreover, the latter product has recently displayed notable effects against both cell division and spore formation in *Clostridium difficile* [83].

Mechanisms of antibacterial activity have been quite thoroughly investigated in the case of **7**. Gene assays and proteomic profiling experiments in *B. subtilis* indicate that the cytoplasmic membrane is the main target of this compound. In *S. aureus*, it was reported to decrease the membrane potential at low doses, and to cause the release of ATP and cytoplasmic proteins. Local membrane damage was confirmed through lipid staining, and the protective effect displayed by saturated fatty acids was explained in terms of counteractive mending. It can be speculated that resistance to **7** by Gram-negative bacteria is due to the reduced penetration of the product through the outer membrane, and its neutralization by lipopolysaccharides. Moreover, interferences in the proteome and metabolome of *Streptococcus pneumoniae* were documented after exposure to **7**, consisting in a reduction of the levels of two enzymes (glycosyltransferase and glucose-1-phosphate uridylyltransferase) and of the uridine diphosphate derivatives of glucose, glucuronic acid, and *N*-acetyl-D-galactosamine, which participate in the synthesis of the capsule; as a matter of fact, a reduction in capsule size was confirmed through a colorimetric assay and electron microscopy [89].

Exposure of MRS*a* to subinhibitory concentrations of **7** revealed a significant modulation of gene expression. Prominent changes involved genes encoding essential proteins for metabolic pathways and processes, such as membrane function, ATP-binding cassette transportation, and metabolism of amino acids, lipoproteins and nucleotides. Although the amino acid content of peptidoglycan in rhodomyrtone-treated MRS*a* did not differ significantly from the control, data gathered on genes involved in the biosynthesis of amino acids and the diaminopimelate pathway indicate that peptidoglycan represents a target for bioactivity of this compound [91]. Moreover, proteome analyses in MRS*a* revealed that exposure to subinhibitory concentrations of **7** affects the expression of several major functional classes of whole cell proteins, which act as surface antigens and virulence factors, or are involved in cell wall biosynthesis, cell division, oxidative stress, and various metabolic pathways. Transmission electron micrographs confirmed that **7** causes morphological and ultrastructural alterations in the bacterial cells, affecting the cell wall with abnormal septum formation and ensuing cell lysis [79].

The protein secretome was also investigated in a representative clinical MRS*a* isolate, where the immunodominant antigen A, the staphylococcal secretory antigen, and other antigenic proteins involved in cell wall hydrolysis were downregulated after treatment with a subinhibitory concentration of **7**. Ribosomal and cytoplasmic proteins, such as glycerol phosphate lipoteichoic acid synthase and the stage V sporulation protein G (SpoVG), were found in the treated sample, while glycerophosphoryl diester phosphodiesterase, and another lipase precursor were absent. Finally, the finding of several cytoplasmic proteins in the supernatant of the treated cultures indicated impairment in the cell wall synthesis [80].

Again, the proteomic approach was followed in assays carried out with *Streptococcus pyogenes*. Various enzymes associated with important metabolic pathways, including alcohol dehydrogenase, glyceraldehyde-3-phosphate dehydrogenase, Xaa-His dipeptidase, ornithine carbamoyltransferase, putative *O*-acetylserine lyase, enolase (2-phosphoglycerate dehydratase), fructose-bisphosphate aldolase, and cysteine synthase, were strongly affected in a clinical isolate treated with **7** at half the MIC. Moreover, a series of known virulence factors, such as glyceraldehyde-3-phosphate dehydrogenase, CAMP factor, and exotoxin C, were downregulated [84]. It was also found that **7** reduces the synthesis of staphyloxanthin, a pigment promoting resistance to reactive oxygen species (ROS) whose shortage increases the bacterial susceptibility to H_2O_2 and singlet oxygen killing [92]. Furthermore, it could reduce biofilm formation by *S. aureus* and *S. epidermidis*, ensuing a reduction in the transcription signals of biofilm-related genes, with a different autolysin profile detectable in the treated cells [77,78]. Besides formation, effects on the disorganization of established biofilms have been recently documented in assays carried out on *Propionibacterium acnes* [86].

Another documented molecular effect of **7** consists in the competitive binding of the tubulin homologue protein FtsZ. In fact, conformational changes in this main bacterial cell division protein were observed in both the (S)- and (R)- binding states of **7**. The compound reduced FtsZ polymerization by 36% and inhibited guanosine triphosphatase activity by up to 45%. However, at inhibitory concentrations, the compound had no effect on FtsZ localization in *B. subtilis*, and cells did not elongate after treatment. Higher concentrations of **7** affected the localization of FtsZ and of its membrane anchor proteins FtsA and SepF, showing that it did not specifically inhibit FtsZ but rather impaired multiple divisome proteins. Cell morphology was sometimes modified to a bean-like shape, possibly implying that the compound may also target cell wall synthesis, or maintenance [93].

At a more applicative level of investigation, it has been observed that subinhibitory concentrations of **7** affect the pathogenicity of oral bacteria (*S. aureus, Streptococcus mutans*) by impairing their adherence to both buccal epithelial cells and a polystyrene support in in vitro assays [84]. Moreover, **7** in a liposomal encapsulated preparation has been proposed for the treatment of bovine mastitis based on observations that were carried out in a bovine udder epidermal tissue model demonstrating remarkable effects against adhesion and invasion of the bacterial agents (*S. aureus* and *S. epidermidis*) [88].

Within a general framework of consistent antibacterial properties, data gathered in Table 3 emphasize that activity may appreciably change among the structural analogues. As an example, the higher efficacy of callistenone A (**27**) in comparison with its B isomer (**28**) is reported to depend on the point of attachment of the isovaleryl side chain [39]. This is confirmed by specific studies based on synthetic analogues, providing indication that the acyl tail of myrtucommulones and related compounds is a prerequisite for the antibacterial properties, and that its affinity for lipids is critical for activity more than its spatial dimension [94,95]. Additional clues in this respect also derive from assays concerning cytotoxicity, which in the case of tomentodiones is potentiated by an isobutyryl chain [50].

5.2. Bioactivities against Other Microorganisms and Viruses

The bioactivity of myrtucommulones and related acylphloroglucinols has been also investigated on fungi, with general negative results deriving from assays on species such as *Candida albicans, Cryptococcus neoformans, Microsporum gypseum,* and *Saccharomyces cerevisiae* [27,39].

Conversely, there are positive indications concerning antimalarial properties. In fact, antiplasmodial activity in the nanomolar range was reported for myrtucommulone A (**1**), and at a lower extent for semimyrtucommulone (**6**) [5,96]. Watsonianone B (**32**) displayed a potent activity against strain 3D7 of *Plasmodium falciparum* (IC_{50} 0.289 μM), particularly against the young ring stages, coupled with selectivity towards a human embryonic kidney cell line (HEK 293) [40]. Strong effects against the chloroquine resistant strain Dd2 (IC_{50} 0.10 μM) and low toxicity towards HEK 293 also characterize rhodomyrtosone F (**35**) [42], while a more moderate activity was assessed for rhodomyrtone A (**7**) against both 3D7 and Dd2 (IC_{50} 1.84 μM and 4.00 μM, respectively) [30]. Moreover, tomentosone A (**23**) inhibited the growth of both chloroquine resistant and sensitive strains (IC_{50} 1.49 μM and 1.0 μM, respectively), while its B analogue (**24**) was significantly less active [37]. Structural comparisons indicated that the syncarpic acid moiety is essential for antiplasmodial activity [97]. Additional data concerning antiplasmodial properties of the above products extracted from the flowers of *Angophora woodsiana* are reported in another paper by the same research group [30].

Antiviral effects were displayed in vitro by the mixture of compounds **3–4**, so far only extracted from two *Kunzea* species, based on the inhibition of the cytopathic effects of *Herpes simplex* type 1 (HSV-1) and Polio type 1 viruses [26]. Moderate effects in similar assays against HSV-1 have been also reported for callistrilones H–I (**38–39**) [49].

5.3. Antioxidant and Anti-Inflammatory Activities

Besides the possible applications in the treatment of infective diseases based on the above-reported effects, multiple observations concerning consistent antioxidant and anti-inflammatory properties have resulted from investigations that were carried out in several laboratories, representing an indication of a potential for a therapeutic use in the treatment of a series of disorders, ranging from allergopathies to cardiovascular diseases. With reference to the latter, myrtucommulone A (**1**) and semimyrtucommulone (**6**) were reported to exert powerful antioxidant properties during the degradation of cholesterol, preserving the LDL form from oxidative damage induced by copper ions, and inhibiting the increase of oxidative products deriving from polyunsaturated fatty acids [98,99]. At micromolar concentrations both compounds suppressed eicosanoid biosynthesis in vitro and in vivo by directly inhibiting cyclooxygenase (COX)-1 and 5-lipoxygenase. Moreover, they were successful in preventing the mobilization of Ca^{2+} in polymorphonuclear leukocytes, mediated by G protein signalling pathways, with the first compound acting at lower concentrations, and suppressed the formation of ROS and the release of elastase [100].

Binding affinity to the thyrotropin-releasing hormone (TRH) receptor-2, which is known to play a role in the phosphoinositide metabolism and is regarded as a potential therapeutic target to treat pain, was also reported for myrtucommulones A, D, and F–I (**1**, **11**, **14–17**) at micromolar concentrations [23]. After experiments carried out on both a human lung adenocarcinoma cell line (A549) and in a cell-free assay based on microsomal preparations of A549 cells stimulated with interleukin (IL)-1β, myrtucommulone resulted to be the first natural product to inhibit microsomal prostaglandin synthase-1 that efficiently suppresses prostaglandin formation without significant inhibition of cyclooxigenases, hence without displaying the typical side effects of non-steroidal anti-inflammatory drugs [101]. Furthermore, **1** exerted anti-inflammatory effects in the pleurisy model. In particular, a reduction was observed in the exudate volume, leukocyte numbers, lung injury, and neutrophil infiltration, and in a series of more specific effects mediated by enzymes and cytokines [102].

The interest for a potential therapeutic use of rhodomyrtone A (**7**) is also based on consistent properties that may prevent or delay the progression of inflammation in skin diseases, such as psoriasis. After stimulating human skin organ cultures with TNF-α and IL-17A to mimic skin inflammation, **7** significantly decreased inflammatory gene expression and the secretion of inflammatory proteins. Particularly, it inhibited TNF-induced extracellular signal–regulated kinases (ERK), c-Jun *N*-terminal kinases (JNK), the mitogen-activated protein (MAP) kinase p38, and phosphorylation of the NF-κB transcription factor p65, suggesting that it acts by modulating MAP kinase and NF-κB signalling pathways. Moreover, it reversed imiquimod-induced skin hyperplasia and epidermal thickening in mice [103]. The potential of **7** as an anti-psoriasis agent is further increased by its property to inhibit proliferation and to induce growth arrest and apoptosis in HaCaT keratinocytes [104]. The expression of pro-inflammatory molecules, including IL-1β, IL-6, TNF-α, and inducible nitric oxide synthase (iNOS) was enhanced in THP-1 monocytes that were stimulated with a high dose of heat-killed MR*Sa*. In contrast, monocytes stimulated with lower doses did not express these cytokines. However, in monocytes stimulated with heat-killed MR*Sa* at low doses, **7** significantly increased the expression of pro-inflammatory mediators, IL-6 and iNOS, and displayed some anti-inflammatory activity by reducing TNF-α expression. Treatment with **7** also significantly upregulated the expression of key pattern recognition receptor proteins (TLR2 and CD14). The ability of **7** to eliminate the resistant bacteria was observed within 24 h after treatment, following enhancement of the expression in monocytes of MR*Sa* recognition receptors, which possibly improved MR*Sa* clearance by modulating pro- and anti-inflammatory cytokine responses [105].

The possible application of the anti-inflammatory and immunomodulatory properties of **7** has been suggested in aquaculture as a result of observations carried out in vitro on head kidney macrophages of the rainbow trout (*Oncorhynchus mykiss*). In fact, exposure to **7** (1 μg mL^{-1}) induced changes in the expression of genes involved in innate immune and inflammatory responses, particularly with reference to pro-inflammatory cytokines (IL-1β, IL-8, TNF-α), anti-inflammatory cytokines (IL-10, TGF-β), the antioxidant enzyme glutathione peroxidase 1, and other inducible enzymes (iNOS, COX-2, arginase). Co-exposure of **7** with lipopolysaccharides led to a downregulation of genes encoding for some of the above inflammation-related products and a reduction in ROS levels [106].

Myrtucommuacetalone (**26**) exhibited a significant inhibitory effect against production of nitric oxide, a ROS generated by NADPH oxidases in human peripheral blood phagocytes whose excess is associated with the pathogenesis of various diseases, such as colitis, diabetes, septic shock, and ischemic neuronal damage. Moreover, it inhibited the proliferation of T-cells, which is a relevant effect for the prevention or treatment of autoimmune disorders, such as Parkinson's disease, rheumatoid arthritis, and diabetes [38]. Additionally, inhibitory effects on nitric oxide production after lipopolysaccharide stimulation in murine macrophage cells (RAW 264.7) have been documented for baefrutones A–D (**64–67**) [52], and for tomentodione T (**61**) and rhodomyrtosones B, G, and I (**19**, **43**, **22**) [50].

5.4. Cytotoxic and Antiproliferative Activities

Apart from the diverse implications arising from the above properties, the observation of a pro-apoptotic effect induced on human cancer cell lines more directly introduces a possible relevance of myrtucommulones and associated acylphloroglucinols as antitumor drugs. Tretiakova et al. [107] first showed that, at micromolar concentrations, myrtucommulone A (**1**) induces apoptosis in several cancer cell lines, such as PC-3 (androgen-independent prostate carcinoma), LNCaP (androgen-dependent prostate carcinoma), KFR (rhabdomyosarcoma), HL-60 (acute promyelocytic leukemia), MM6 (acute monocytic leukemia), H9 (cutaneous T-cell lymphoma), DLD-1 (colorectal adenocarcinoma), and Jurkat (acute T-cell leukemia). Cell death occurred through the mitochondrial pathway involving the activation of caspase-3, -8, and -9, cleavage of poly(ADP-ribose)polymerase (PARP), release of nucleosomes into the cytosol, and DNA fragmentation. A lower cytotoxic effect was displayed on non-transformed human peripheral blood mononuclear cells and foreskin fibroblasts. Apoptosis appeared to be mediated by the intrinsic pathway, with the loss of the mitochondrial membrane potential in MM6 cells and the release of cytochrome c from mitochondria. Interestingly, Jurkat cells deficient in caspase-9 were resistant to apoptosis, and no processing of PARP or caspase-8 was evident. Conversely, in cell lines that were deficient in either CD95 signalling or caspase-8, myrtucommulone was still able to induce cell death and PARP cleavage.

A more direct indication that **1** induces apoptosis by triggering the intrinsic pathway and directly disrupting the mitochondrial functions resulted in assays carried out on HL-60 cells. In these cells, the compound caused the loss of the mitochondrial membrane potential and suppressed mitochondrial ATP synthesis, consequently inducing the adenosine monophosphate-activated protein kinase (AMPK), an energy sensor involved in apoptosis of cancer cells. More in detail, **1** acts as a protonophore that primarily dissipates the mitochondrial membrane potential through a direct structural interaction, and suppresses the proton motive force that impairs mitochondrial viability and activates AMPK due to lowered ATP levels [108]. The chaperonin heat-shock protein 60 (HSP60) also represents a molecular target of myrtucommulone A, which binds the protein and modulates its mitochondrial functions. Particularly, in a protein refolding assay the compound was found to prevent HSP60-mediated reactivation of denatured malate-dehydrogenase [109].

Myrtucommulone A (**1**) also induced apoptosis in several chronic myelogenous leukemia cell lines (K-562, MEG-01, KBM-5) in consequence of downregulation of anti-apoptotic proteins, as evidenced by nuclear fragmentation and PARP cleavage. Interestingly, the compound displayed differential toxicity, since peripheral blood mononuclear cells from healthy donors that were used as control were unaffected [110].

In further studies carried out on murine breast cancer cells (4T1), **1** was found to trigger apoptosis at micromolar concentrations through both the intrinsic and extrinsic apoptotic pathways. The compound mediated an increased expression of several apoptotic genes, such as Fas, FasL, Gadd45a, Tnf, Tnfsf12, Trp53, and caspase-4. Moreover, the results of a wound healing experiment showed that it is also able to inhibit cancer cell migration [111]. All these effects were enhanced when treatment was operated in combination with epirubicin or cisplatin, evidencing a synergistic effect that could be exploited for setting more effective therapeutic schemes [112].

Another relevant effect of **1** that may contrast tumor development consists in a reduction of the expression of endoglin, a membrane glycoprotein that has a crucial role in angiogenesis. Treatment with this product reduced the chondrogenic potential in human mesenchymal stem cell (hMSC) lines, possibly as a consequence of the NF-κB p65 activation, while the adipogenic or osteogenic differentiation was not dramatically affected. The exploitation of these properties could be useful in targeted differentiation studies [113].

It is known that hMSCs can be observed in tissues surrounding tumors, where they could play a role in regulating cancer cell behaviour through paracrine signalling. Therefore, the modulation of their secretome is highly significant in view of attempts to control the disease. **1** was effective in modulating cytokine expression in hMSCs with a decrease of TNF-α, IL-6, IL-8, the vascular endothelial growth factor (VEGF), and the basic fibroblast growth factor (FGF-2), and in reducing the proliferation, migration, and clonogenicity of human bladder (HTB-9) and 4T1 cancer cells [114]. Moreover, when considering the prominent role that is played by the epithelial-mesenchymal transition in cancer progression and metastasis, the ability of **1** to interfere in this process again by modulating signalling pathways and inhibiting phosphorylation of multiple proteins represents a notable therapeutic property [115]. Besides reducing the viability and proliferation in HTB-9 cells, in cancer stem cells the compound downregulated the expression of markers associated to pluripotency and multipotency (i.e., NANOG, OCT-4, SOX-2, SSEA-4, TRA-1-60, CD90, CD73, and CD44), and decreased sphere-forming ability [116].

As an integration to data concerning efficacy, perspectives for a possible pharmaceutical use of myrtucommulones are corroborated by the absence of substantial cytotoxicity for non-malignant cells [107], and evidence of molecular stability in human and rat plasma [117].

Additional indications concerning the inhibitory effects on tumor cells have resulted in several laboratories from assays that were carried out with a number of compounds of the myrtucommulone series. Particularly, antiproliferative activity has been reported for bullataketals A and B (**8–9**) on murine leukemia cells (P388) [31], for myrtucommulones A and J (**1, 21**) on another prostate cancer cell line (DU145), HEpG2 (human liver carcinoma) and MT-4 (lymphocytic leukemia) [35], on PC3 and DU145 treated with a mixture of **1** and myrtucommulone D (**11**) [73], on HCT116 (human colorectal carcinoma) treated with **11** and isomyrtucommulone B (**5**) [25], on HeLa (human cervix uteri carcinoma) for tomentodiones S and T (**60–61**), rhodomyrtone A (**7**), and rhodomyrtosones A, B, G, and I (**18, 19, 43, 22**) [50].

Rhodomyrtone A (**7**) is cytotoxic for several types of eukaryotic cells, and eryptosis induced in human erythrocytes progresses along with cell shrinkage, membrane blebbing, and phosphatidylserine translocation to the cell surface. The distinctive interactions with the cytoplasmic membrane assimilate **7** to amphipathic products, introducing it as a useful tool in studies on membrane physiology [118]. While confirming that it is responsible for membrane invaginations that form intracellular vesicles trapping a broad range of membrane proteins, Saeloh et al. [119] observe that **7** does not behave as a typical membrane-inserting molecule; in fact, it transiently binds to phospholipid head groups and causes a distortion of lipid packing, which explains membrane fluidization and curvature. However, both the transient binding mode and the ability to form protein-trapping vesicles are unique properties, which are possibly indicative of a peculiar mechanism of action.

More detailed investigations provide further insight in the antitumor properties of **7**, which has been recently characterized as an antimetastatic agent for the treatment of skin cancer cells after studies that were carried out on an epidermoid carcinoma cell line (A431). In fact, at subcytotoxic concentrations the compound reduced migration of tumor cells, as well as their adhesive and invasive ability. At the molecular level it was able to inhibit the focal adhesion kinase (FAK) and phosphorylation of protein kinase B (AKT), c-Raf, ERK1/2, and p38 involved in the downregulation of enzyme activities and the expression of matrix metalloproteinase (MMP)-2 and -9. Moreover, the compound increased the expression of TIMP-1 and TIMP-2, which are inhibitors of MMP-9 and MMP-2, respectively, and inhibited the expression and phosphorylation of NF-κB in a dose-dependent manner [120].

5.5. Other Pharmacological Perspectives

The α-glucosidase-inhibitory activity displayed by myrtucommulones C–E (**10**–**12**) could introduce therapeutic application of these compounds in another widespread disease, such as diabetes [33]. Moreover, rhodomyrtosone E (**33**) showed weak effects on the translocation to the plasma membrane of the insulin-responsive glucose transporter 4 (GLUT-4) protein, representing another attractive target for anti-diabetic drug development [41]. Finally, the strong inhibitory activity towards soluble epoxyde hydrolases that is exhibited by myrtucommulone B (**2**) and callistenone B (**28**) could be exploited for the treatment of a variety of pathological conditions [24].

Besides data resulting from specific assays carried out with the purified compounds, the widespread use in ethnomedicine of extracts of plant species in the Myrtaceae represents a reliable guide for the possible pharmaceutical applications of these products [121–123]. Based on its consistent antibacterial properties, the use of myrtle extracts has been proposed for the treatment of mild bacterial disorders such as vaginosis [124], and an ethanolic myrtle extract (Myrtacine®) has been registered for the treatment of acne lesions whose active principles are claimed to be myrtucommulones A and B [125,126]. For the latter dermatological use, rhodomyrtone A has been proposed in an innovative liposome encapsulation in order to overcome problems deriving from the poor water solubility [127]. The same kind of preparation has also been tested for zootechnical use [88].

6. Conclusions

Structures and properties of 69 myrtucommulone-related products that were extracted from plant species belonging to the Myrtaceae during the past 45 years have been reviewed in this paper. Considering that about half of these compounds have been discovered in the last two years, and that attention of the scientific community to the exploitation of natural resources of bioactive products is increasing more and more, it is predictable that their number is going to increase quite quickly in the future. Furthermore, in the aim to study more in detail aspects concerning the relationships between structure and bioactivity of these products, synthetic studies have been recently started in a few laboratories from which quite a high number of analogues have been obtained, indicating that these structural models may be used for the synthesis of novel variants with improved effects [94,95,128,129].

Advances in the exploitation of these valuable products require a more refined capacity to detect their occurrence in plants [130–132], and to perform a careful extraction and purification of the effective analogues. Alternative opportunities to obtain these products through controlled fermentation also represent a basic investigational line that is likely to attract the attention of leading actors in the field of drug discovery and development in the near future.

Author Contributions: Conceptualization, R.N. and A.A.; experimental data, P.F.; literature search, M.M.S.; writing—review and editing, R.N., M.M.S., A.A.

Funding: Financial support was provided by Ministero Italiano dell'Istruzione, dell'Università e della Ricerca (MIUR) through Finanziamento delle Attività Base della Ricerca (FFABR) 2017.

Conflicts of Interest: The authors declare no conflict of interest.

References

1. Zilkah, S.; Goldschdmidt, E.E. Myrtle (*Myrtus communis* L.)—A native Mediterranean and cultured crop species. In *Medicinal and Aromatic Plants of the Middle-East. Medicinal and Aromatic Plants of the World*; Yaniv, Z., Dudai, N., Eds.; Springer: Dordrecht, The Netherlands, 2014; Volume 2.
2. Alipour, G.; Dashti, S.; Hosseinzadeh, H. Review of pharmacological effects of *Myrtus communis* L. and its active constituents. *Phytother. Res.* **2014**, *28*, 1125–1136. [CrossRef] [PubMed]

3. Aleksic, V.; Knezevic, P. Antimicrobial and antioxidative activity of extracts and essential oils of *Myrtus communis* L. *Microb. Res.* **2014**, *169*, 240–254. [CrossRef] [PubMed]
4. Tuberoso, C.I.G.; Rosa, A.; Bifulco, E.; Melis, M.P.; Atzeri, A.; Pirisi, F.M.; Dessì, M.A. Chemical composition and antioxidant activities of *Myrtus communis* L. berries extracts. *Food Chem.* **2010**, *123*, 1242–1251. [CrossRef]
5. Appendino, G.; Bianchi, F.; Minassi, A.; Sterner, O.; Ballero, M.; Gibbons, S. Oligomeric acylphloroglucinols from myrtle (*Myrtus communis*). *J. Nat. Prod.* **2002**, *65*, 334–338. [CrossRef]
6. Yoshimura, M.; Amakura, Y.; Tokuhara, M.; Yoshida, T. Polyphenolic compounds isolated from the leaves of *Myrtus communis*. *J. Nat. Med.* **2008**, *62*, 366–368. [CrossRef]
7. Piras, F.M.; Dettori, M.F.; Magnani, A. ToF-SIMS PCA analysis of *Myrtus communis* L. *Appl. Surface Sci.* **2009**, *255*, 7805–7811. [CrossRef]
8. Barboni, T.; Cannac, M.; Massi, L.; Perez-Ramirez, Y.; Chiaramonti, N. Variability of polyphenol compounds in *Myrtus communis* L. (Myrtaceae) berries from Corsica. *Molecules* **2010**, *15*, 7849–7860. [CrossRef] [PubMed]
9. Usai, M.; Mulas, M.; Marchetti, M. Chemical composition of essential oils of leaves and flowers from five cultivars of myrtle (*Myrtus communis* L.). *J. Essent. Oil Res.* **2015**, *27*, 465–476. [CrossRef]
10. Hamdy, A.A.; Kassem, H.A.; Awad, G.E.A.; El-Kady, S.M.; Benito, M.T.; Doyagüez, E.G.; Jimeno, M.L.; Lall, N.; Hussein, A.A. *In-vitro* evaluation of certain Egyptian traditional medicinal plants against *Propionibacterium acnes*. *S. Afr. J. Bot.* **2017**, *109*, 90–95. [CrossRef]
11. Christenhusz, M.J.M.; Byng, J.W. The number of known plants species in the world and its annual increase. *Phytotaxa* **2016**, *261*, 201–217. [CrossRef]
12. de Oliveira Bernardes, C.; Tuler, A.C.; Ferreira, A.; Carvalho, M.S.; Nogueira, A.M.; da Silva Ferreira, M.F. Transferability of *Psidium microsatellite* loci in Myrteae (Myrtaceae): A phylogenetic signal. *Euphytica* **2018**, *214*, 105. [CrossRef]
13. Stefanello, M.E.A.; Pascoal, A.C.; Salvador, M.J. Essential oils from neotropical Myrtaceae: Chemical diversity and biological properties. *Chem. Biodivers.* **2011**, *8*, 73–94. [CrossRef] [PubMed]
14. Moraes Cascaes, M.; Skelding Pinheiro Guilhon, G.M.; de Aguiar Andrade, E.H.; Bichara Zoghbi, M.G.; da Silva Santos, L. Constituents and pharmacological activities of *Myrcia* (Myrtaceae): A review of an aromatic and medicinal group of plants. *Int. J. Mol. Sci.* **2015**, *16*, 23881–23904. [CrossRef] [PubMed]
15. de Souza, A.M.; Freitas de Oliveira, C.; Bednarczuk de Oliveira, V.; Martins Betim, F.C.; Gomes Miguel, O.; Dallarmi Miguel, M. Traditional uses, phytochemistry, and antimicrobial activities of *Eugenia* species—A review. *Planta Med.* **2018**, *84*, 1232–1248. [CrossRef] [PubMed]
16. Prosser, J.A.; Woods, R.R.; Horswell, J.; Robinson, B.H. The potential in-situ antimicrobial ability of Myrtaceae plant species on pathogens in soil. *Soil Biol. Biochem.* **2016**, *96*, 1–3. [CrossRef]
17. Kashman, Y.; Rotstein, A.; Lifshitz, A. The structure determination of two new acylphloroglucinols from *Myrtus communis* L. *Tetrahedron* **1974**, *30*, 991–997. [CrossRef]
18. Bianchi, A. The Mediterranean aromatic plants and their culinary use. *Nat. Prod. Res.* **2015**, *29*, 201–206. [CrossRef]
19. Montoro, P.; Tuberoso, C.I.; Piacente, S.; Perrone, A.; De Feo, V.; Cabras, P.; Pizza, C. Stability and antioxidant activity of polyphenols in extracts of *Myrtus communis* L. berries used for the preparation of myrtle liqueur. *J. Pharm. Biomed. Anal.* **2006**, *41*, 1614–1619. [CrossRef]
20. Craven, L.A. New combinations in Melaleuca for Australian species of Callistemon (Myrtaceae). *Novon* **2006**, *16*, 468–478. [CrossRef]
21. Singh, I.P.; Bharate, S.B. Phloroglucinol compounds of natural origin. *Nat. Prod. Rep.* **2006**, *23*, 558–591. [CrossRef]
22. Lounasmaa, M.; Puri, H.S.; Widén, C.-J. Phloroglucinol derivatives of *Callistemon lanceolatus* leaves. *Phytochemistry* **1977**, *16*, 1851–1852. [CrossRef]
23. Carroll, A.R.; Lamb, J.; Moni, R.; Guymer, G.P.; Forster, P.I.; Quinn, R.J. Myrtucommulones F–I, phloroglucinols with thyrotropin-releasing hormone receptor-2 binding affinity from the seeds of *Corymbia scabrida*. *J. Nat. Prod.* **2008**, *71*, 1564–1568. [CrossRef] [PubMed]
24. Khanh, P.N.; Duc, H.V.; Huong, T.T.; Son, N.T.; Ha, V.T.; Van, D.T.; Tai, B.H.; Kim, J.E.; Jo, A.R.; Kim, Y.H.; et al. Alkylphloroglucinol derivatives and triterpenoids with soluble epoxide hydrolase inhibitory activity from *Callistemon citrinus*. *Fitoterapia* **2016**, *109*, 39–44. [CrossRef]

25. Qin, X.J.; Liu, H.; Yu, Q.; Yan, H.; Tang, J.F.; An, L.K.; Khan, A.; Chen, Q.R.; Hao, X.J.; Liu, H.Y. Acylphloroglucinol derivatives from the twigs and leaves of *Callistemon salignus*. *Tetrahedron* **2017**, *73*, 1803–1811. [CrossRef]
26. Bloor, S.J. Antiviral phloroglucinols from New Zealand *Kunzea* species. *J. Nat. Prod.* **1992**, *55*, 43–47. [CrossRef] [PubMed]
27. Kaneshima, T.; Myoda, T.; Toeda, K.; Fujimori, T.; Nishizawa, M. Antimicrobial constituents of peel and seeds of camu-camu (*Myrciaria dubia*). *Biosci. Biotechnol. Biochem.* **2017**, *81*, 1461–1465. [CrossRef]
28. Salni, D.; Sargent, M.V.; Skelton, B.W.; Soediro, I.; Sutisna, M.; White, A.H.; Yulinah, E. Rhodomyrtone, an antibotic from *Rhodomyrtus tomentosa*. *Aus. J. Chem.* **2002**, *55*, 229–232. [CrossRef]
29. Mohamed, G.A.; Ibrahim, S.R. Eucalyptone G, a new phloroglucinol derivative and other constituents from *Eucalyptus globulus* Labill. *Arkivoc* **2007**, *15*, 281–291.
30. Senadeera, S.P.; Duffy, S.; Avery, V.M.; Carroll, A.R. Antiplasmodial β-triketones from the flowers of the Australian tree *Angophora woodsiana*. *Bioorg. Med. Chem. Lett.* **2017**, *27*, 2602–2607. [CrossRef]
31. Larsen, L.; Benn, M.H.; Parvez, M.; Perry, N.B. A cytotoxic triketone–phloroglucinol–bullatenone hybrid from *Lophomyrtus bullata*. *Org. Biomol. Chem.* **2005**, *3*, 3236–3241. [CrossRef]
32. Woollard, J.M.R.; Perry, N.B.; Weavers, R.T.; van Klink, J.W. Bullatenone, 1, 3-dione and sesquiterpene chemotypes of *Lophomyrtus* species. *Phytochemistry* **2008**, *69*, 1313–1318. [CrossRef]
33. Shaheen, F.; Ahmad, M.; Khan, S.N.; Hussain, S.S.; Anjum, S.; Tashkhodjaev, B.; Turguniv, K.; Sultankhodzhaev, M.N.; Choudhary, M.I.; Ur-Rahman, A. New α-glucosidase inhibitors and antibacterial compounds from *Myrtus communis* L. *Eur. J. Org. Chem.* **2006**, *2016*, 2371–2377. [CrossRef]
34. Hiranrat, A.; Mahabusarakam, W. New acylphloroglucinols from the leaves of *Rhodomyrtus tomentosa*. *Tetrahedron* **2008**, *64*, 11193–11197. [CrossRef]
35. Cottiglia, F.; Casu, L.; Leonti, M.; Caboni, P.; Floris, C.; Busonera, B.; Farci, P.; Ouhtit, A.; Sanna, G. Cytotoxic phloroglucinols from the leaves of *Myrtus communis*. *J. Nat. Prod.* **2012**, *75*, 225–229. [CrossRef] [PubMed]
36. Hiranrat, A.; Chitbankluoi, W.; Mahabusarakam, W.; Limsuwan, S.; Voravuthikunchai, S.P. A new flavellagic acid derivative and phloroglucinol from *Rhodomyrtus tomentosa*. *Nat. Prod. Res.* **2012**, *26*, 1904–1909. [CrossRef]
37. Hiranrat, A.; Mahabusarakam, W.; Carroll, A.R.; Duffy, S.; Avery, V.M. Tomentosones A and B, hexacyclic phloroglucinol derivatives from the Thai shrub *Rhodomyrtus tomentosa*. *J. Org. Chem.* **2012**, *77*, 680–683. [CrossRef]
38. Choudhary, M.I.; Khan, N.; Ahmad, M.; Yousuf, S.; Fun, H.-K.; Soomro, S.; Asif, M.; Mesaik, M.A.; Shaheen, F. New inhibitors of ROS generation and T-cell proliferation from *Myrtus communis*. *Org. Lett.* **2013**, *15*, 1862–1865. [CrossRef]
39. Rattanaburi, S.; Mahabusarakam, W.; Phongpaichit, S.; Carroll, A.R. Acylphloroglucinols from *Callistemon lanceolatus* DC. *Tetrahedron* **2013**, *69*, 6070–6075. [CrossRef]
40. Carroll, A.R.; Avery, V.M.; Duffy, S.; Forster, P.I.; Guymer, G.P. Watsonianone A-C, anti-plasmodial β-triketones from the Australian tree, *Corymbia watsoniana*. *Org. Biomol. Chem.* **2013**, *21*, 453–458. [CrossRef] [PubMed]
41. Wang, C.; Yang, J.; Zhao, P.; Zhou, Q.; Mei, Z.; Yang, G.; Yang, X.; Feng, Y. Chemical constituents from *Eucalyptus citriodora* Hook leaves and their glucose transporter 4 translocation activities. *Bioorg. Med. Chem. Lett.* **2014**, *24*, 3096–3099. [CrossRef]
42. Hans, M.; Charpentier, M.; Huch, V.; Jauch, J.; Bruhn, T.; Bringmann, G.; Quandt, D. Stereoisomeric composition of natural myrtucommulone A. *J. Nat. Prod.* **2015**, *78*, 2381–2389. [CrossRef] [PubMed]
43. Su, Q.; Dalal, S.; Goetz, M.; Cassera, M.B.; Kingston, D.G.I. Antiplasmodial phloroglucinol derivatives from *Syncarpia glomulifera*. *Bioorg. Med. Chem.* **2016**, *24*, 2544–2548. [CrossRef] [PubMed]
44. Liu, H.-X.; Chen, Y.-C.; Liu, Y.; Zhang, W.-M.; Wu, J.-W.; Tan, H.-B.; Qiu, S.-X. Acylphloroglucinols from the leaves of *Callistemon viminalis*. *Fitoterapia* **2016**, *114*, 40–44. [CrossRef] [PubMed]
45. Cao, J.Q.; Huang, X.J.; Li, Y.T.; Wang, Y.; Wang, L.; Jiang, R.W.; Ye, W.C. Callistrilones A and B, triketone–phloroglucinol–monoterpene hybrids with a new skeleton from *Callistemon rigidus*. *Org. Lett.* **2016**, *18*, 120–123. [CrossRef] [PubMed]
46. Liu, H.-X.; Tan, H.-B.; Qiu, S.-X. Antimicrobial acylphloroglucinols from the leaves of *Rhodomyrtus tomentosa*. *J. Asian Nat. Prod. Res.* **2016**, *18*, 535–541. [CrossRef] [PubMed]

47. Hiranrat, W.; Hiranrat, A.; Mahabusarakam, W. Rhodomyrtosones G and H, minor phloroglucinols from the leaves of *Rhodomyrtus tomentosa*. *Phytochem. Lett.* **2017**, *21*, 25–28. [CrossRef]
48. Wu, L.; Zhang, Y.; Wang, X.; Liu, R.; Yang, M.; Kong, L.; Luo, J. Acylphloroglucinols from the fruits of *Callistemon viminalis*. *Phytochem. Lett.* **2017**, *20*, 61–65. [CrossRef]
49. Cao, J.Q.; Wu, Y.; Zhong, Y.L.; Li, N.P.; Chen, M.; Li, M.M.; Ye, W.C.; Wang, L. Antiviral triketone-phloroglucinol-monoterpene adducts from *Callistemon rigidus*. *Chem. Biodiv.* **2018**, *15*, e1800172. [CrossRef]
50. Zhang, Y.B.; Li, W.; Jiang, L.; Yang, L.; Chen, N.H.; Wu, Z.N.; Li, Y.L.; Wang, G.C. Cytotoxic and anti-inflammatory active phloroglucinol derivatives from *Rhodomyrtus tomentosa*. *Phytochemistry* **2018**, *153*, 111–119. [CrossRef]
51. Tanaka, N.; Jia, Y.; Niwa, K.; Imabayashi, K.; Tatano, Y.; Yagi, H.; Kashiwada, Y. Phloroglucinol derivatives and a chromone glucoside from the leaves of *Myrtus communis*. *Tetrahedron* **2018**, *74*, 117–123. [CrossRef]
52. Hou, J.Q.; Wang, B.L.; Han, C.; Xu, J.; Wang, Z.; He, Q.W.; Zhang, P.L.; Zhao, S.M.; Pei, X.; Wang, H. Atropisomeric meroterpenoids with rare triketone-phloroglucinol-terpene hybrids from *Baeckea frutescens*. *Org. Biomol. Chem.* **2018**, *16*, 8513–8524. [CrossRef] [PubMed]
53. Dewick, P.M. *Medicinal Natural Products: A Biosynthetic Approach*, 3rd ed.; John Wiley & Sons, Ltd.: New York, NY, USA, 2009; ISBN 0471496413.
54. Appendino, G.; Maxia, L.; Bettoni, P.; Locatelli, M.; Valdivia, C.; Ballero, M.; Stavri, M.; Gibbons, S.; Sterner, O. Antibacterial galloylated alkylphloroglucinol glucosides from myrtle (*Myrtus communis*). *J. Nat. Prod.* **2006**, *69*, 251–254. [CrossRef] [PubMed]
55. Liu, F.; Lu, W.J.; Li, N.P.; Liu, J.W.; He, J.; Ye, W.C.; Wang, L. Four new cinnamoyl-phloroglucinols from the leaves of *Xanthostemon chrysanthus*. *Fitoterapia* **2018**, *128*, 93–96. [CrossRef] [PubMed]
56. Su, J.C.; Wang, S.; Cheng, W.; Huang, X.J.; Li, M.M.; Jiang, R.W.; Li, Y.L.; Wang, L.; Ye, W.C.; Wang, Y. Phloroglucinol derivatives with unusual skeletons from *Cleistocalyx operculatus* and their in vitro antiviral activity. *J. Org. Chem.* **2018**, *83*, 8522–8532. [CrossRef] [PubMed]
57. Liu, C.; Ang, S.; Huang, X.J.; Tian, H.Y.; Deng, Y.Y.; Zhang, D.M.; Wang, Y.; Ye, W.C.; Wang, L. Meroterpenoids with new skeletons from *Myrtus communis* and structure revision of myrtucommulone K. *Org. Lett.* **2016**, *18*, 4004–4007. [CrossRef] [PubMed]
58. Wu, L.; Wang, X.B.; Li, R.J.; Zhang, Y.L.; Yang, M.H.; Luo, J.; Kong, L.Y. Callistiviminenes AO: Diverse adducts of β-triketone and sesqui-or monoterpene from the fruits of *Callistemon viminalis*. *Phytochemistry* **2016**, *131*, 140–149. [CrossRef]
59. Ito, H.; Iwamori, H.; Kasajima, N.; Kaneda, M.; Yoshida, T. Kunzeanones A, B, and C: Novel alkylated phloroglucinol metabolites from *Kunzea ambigua*. *Tetrahedron* **2004**, *60*, 9971–9976. [CrossRef]
60. Singh, I.P.; Sidana, J.; Bharate, S.B.; Foley, W.J. Phloroglucinol compounds of natural origin: Synthetic aspects. *Nat. Prod. Rep.* **2010**, *27*, 393–416. [CrossRef]
61. Müller, H.; Paul, M.; Hartmann, D.; Huch, V.; Blaesius, D.; Koeberle, A.; Werz, O.; Jauch, J. Total synthesis of myrtucommulone A. *Angewandte Chemie* **2010**, *49*, 2045–2049. [CrossRef]
62. Charpentier, M.; Hans, M.; Jauch, J. Enantioselective synthesis of myrtucommulone A. *Eur. J. Org. Chem.* **2013**, *19*, 4078–4084. [CrossRef]
63. Charpentier, M.; Jauch, J. Metal catalysed versus organocatalysed stereoselective synthesis: The concrete case of myrtucommulones. *Tetrahedron* **2017**, *73*, 6614–6623. [CrossRef]
64. Morkunas, M.; Dube, L.; Götz, F.; Maier, M.E. Synthesis of the acylphloroglucinols rhodomyrtone and rhodomyrtosone B. *Tetrahedron* **2013**, *69*, 8559–8563. [CrossRef]
65. Morkunas, M.; Maier, M.E. Alternative routes to the acylphloroglucinol rhodomyrtone. *Tetrahedron* **2015**, *71*, 9662–9666. [CrossRef]
66. Gervais, A.; Lazarski, K.E.; Porco, J.A., Jr. Divergent total syntheses of rhodomyrtosones A and B. *J. Org. Chem.* **2015**, *80*, 9584–9591. [CrossRef] [PubMed]
67. Liu, H.; Huo, L.; Yang, B.; Yuan, Y.; Zhang, W.; Xu, Z.; Qiu, S.; Tan, H. Biomimetic-inspired syntheses of myrtucommuacetalone and myrtucommulone J. *Org. Lett.* **2017**, *19*, 4786–4789. [CrossRef] [PubMed]
68. Cheng, M.J.; Cao, J.Q.; Yang, X.Y.; Zhong, L.P.; Hu, L.J.; Lu, X.; Hou, B.L.; Hu, Y.J.; Wang, Y.; You, X.F.; et al. Catalytic asymmetric total syntheses of myrtucommuacetalone, myrtucommuacetalone B, and callistrilones A, C, D and E. *Chem. Sci.* **2018**, *9*, 1488–1495. [CrossRef] [PubMed]

69. Dethe, D.H.; Dherange, B.D.; Das, S. Biomimetic total syntheses of callistrilones A, B, and D. *Org. Lett.* **2018**, *20*, 680–683. [CrossRef]
70. Guo, Y.; Zhang, Y.; Xiao, M.; Xie, Z. Biomimetic syntheses of callistrilones A–E via an oxidative [3 + 2] cycloaddition. *Org. Lett.* **2018**, *20*, 2509–2512. [CrossRef] [PubMed]
71. Hu, L.J.; Cheng, M.J.; Cao, J.Q.; Zhong, L.P.; Hu, Y.J.; Wang, Y.; Wang, L.; Ye, W.C.; Li, C.C. Asymmetric total syntheses of callistrilones B, G and J. *Org. Chem. Front.* **2018**, *5*, 1506–1510. [CrossRef]
72. Nicoletti, R.; Fiorentino, A. Plant bioactive metabolites and drugs produced by endophytic fungi of Spermatophyta. *Agriculture* **2015**, *5*, 918–970. [CrossRef]
73. Nicoletti, R.; Ferranti, P.; Caira, S.; Misso, G.; Castellano, M.; Di Lorenzo, G.; Caraglia, M. Myrtucommulone production by a strain of *Neofusicoccum australe* endophytic in myrtle (*Myrtus communis*). *World J. Microbiol. Biotechnol.* **2014**, *30*, 1047–1052. [CrossRef] [PubMed]
74. Massaro, C.F.; Smyth, T.J.; Smyth, W.F.; Heard, T.; Leonhardt, S.D.; Katouli, M.; Wallace, H.M.; Brooks, P. Phloroglucinols from anti-microbial deposit-resins of Australian stingless bees (*Tetragonula carbonaria*). *Phytother. Res.* **2015**, *29*, 48–58. [CrossRef] [PubMed]
75. Rotstein, A.; Lifshitz, A.; Kashman, Y. Isolation and antibacterial activity of acylphloroglucinols from *Myrtus communis*. *Antimicrob. Agents Chemother.* **1974**, *6*, 539–542. [CrossRef]
76. Limsuwan, S.; Trip, E.N.; Kouwen, T.R.; Piersma, S.; Hiranrat, A.; Mahabusarakam, W.; Voravuthikunchai, S.P.; Van Dijl, J.M.; Kayser, O. Rhodomyrtone: A new candidate as natural antibacterial drug from *Rhodomyrtus tomentosa*. *Phytomedicine* **2009**, *16*, 645–651. [CrossRef]
77. Saising, J.; Ongsakul, M.; Voravuthikunchai, S.P. *Rhodomyrtus tomentosa* (Aiton) Hassk. ethanol extract and rhodomyrtone: A potential strategy for the treatment of biofilm-forming staphylococci. *J. Med. Microbiol.* **2011**, *60*, 1793–1800. [CrossRef]
78. Saising, J.; Götz, F.; Dube, L.; Ziebandt, A.K.; Voravuthikunchai, S.P. Inhibition of staphylococcal biofilm-related gene transcription by rhodomyrtone, a new antibacterial agent. *Ann. Microbiol.* **2015**, *65*, 659–665. [CrossRef]
79. Sianglum, W.; Srimanote, P.; Wonglumsom, W.; Kittiniyom, K.; Voravuthikunchai, S.P. Proteome analyses of cellular proteins in methicillin-resistant *Staphylococcus aureus* treated with rhodomyrtone, a novel antibiotic candidate. *PLoS ONE* **2011**, *6*, e16628. [CrossRef]
80. Visutthi, M.; Srimanote, P.; Voravuthikunchai, S.P. Responses in the expression of extracellular proteins in methicillin-resistant *Staphylococcus aureus* treated with rhodomyrtone. *J. Microbiol.* **2011**, *49*, 956–964. [CrossRef]
81. Limsuwan, S.; Homlaead, S.; Watcharakul, S.; Chusri, S.; Moosigapong, K.; Saising, J.; Voravuthikunchai, S.P. Inhibition of microbial adhesion to plastic surface and human buccal epithelial cells by *Rhodomyrtus tomentosa* leaf extract. *Arch. Oral Biol.* **2014**, *59*, 1256–1265. [CrossRef]
82. Voravuthikunchai, S.P.; Dolah, S.; Charernjiratrakul, W. Control of *Bacillus cereus* in foods by *Rhodomyrtus tomentosa* (Ait.) hassk. leaf extract and its purified compound. *J. Food Prot.* **2010**, *73*, 1907–1912. [CrossRef]
83. Srisuwan, S.; Tongtawe, P.; Srimanote, P.; Voravuthikunchai, S.P. Rhodomyrtone modulates innate immune responses of THP-1 monocytes to assist in clearing methicillin-resistant *Staphylococcus aureus*. *PLoS ONE* **2014**, *9*, e110321. [CrossRef] [PubMed]
84. Limsuwan, S.; Hesseling-Meinders, A.; Voravuthikunchai, S.P.; Van Dijl, J.M.; Kayser, O. Potential antibiotic and anti-infective effects of rhodomyrtone from *Rhodomyrtus tomentosa* (Aiton) Hassk. on *Streptococcus pyogenes* as revealed by proteomics. *Phytomedicine* **2011**, *18*, 934–940. [CrossRef] [PubMed]
85. Saising, J.; Voravuthikunchai, S.P. Anti *Propionibacterium acnes* activity of rhodomyrtone, an effective compound from *Rhodomyrtus tomentosa* (Aiton) Hassk. leaves. *Anaerobe* **2012**, *18*, 400–404. [CrossRef] [PubMed]
86. Wunnoo, S.; Saising, J.; Voravuthikunchai, S.P. Rhodomyrtone inhibits lipase production, biofilm formation, and disorganizes established biofilm in *Propionibacterium acnes*. *Anaerobe* **2017**, *43*, 61–68. [CrossRef] [PubMed]
87. Leejae, S.; Taylor, P.W.; Voravuthikunchai, S.P. Antibacterial mechanisms of rhodomyrtone against important hospital-acquired antibiotic-resistant pathogenic bacteria. *J. Med. Microbiol.* **2013**, *62*, 78–85. [CrossRef] [PubMed]

88. Mordmuang, A.; Shankar, S.; Chethanond, U.; Voravuthikunchai, S.P. Effects of *Rhodomyrtus tomentosa* leaf extract on staphylococcal adhesion and invasion in bovine udder epidermal tissue model. *Nutrients* **2015**, *7*, 8503–8517. [CrossRef]
89. Mitsuwan, W.; Olaya-Abril, A.; Calderon-Santiago, M.; Jimenez-Munguia, I.; Gonzalez-Reyes, J.A.; Priego-Capote, F.; Voravuthikunchai, S.P.; Rodriguez-Ortega, M.J. Integrated proteomic and metabolomic analysis reveals that rhodomyrtone reduces the capsule in *Streptococcus pneumoniae*. *Sci. Rep.* **2017**, *7*, 1. [CrossRef]
90. Zhao, L.Y.; Liu, H.X.; Wang, L.; Xu, Z.F.; Tan, H.B.; Qiu, S.X. Rhodomyrtosone B, a membrane-targeting anti-MRSA natural acylgphloroglucinol from *Rhodomyrtus tomentosa*. *J. Ethnopharmacol.* **2019**, *228*, 50–57. [CrossRef]
91. Sianglum, W.; Srimanote, P.; Taylor, P.W.; Rosado, H.; Voravuthikunchai, S.P. Transcriptome analysis of responses to rhodomyrtone in methicillin-resistant *Staphylococcus aureus*. *PLoS ONE* **2012**, *7*, e45744. [CrossRef]
92. Leejae, S.; Hasap, L.; Voravuthikunchai, S.P. Inhibition of staphyloxanthin biosynthesis in *Staphylococcus aureus* by rhodomyrtone, a novel antibiotic candidate. *J. Med. Microbiol.* **2013**, *62*, 421–428. [CrossRef]
93. Saeloh, D.; Wenzel, M.; Rungrotmongkol, T.; Hamoen, L.W.; Tipmanee, V.; Voravuthikunchai, S.P. Effects of rhodomyrtone on Gram-positive bacterial tubulin homologue FtsZ. *PeerJ* **2017**, e2962. [CrossRef]
94. Tan, H.; Liu, H.; Zhao, L.; Yuan, Y.; Li, B.; Jiang, Y.; Gong, L.; Qiu, S. Structure-activity relationships and optimization of acyclic acylphloroglucinol analogues as novel antimicrobial agents. *Eur. J. Med. Chem.* **2017**, *125*, 492–499. [CrossRef] [PubMed]
95. Zhao, L.; Liu, H.; Huo, L.; Wang, M.; Yang, B.; Zhang, W.; Xu, Z.; Tan, H.; Qiu, S.X. Structural optimization and antibacterial evaluation of rhodomyrtosone B analogues against MRSA strains. *Med. Chem. Comm.* **2018**, *9*, 1698–1707. [CrossRef] [PubMed]
96. Verotta, L. Are acylphloroglucinols lead structures for the treatment of degenerative diseases? *Phytochem. Rev.* **2003**, *1*, 389–407. [CrossRef]
97. Senadeera, S.P.; Lucantoni, L.; Duffy, S.; Avery, V.M.; Carroll, A.R. Antiplasmodial β-triketone-flavanone hybrids from the flowers of the Australian tree *Corymbia torelliana*. *J. Nat. Prod.* **2018**, *81*, 1588–1597. [CrossRef]
98. Rosa, A.; Deiana, M.; Casu, V.; Corona, G.; Appendino, G.; Bianchi, F.; Ballero, M.; Dessì, M.A. Antioxidant activity of oligomeric acylphloroglucinols from *Myrtus communis* L. *Free Radic. Res.* **2003**, *37*, 1013–1019.
99. Rosa, A.; Melis, M.P.; Deiana, M.; Atzeri, A.; Appendino, G.; Corona, G.; Incani, A.; Loru, D.; Dessì, M.A. Protective effect of the oligomeric acylphloroglucinols from *Myrtus communis* on cholesterol and human low density lipoprotein oxidation. *Chem. Phys. Lipids* **2008**, *155*, 16–23. [CrossRef] [PubMed]
100. Feißt, C.; Franke, L.; Appendino, G.; Werz, O. Identification of molecular targets of the oligomeric nonprenylated acylphloroglucinols from *Myrtus communis* and their implication as anti-inflammatory compounds. *J. Pharmacol. Exp. Ther.* **2005**, *315*, 389–396. [CrossRef] [PubMed]
101. Koeberle, A.; Pollastro, F.; Northoff, H.; Werz, O. Myrtucommulone, a natural acylphloroglucinol, inhibits microsomal prostaglandin E_2 synthase-1. *Br. J. Pharmacol.* **2009**, *156*, 952–961. [CrossRef]
102. Rossi, A.; Di Paola, R.; Mazzon, E.; Genovese, T.; Caminiti, R.; Bramanti, P.; Pergola, C.; Koeberle, A.; Werz, O.; Sautebin, L.; et al. Myrtucommulone from *Myrtus communis* exhibits potent anti-inflammatory effectiveness in vivo. *J. Pharmacol. Exp. Ther.* **2009**, *329*, 76–86. [CrossRef]
103. Chorachoo, J.; Lambert, S.; Furnholm, T.; Roberts, L.; Reingold, L.; Auepemkiate, S.; Voravuthikunchai, S.P.; Johnston, A. The small molecule rhodomyrtone suppresses TNF-α and IL-17A-induced keratinocyte inflammatory responses: A potential new therapeutic for psoriasis. *PLoS ONE* **2018**, *13*, e0205340. [CrossRef] [PubMed]
104. Chorachoo, J.; Saeloh, D.; Srichana, T.; Amnuaikit, T.; Musthafa, K.S.; Sretrirutchai, S.; Voravuthikunchai, S.P. Rhodomyrtone as a potential anti-proliferative and apoptosis inducing agent in HaCaT keratinocyte cells. *Eur. J. Pharmacol.* **2016**, *772*, 144–151. [CrossRef] [PubMed]
105. Srisuwan, S.; Mackin, K.E.; Hocking, D.; Lyras, D.; Bennett-Wood, V.; Voravuthikunchai, S.P.; Robins-Browne, R.M. Antibacterial activity of rhodomyrtone on *Clostridium difficile* vegetative cells and spores in vitro. *Int. J. Antimicrob. Agents* **2018**, *52*, 724–729. [CrossRef]

106. Na-Phatthalung, P.; Teles, M.; Voravuthikunchai, S.P.; Tort, L.; Fierro-Castro, C. Immunomodulatory effects of *Rhodomyrtus tomentosa* leaf extract and its derivative compound, rhodomyrtone, on head kidney macrophages of rainbow trout (*Oncorhynchus mykiss*). *Fish Physiol. Biochem.* **2018**, *44*, 543–555. [CrossRef] [PubMed]
107. Tretiakova, I.; Blaesius, D.; Maxia, L.; Wesselborg, S.; Schulze-Osthoff, K.; Cinatl, J.; Michaelis, M.; Werz, O. Myrtucommulone from *Myrtus communis* induces apoptosis in cancer cells via the mitochondrial pathway involving caspase-9. *Apoptosis* **2008**, *13*, 119–131. [CrossRef] [PubMed]
108. Wiechmann, K.; Müller, H.; Fischer, D.; Jauch, J.; Werz, O. The acylphloroglucinols hyperforin and myrtucommulone A cause mitochondrial dysfunctions in leukemic cells by direct interference with mitochondria. *Apoptosis* **2015**, *20*, 1508–1517. [CrossRef] [PubMed]
109. Wiechmann, K.; Müller, H.; König, S.; Wielsch, N.; Svatoš, A.; Jauch, J.; Werz, O. Mitochondrial chaperonin HSP60 is the apoptosis-related target for myrtucommulone. *Cell Chem. Biol.* **2017**, *24*, 614–623. [CrossRef] [PubMed]
110. Grandjenette, C.; Schnekenburger, M.; Morceau, F.; Mack, F.; Wiechmann, K.; Werz, O.; Dicato, M.; Diederich, M. Dual induction of mitochondrial apoptosis and senescence in chronic myelogenous leukemia by myrtucommulone A. *Anti-Cancer Agents Med. Chem.* **2015**, *15*, 363–373. [CrossRef]
111. Izgi, K.; Iskender, B.; Jauch, J.; Sezen, S.; Cakir, M.; Charpentier, M.; Canatan, H.; Sakalar, C. Myrtucommulone-A induces both extrinsic and intrinsic apoptotic pathways in cancer cells. *J. Biochem. Mol. Toxicol.* **2015**, *29*, 432–439. [CrossRef]
112. Izgi, K.; Iskender, B.; Sakalar, C.; Arslanhan, A.; Yüksek, E.H.; Hizar, E.; Canatan, H. Effects of epirubicin and cisplatin against 4T1 breast cancer cells are enhanced by myrtucommulone-A. *Anti-Cancer Agents Med. Chem.* **2017**, *17*, 404–414. [CrossRef]
113. Izgi, K.; Sonmez, M.F.; Canatan, H.; Iskender, B. Long term exposure to myrtucommulone-A changes CD105 expression and differentiation potential of mesenchymal stem cells. *Tissue Engin. Regenener. Med.* **2017**, *14*, 113–121. [CrossRef]
114. Iskender, B.; Izgi, K.; Sakalar, C.; Canatan, H. Priming hMSCs with a putative anti-cancer compound, myrtucommulone-a: A way to harness hMSC cytokine expression via modulating PI3K/Akt pathway? *Tumor Biol.* **2016**, *37*, 1967–1981. [CrossRef] [PubMed]
115. Iskender, B.; Izgi, K.; Canatan, H. Novel anti-cancer agent myrtucommulone-A and thymoquinone abrogate epithelial–mesenchymal transition in cancer cells mainly through the inhibition of PI3K/AKT signalling axis. *Mol. Cell. Biochem.* **2016**, *416*, 71–84. [CrossRef]
116. Iskender, B.; Izgi, K.; Karaca, H.; Canatan, H. Myrtucommulone-A treatment decreases pluripotency-and multipotency-associated marker expression in bladder cancer cell line HTB-9. *J. Nat. Med.* **2015**, *69*, 543–554. [CrossRef]
117. Gerbeth, K.; Meins, J.; Werz, O.; Schubert-Zsilavecz, M.; Abdel-Tawab, M. Determination of myrtucommulone from *Myrtus communis* in human and rat plasma by liquid chromatography/tandem mass spectrometry. *Planta Med.* **2011**, *77*, 450–454. [CrossRef]
118. Saising, J.; Nguyen, M.-T.; Haertner, T.; Ebner, P.; Al Mamun Bhuyan, A.; Berscheid, A.; Muehlenkamp, M.; Schaekermann, S.; Kumari, N.; Maier, M.E.; et al. Rhodomyrtone (Rom) is a membrane-active compound. *Biochim. Biophys. Acta Biomembr.* **2018**, *1860*, 1114–1124. [CrossRef] [PubMed]
119. Saeloh, D.; Tipmanee, V.; Jim, K.K.; Dekker, M.P.; Bitter, W.; Voravuthikunchai, S.P.; Wenzel, M.; Hamoen, L.W. The novel antibiotic rhodomyrtone traps membrane proteins in vesicles with increased fluidity. *PLoS Pathog.* **2018**, *14*, e1006876. [CrossRef]
120. Tayeh, M.; Nilwarangoon, S.; Mahabusarakum, W.; Watanapokasin, R. Anti-metastatic effect of rhodomyrtone from *Rhodomyrtus tomentosa* on human skin cancer cells. *Int. J. Oncol.* **2017**, *50*, 1035–1043. [CrossRef]
121. Romani, A.; Coinu, R.; Carta, S.; Pinelli, P.; Galardi, C.; Vincieri, F.F.; Franconi, F. Evaluation of antioxidant effect of different extracts of *Myrtus communis* L. *Free Radic. Res.* **2004**, *38*, 97–103. [CrossRef]
122. Geetha, K.M.; Sridhar, C.; Murugan, V. Antioxidant and healing effect of aqueous alcoholic extract of *Rhodomyrtus tomentosa* (Ait.) Hassk on chronic gastric ulcers in rats. *J. Pharm. Res.* **2010**, *3*, 2860–2862.
123. Jeong, D.; Yang, W.S.; Yang, Y.; Nam, G.; Kim, J.H.; Yoon, D.H.; Noh, H.J.; Lee, S.; Kim, T.W.; Sung, G.H.; et al. In vitro and in vivo anti-inflammatory effect of *Rhodomyrtus tomentosa* methanol extract. *J. Ethnopharmacol.* **2013**, *146*, 205–213. [CrossRef] [PubMed]

124. Masoudi, M.; Kopaei, M.R.; Miraj, S. A comparison of the efficacy of metronidazole vaginal gel and *Myrtus* (*Myrtus communis*) extract combination and metronidazole vaginal gel alone in the treatment of recurrent bacterial vaginosis. *Avicenna J. Phytomed.* **2017**, *7*, 129–136. [PubMed]
125. Fiorini-Puybaret, C.; Aries, M.F.; Fabre, B.; Mamatas, S.; Luc, J.; Degouy, A.; Ambonati, M.; Mejean, C.; Poli, F. Pharmacological properties of Myrtacine® and its potential value in acne treatment. *Planta Med.* **2011**, *77*, 1582–1589. [CrossRef] [PubMed]
126. Feuillolay, C.; Pecastaings, S.; Le Gac, C.; Fiorini-Puybaret, C.; Luc, J.; Joulia, P.; Roques, C. A *Myrtus communis* extract enriched in myrtucummulones and ursolic acid reduces resistance of *Propionibacterium acnes* biofilms to antibiotics used in acne vulgaris. *Phytomedicine* **2016**, *23*, 307–315. [CrossRef] [PubMed]
127. Chorachoo, J.; Amnuaikit, T.; Voravuthikunchai, S.P. Liposomal encapsulated rhodomyrtone: A novel antiacne drug. *Evid.-Based Complement. Altern. Med.* **2013**, 157635. [CrossRef] [PubMed]
128. Leejae, S.; Yingyongnarongkul, B.E.; Suksamrarn, A.; Voravuthikunchai, S.P. Synthesis and structure–activity relationship of rhodomyrtone derivatives as antibacterial agent. *Chin. Chem. Lett.* **2012**, *23*, 1011–1014. [CrossRef]
129. Wiechmann, K.; Müller, H.; Huch, V.; Hartmann, D.; Werz, O.; Jauch, J. Synthesis and biological evaluation of novel myrtucommulones and structural analogues that target mPGES-1 and 5-lipoxygenase. *Eur. J. Med. Chem.* **2015**, *101*, 133–149. [CrossRef] [PubMed]
130. Pereira, P.; Cebola, M.J.; Oliveira, M.C.; Bernardo-Gil, M.G. Supercritical fluid extraction vs conventional extraction of myrtle leaves and berries: Comparison of antioxidant activity and identification of bioactive compounds. *J. Supercrit. Fluids* **2016**, *113*, 1–9. [CrossRef]
131. Díaz-de-Cerio, E.; Arráez-Román, D.; Segura-Carretero, A.; Ferranti, P.; Nicoletti, R.; Perrotta, G.M.; Gómez-Caravaca, A.M. Establishment of pressurized-liquid extraction by response surface methodology approach coupled to HPLC-DAD-TOF-MS for the determination of phenolic compounds of myrtle leaves. *Anal. Bioanal. Chem.* **2018**, *410*, 3547–3557. [CrossRef]
132. González de Peredo, A.V.; Vázquez-Espinosa, M.; Espada-Bellido, E.; Jiménez-Cantizano, A.; Ferreiro-González, M.; Amores-Arrocha, A.; Palma, M.; Barroso, C.G.; Barbero, F.G. Development of new analytical microwave-assisted extraction methods for bioactive compounds from myrtle (*Myrtus communis* L.). *Molecules* **2018**, *23*, 2992. [CrossRef]

molecules

MDPI

Article

Chlamyphilone, a Novel *Pochonia chlamydosporia* Metabolite with Insecticidal Activity

Federica Lacatena [1,†], Roberta Marra [1,†], Pierluigi Mazzei [2,3], Alessandro Piccolo [1,3], Maria Cristina Digilio [1], Massimo Giorgini [4], Sheridan L. Woo [4,5], Pierpaolo Cavallo [6,7], Matteo Lorito [1,4] and Francesco Vinale [1,4,*]

1 Dipartimento di Agraria, Università degli Studi di Napoli Federico II, 80055 Portici (NA), Italy; federica.lacatena@unina.it (F.L.); robmarra@unina.it (R.M.); alessandro.piccolo@unina.it (A.P.); digilio@unina.it (M.C.D.); lorito@unina.it (M.L.)
2 Dipartimento di Farmacia (DIFARMA), Università degli Studi di Salerno, 84084 Fisciano (SA), Italy; pmazzei@unisa.it
3 Centro Interdipartimentale di Ricerca sulla Spettroscopia di Risonanza Magnetica Nucleare, per l'Ambiente, l'Agro-Alimentare ed i Nuovi Materiali (CERMANU), Università degli Studi di Napoli Federico II, 80055 Portici (NA), Italy
4 Istituto per la Protezione Sostenibile delle Piante, Consiglio Nazionale delle Ricerche (IPSP-CNR), 80055 Portici (NA), Italy; massimo.giorgini@ipsp.cnr.it (M.G.); woo@unina.it (S.L.W.)
5 Dipartimento di Farmacia, Università degli Studi di Napoli Federico II, 80131 Napoli, Italy
6 Dipartimento di Fisica "E.R. Caianiello", Università degli Studi di Salerno, 84084 Fisciano (SA), Italy; pcavallo@unisa.it
7 Istituto Sistemi Complessi, Consiglio Nazionale delle Ricerche (ISC-CNR), 00185 Rome, Italy
* Correspondence: francesco.vinale@ipsp.cnr.it; Tel.: +39-081-253-9338
† These authors contributed equally to this work.

Academic Editor: Josphat Matasyoh
Received: 14 January 2019; Accepted: 14 February 2019; Published: 19 February 2019

Abstract: Metabolites from a collection of selected fungal isolates have been screened for insecticidal activity against the aphid *Acyrthosiphon pisum*. Crude organic extracts of culture filtrates from six fungal isolates (*Paecilomyces lilacinus*, *Pochonia chlamydosporia*, *Penicillium griseofulvum*, *Beauveria bassiana*, *Metarhizium anisopliae* and *Talaromyces pinophilus*) caused mortality of aphids within 72 h after treatment. In this work, bioassay-guided fractionation has been used to characterize the main bioactive metabolites accumulated in fungal extracts. Leucinostatins A, B and D represent the bioactive compounds produced by *P. lilacinus*. From *P. griseofulvum* and *B. bassiana* extracts, griseofulvin and beauvericin have been isolated, respectively; 3-*O*-Methylfunicone and a mixture of destruxins have been found in the active fractions of *T. pinophilum* and *M. anisopliae*, respectively. A novel azaphilone compound, we named chlamyphilone, with significant insecticidal activity, has been isolated from the culture filtrate of *P. chlamydosporia*. Its structure has been determined using extensive spectroscopic methods and chemical derivatization.

Keywords: secondary metabolites; beneficial microbes; pea aphid; azaphilones

1. Introduction

Alarm over the impact of pesticides on the environment and human health is increasing year after year, and rigorous pesticide registration procedures have been introduced. Through Directive 2009/128/EC, the European Community has severely restricted the use of synthetic pesticides in plant protection, and the new regulations have reduced the number of chemicals available in agriculture [1]. Novel pesticides, including natural product-based formulations, have been developed to counteract the evolution of resistance among plant pathogen and pest populations [2].

Microorganisms biosynthesize thousands of compounds displaying different biological activities, e.g., acting as antibiotics, therapeutic agents, toxins, hormones, etc. [3]. Thus, the exploitation of the microbial metabolome has become an important area of research to isolate novel natural products potentially useful for agricultural applications [3]. The improvement of screening technologies can be used to extend the search of new microbial active metabolites, i.e., through (i) upgrading of fermentation techniques, (ii) the development of new methods for detection, and (iii) genetic manipulation to create mutants able to produce qualitatively or quantitatively different metabolites [4].

Microorganisms have been extensively screened for antibiotic production; the main producers are fungi and bacteria, particularly the actinomycetes *Streptomyces* spp. [5]. A large number of manuscripts deal with microbial metabolites, which have demonstrated efficacy as crop protection agents, but only a few compounds have been marketed so far [2]. Examples of commercially available microbial metabolites with insecticide activity are: (1) abamectin and anthelmintic produced by the soil-dwelling actinomycete *S. avermitilis*; (2) milbemycin (also known as milbemectin), an insecticide and acaricide from *S. hygroscopicus* subsp. *aureolacrimosus*; (3) polynactins, secondary metabolites from the actinomycete *S. aureus*, isolated and applied as a mixture of tetranactin, trinactin and dynactin; (4) spinosad, a secondary metabolite from the soil actinomycete *Saccharopolyspora spinosa* [2].

The interaction between some fungal strains and the plant establish a molecular cross-talk in which fungal metabolites can act as elicitors that activate the expression of genes involved in plant defence response, and promote the growth of the plant [6]. Identification of new bioactive compounds may be obtained with a good microbial collection or isolating metabolites not expressed under standard laboratory conditions [6–8].

In this work, we show the results of a screening aimed to isolate microbial metabolites with insecticidal activity. Various genera of fungi, including some entomopathogenic species, were grown in liquid culture and their major secondary metabolites were investigated. Bioassay-guided fractionation used to isolate the bioactive metabolites allowed the identification of seven known compounds and a new metabolite named chlamyphilone. This compound was fully characterized by spectrometric analysis and chemical derivatization. The fungal metabolites have been screened for insecticidal activity against *Acyrthosiphon pisum* (Hemiptera: Aphididae), the pea aphid, whose high rate of increase makes it a useful model for screening [9].

2. Results

The microbes used in the present study includes various genera of fungi identified according to morphological features and molecular analyses (rDNA-ITS and β-tubulin gene sequencing). In particular, 5 isolates belong to the *Trichoderma* genus (*T. tomentosum* F19, *T. asperellum* CINO1, *T. harzianum* M10, *T. harzianum* F53, *T. velutinum* F28) and 4 to *Penicillium* (*P. chrysogenum* F5, *P. decumbens* F29, *P. griseofulvum* F11, *P. restrictum* F55). *Beauveria bassiana* BB1, *Chloridium virescens* F57, *Metarhizium anisopliae* MA3, *Paecilomyces lilacinus* 100379, *Pochonia chlamydosporia* B and *Talaromyces pinophilus* F36CF are present as single species.

All fungi have been grown in static conditions and the culture filtrates were extracted exhaustively with ethyl acetate. The crude extracts were fractionated by column chromatography to provide a total of 79 organic fractions. The extracts and all the fractions were tested at different concentrations on *Acyrthosiphon pisum*. Table 1 reports the percentage of mortality at 72h of the active extracts/fractions obtained from different fungal isolates. Only extracts obtained from *P. lilacinus*, *P. griseofulvum*, *P. chlamydosporia*, *B. bassiana*, *M. anisopliae* and *T. pinophilus* showed significant insecticidal activity (data obtained from the other fungal strains are not showed). Moreover, the bioassay-guided fractionation indicated that only one fraction per strain was responsible of a significant aphid mortality after 72 h exposure (Table 1).

Chemical characterization of the active fractions was obtained using NMR and/or LC-MS qTOF analyses. Fraction No. 3 of *P. lilacinus* (showing 30% aphid mortality; Table 1) was constituted by a mixture of leucinostatins A, B and D (**1**, **2** and **3**, respectively; Figure 1), with the molecular weights (MW) 1218.634 g/mol, 1204.607 g/mol and 1104.4799 g/mol, respectively (Figure S1).

Table 1. Insecticidal activity of fungal organic extracts and active fractions as mortality (%) of the pea aphid *Acyrthosiphon pisum* at 72 h after exposure. Different lower letters refer to significant differences ($p < 0.05$) among treatments at the same incubation time. Extracts from *T. tomentosum* F19, *T. asperellum* CINO1, *T. harzianum* M10, *T. harzianum* F53, *T. velutinum* F28, *P. chrysogenum* F5, *P. decumbens* F29, *P. restrictum* F55 and *C. virescens* F57 did not show significant insecticidal activity.

Microbes	Aphid Mortality (%) at 72 h after Exposure to Organic Extract (10 mg/mL)	Standard Deviation	Active Fraction (Number)	Aphid Mortality (%) at 72 h after Exposure to Active Fraction (0.5 mg/mL)	Standard Deviation
Paecilomyces lilacinus	6.0 [a]	±3.6	3	30.0 [a]	±18.0
Penicillium griseofulvum	63.3 [b]	±21.0	7	73.3 [b]	±8.0
Pochonia chlamydosporia	50.0 [b]	±13.0	2	100.0 [c]	±0.0
Beuveria bassiana	40.0 [b]	±17.0	3	40.0 [a]	±22.9
Metarhizium anisopliae	53.3 [b]	±5.7	5	60.0 [a]	±10.0
Talaromyces pinophilus	65.0 [b]	±15.0	2	48.3 [a]	±5.8

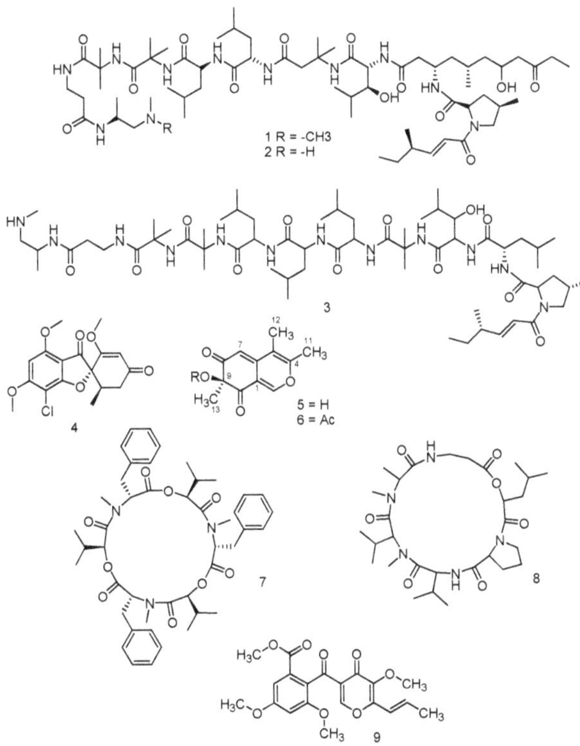

Figure 1. Chemical structures of leucinostatin A, **1**; leucinostatin B, **2**; leucinostatin D, **3**; griseofulvin, **4**; chlamyphilone, **5**; Ac-chlamyphilone, **6**; beauvericin, **7**; dextrusin B2, **8**; 3-*O*-Methylfunicone, **9**.

The mycotoxin griseofulvin (**4** in Figure 1) was determined as the main metabolite of *P. griseofulvum* and was isolated in the active fraction number 7 (showing 73,3% aphid mortality; Table 1). Figure S3 reports the mass spectrum of the isolated compound.

The residue recovered after the organic extraction (350 mg) of *P. chlamydosporia* culture filtrate was subjected to flash column chromatography, eluting with CH_2Cl_2/MeOH (90:10 *v*/*v*). Fractions showing similar thin-layer chromatography (TLC) profiles were combined and further purified by using preparative TLC separation (Si gel; CH_2Cl_2/MeOH 90:10 *v*/*v*). Both fractions and pure compounds were tested for insecticidal activity. Twelve milligrams of a novel compound, named chlamyphilone (**5** in Figure 1; 5 mg/L), were obtained as a yellow amorphous solid in pure form after TLC (Rf in CH_2Cl_2/MeOH 90:10 *v*/*v* = 0.8) and has: $[\alpha]^{25}{}_{D}$ −21.9° (c 1; CH_2Cl_2); UV (CH_2Cl_2) λ max (log ε) 219 (3.57), 332 (4.23). ^1H and ^{13}C NMR spectral data of chlamyphilone (in $CDCl_3$) are presented in Table 2. LC-MS qTOF analysis detected the precursor ions at *m*/*z* 221.0814 (pseudomolecular ion $[M + H]^+$); 463.1360 $[M_2 + Na]^+$; 243.0916 $[M + Na]^+$; 221.0814 $[M + H]^+$ (calcd. 221.0814); 203.1313 $[M + H - H_2O]^+$ (Figure S4). The results obtained by ^{13}C NMR (Table 2 and Figure S6) and LC-MS qTOF analyses (Figure S4) were consistent with a compound having a molecular weight of 220.0814, corresponding to the molecular formula $C_{12}H_{12}O_4$ with seven unsaturations.

Table 2. ^1H and ^{13}C NMR spectral data of chlamyphilone (in $CDCl_3$).

Position	δ_C Mult.	δ_H (*J* in Hz)	HMBC *
1	115.3 qC		
2	151.7 CH	7.91 d (1.37)	1, 10, 11
4	155.1 qC		
5	114.4 qC		
6	145.2 qC		
7	102.9 CH	5.55 d (1.37)	6, 9
8	196.3 qC		
9	83.1 qC		
10	196.5 qC		
11	17.5 CH_3	2.26 d (0.6)	2
12	12.7 CH_3	1.93 d (0.6)	4, 6
13	28.6 CH_3	1.55 s	9, 10, 8
OH		3.9	8

* All correlations represent 2 or 3 bond couplings. Abbreviation, s: singlet, d: doublet, t: triplet, dd: doublet of doublets.

The UV spectrum showed absorption peaks characteristic of the azaphilone family, which are natural products containing a 6*H*-isochromene-6,8(7*H*)-dione. Six carbon signals at 102.9, 114.4, 115.3, 145.2, 151.7, and 155.1 ppm revealed three double bonds, whereas signals at 196.3 and 196.5 ppm indicated the presence of two ketones. The other four signals in the ^{13}C spectrum were all shifted upfield in the 12.7–83.11 ppm range. The DEPT data demonstrated that one of the protons in the molecule is bound to oxygen, and that in the molecule there are three CH_3, two CH, and seven fully substituted C atoms. The ^1H-^1H COSY, HSQC analysis, and the chemical shift evaluation (Table 2) allowed the identification of the structural fragment. In Table 2 all the compound signals are reported.

The connectivity of the spin systems was deduced by a long-range ^1H-^{13}C heterocorrelated experiment that was obtained with the HMBC (Table 2). In particular, the most relevant HMBC correlations are reported in Figure 2, which implied that the structure of the metabolite **5** is 7-hydroxy-3,4,7-trimethyl-isochromene-6,8-dione. The configuration of chlamyphilone was evident from NOESY experiment. The MM-2 energy calculation (16.0 Kcal/mol) was run to find the most stable conformational model (Figure S12).

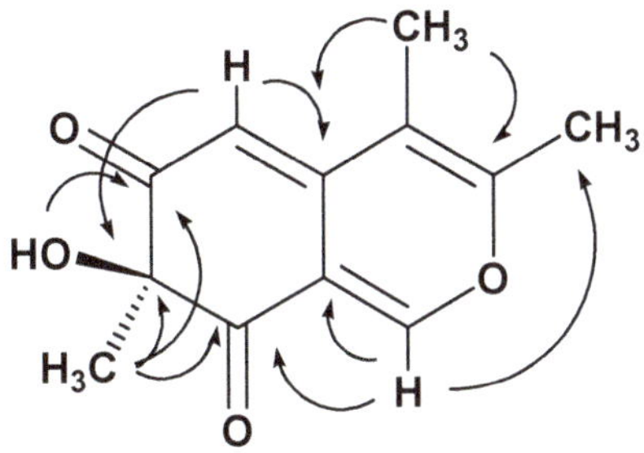

Figure 2. HMBC correlation of chlamyphilone.

To confirm the proposed structure, a sample of chlamyphilone was acetylated using acetic anhydride/pyridine (**6** in Figure 1). The ^{1}H-NMR spectrum of the product (acetic acid 3,4,7-trimethyl-6,8-dioxo-7,8-dihydro-6H-isochromen-7-yl ester) contained one acetate resonance, confirming the presence of the hydroxyl group bound to the carbon at 83.1. The ^{13}C-NMR spectrum contained an additional signal for the acetyl carbonyl resonance at 170.8 and for the acetate methyl carbon at 18.4 ppm.

Chlamyphilone (**5** in Figure 1) showed the highest insecticidal activity against *Acyrthosiphon pisum* with a median Lethal Dose (LD50) of 175 μg/mL and a minimal inhibitory concentration (MIC) of 150 μg/mL (Table 3).

Table 3. Insecticidal activity of chlamyphilone **5** at different concentrations. Treatments were evaluated as mortality (%) of the pea aphid *Acyrthosiphon pisum* at 72 h after exposure to **5**.

Concentrations (μg/mL)	Aphid Mortality (%)	Standard Deviation
500	100.0	±0.0
200	100.0	±0.0
190	96.7	±5.8
180	51.7	±2.9
170	30.0	±5.0
160	8.3	±2.9
150	3.3	±2.8

The insecticidal activity of the *B. bassiana* strain used in the present work was mainly dependent on the presence of beauvericin (**7** in Figure 1). This metabolite is a cyclohexadepsipeptide mycotoxin with the molecular formula $C_{45}H_{57}N_3O_9$ (Figure S13).

From *M. anisopliae* extract, destruxin B2 has been characterized (**8** in Figure 1). Destruxins are cyclic hexadepsipeptides composed of an α-hydroxy acid and five amino acid residues. Individual destruxins differ in terms of the hydroxy acid, *N*-methylation, and R group of the amino acid residues. Destruxin B2 has the molecular formula $C_{29}H_{49}N_5O_7$, as revealed by the MS spectrum (Figure S14).

Finally, the major active metabolite produced by *T. pinophilus* strain F36CF was 3-*O*-methylfunicone (OMF—**9** in Figure 1), a well-known γ-pyrone derivative, previously isolated from *Talaromyces* spp.

3. Discussion

In this work, a functional screening of fungal strains, preliminarily selected because of their ability to synthetize active metabolites, has been carried out. Bioassay-guided fractionation has been used to isolate microbial metabolites with insecticidal activity against the pea aphid *A. pisum* (Hemiptera: Aphididae), long used as a model to study plant-insect interactions [9]. This aphid is an important pest of several plants, is a phloem-feeding insect determining direct negative effects on plants in terms of nutritive subtraction and injection of toxic saliva. Moreover, it has been involved in the transmission of diverse plant viruses [9].

The active fraction obtained from *P. lilacinus* was constituted by a mixture of leucinostatins A, B, and D. Leucinostatins are peptides containing an unsaturated fatty acid, nine amino acid residues, and a basic component joined together by amide linkages [10]. These antibiotic metabolites also showed antitumor and nematocidal activities and an uncoupling effect on rat liver mitochondrial function [10,11]. To the best of our knowledge, our work is the first report on the insecticidal activity of leucinostatins against *A. pisum*.

Griseofulvin was determined to be the main metabolite of *P. griseofulvum*. The insecticidal activity of this metabolite has been previously demonstrated in the corn earworm, *Helicoverpa zea* Boddie, and the fall armyworm, *Spodoptera frugiperda* (J. E. Smith), by oral administration in an artificial diet (250 ppm) [12–14]. In the case of soft-tegument aphids like *A. pisum*, the activity is probably topical as oral administration may be excluded because of their peculiar feeding pattern. In fact, aphids use their piercing, sucking mouthparts to feed on plant fluid [9].

The extract of *P. chlamydosporia* culture filtrate yielded 12 mg of a novel compound named chlamyphilone. This metabolite belongs to the class of azaphilones, which contain a 6*H*-isochromene-6,8(7*H*)-dione or an isoquinoline-6,8(2*H*,7*H*)-dione skeleton (and its substituted derivatives thereof). This is the first report on the isolation of an azaphilone metabolite from *P. chlamydosporia*. A similar azaphilone, named myxostiol, with plant growth regulating activity, has been previously isolated from *Myxotrichum stipitatum* [15]. Over 170 different azaphilone compounds, classified into 10 structural groups, have been isolated from fungi belonging to 23 genera [16]. Azaphilones display a wide range of biological activities, such as antimicrobial, antifungal, antiviral, antioxidant, cytotoxic, nematocidal, and anti-inflammatory [16]. Many of these properties may be explained by the reactions of azaphilones with amino groups, such as those found in amino acids, proteins, and nucleic acids, resulting in the formation of vinylogous c-pyridones [16–18]. To the best of our knowledge, this is the first evidence of the insecticidal activity of a metabolite belonging to the class of azaphilones. Recently, the involvement of small molecules (e.g., aurovertins) in the interaction between nematodes and *P. chlamydosporia* has been demonstrated [19].

An active beauvericin has been extracted from culture filtrate of *B. bassiana*. This metabolite is a mycotoxin whose insecticidal properties have been previously reported (e.g., against the wheat aphid *Schizaphis graminum* at 0.5 mg/mL) [20,21]. This molecule contains three residues, each with D-2-hydroxyisovaleric acid (Hiv) and L-*N*-methylphenylalanine linked alternately [21].

Destruxin B2 is the active molecule isolated from *M. anisopliae* extract. Destruxins exhibited a wide variety of biological activities, but are best known for their insecticidal and phytotoxic properties [22,23]. However, the effect of destruxin B2 against *Acyrthosiphon pisum* is reported here for the first time.

T. pinophilus strain F36CF produced 3-*O*-methylfunicone. This compound was downregulated in the presence of *T. harzianum* M10 [8], exhibited notable antibiotic and antitumor properties, and recently exhibited an insecticidal effect, thus expanding the biological activities of this compound [24]. Recently, talarodiolide, a new 12-membered macrodiolide, was isolated and characterized from the culture filtrate of a *T. pinophilus* strain [25]. This metabolite did not show insecticidal activity, highlighting that the strain collection is important to select the best producers of bioactive compounds.

4. Materials and Methods

Fungal strains. The microbes used in the present study were present in the fungal collection at Department of Agricultural Sciences, University of Naples Federico II (UNINA Collection) or isolated from different sources (Table 4).

Table 4. Fungal strains used in the present study for the isolation of metabolites with insecticidal activity against the pea aphid *A. pisum*.

Microbes	Strain	Division	Class	Order	Family	Habitat/Source *
Beauveria bassiana	BB1	*Ascomycota*	*Sordariomycetes*	*Hypocreales*	*Cordycipitaceae*	UNINA Collection
Chloridium virescens	F57	*Ascomycota*	*Sordariomycetes*	*Chaetosphaeriales*	*Chaetosphaeriaceae*	Pasture
Metarhizium anisopliae	MA3	*Ascomycota*	*Sordariomycetes*	*Hypocreales*	*Clavicipitaceae*	UNINA Collection
Paecilomyces lilacinus	100379	*Ascomycota*	*Eurotiomycetes*	*Eurotiales*	*Trichocomaceae*	UNINA Collection
Penicillium chrysogenum	F5	*Ascomycota*	*Eurotiomycetes*	*Eurotiales*	*Trichocomaceae*	Mediterranean area
Penicillium decumbens	F29	*Ascomycota*	*Eurotiomycetes*	*Eurotiales*	*Trichocomaceae*	Mediterranean area
Penicillium griseofulvum	F11	*Ascomycota*	*Eurotiomycetes*	*Eurotiales*	*Trichocomaceae*	Mediterranean area
Penicillium restrictum	F55	*Ascomycota*	*Eurotiomycetes*	*Eurotiales*	*Trichocomaceae*	Mediterranean area
Pochonia chlamydosporia	B	*Ascomycota*	*Sordariomycetes*	*Hypocreales*	*Clavicipitaceae*	UNINA Collection
Talaromyces pinophilus	F36CF	*Ascomycota*	*Eurotiomycetes*	*Eurotiales*	*Trichocomaceae*	*Arbutus unedo*
Trichoderma tomentosum	F19	*Ascomycota*	*Sordariomycetes*	*Hypocreales*	*Hypocreaceae*	Woodland (Oak)
Trichoderma asperellum	CINO1	*Ascomycota*	*Sordariomycetes*	*Hypocreales*	*Hypocreaceae*	UNINA Collection
Trichoderma harzianum	M10	*Ascomycota*	*Sordariomycetes*	*Hypocreales*	*Hypocreaceae*	UNINA Collection
Trichoderma harzianum	F53	*Ascomycota*	*Sordariomycetes*	*Hypocreales*	*Hypocreaceae*	Mediterranean area
Trichoderma velutinum	F28	*Ascomycota*	*Sordariomycetes*	*Hypocreales*	*Hypocreaceae*	Woodland (Oak)

* UNINA Collection: fungal collection available at the Department of Agricultural Sciences/University of Naples Federico II, Naples, Italy.

The fungi were identified according to morphological features and molecular analyses. Briefly, an amount of 10^6 spores/mL was inoculated in a 250-mL Erlenmeyer Flasks containing 100 mL of sterile Potato Dextrose Broth (PDB, HIMEDIA, Mumbai, India). Each flask was incubated at 25 °C in an orbital shaker (150 r.p.m.) and fungi were left to grow for seven days. Mycelium was recovered, ground to a powder in liquid nitrogen, and used to perform the DNA extraction. Genetic analysis was carried out through PCR and sequencing of the rDNA Internal Transcribed Spacer (ITS) and β-tubulin gene, using the most common primers to identify fungal strains: ITS1 and ITS4 [26], as well as tub2 and BenA [27]. The PCR products were subjected to gel electrophoresis, excised, and sequenced [26,27]. Analysis of the ITS gave 99% of identity with GenBank sequences of the fungi reported in Table 4, confirming the identity of the microbes.

Secondary metabolites production. The fungi were maintained on potato dextrose agar (PDA, HIMEDIA) at room temperature and sub-cultured bimonthly. Liquid cultures were prepared in 5000 L-Erlenmeyer flasks containing 1 L of PDB and inoculated with five mycelial plugs (7 mm^2) from a fresh PDA culture of each fungal strain. After 30 days of incubation at 25 °C in static conditions, the cultures were filtered through filter paper (Whatman No. 4) and the filtrate was analyzed by LC-MS qTOF [28].

Extraction and isolation of fungal secondary metabolites. The culture filtrates were acidified to pH 4 with 5 M HCl and extracted exhaustively with ethyl acetate (EtOAc). The combined organic extracts were dried (Na_2SO_4) and solvent eliminated under reduced pressure at 35 °C (Rotavapor RV 10 IKA® - Werke GmbH & Co. KG, Staufen, Germany). The residues recovered were fractionated by column chromatography (silica gel; 200 g) eluted with different eluents (Table 5) and the homogeneous fractions were collected as reported in Table 5. For thin layer chromatography (TLC) the following solvents were used: dichloromethane (CH_2Cl_2)/methanol (MeOH) 90:10; 80:20 (v/v), chloroform ($CHCl_3$)/MeOH 90:10; 80:20 (v/v); EtOAc/petroleum ether 90:10; 80:20 (v/v) [28]. All fractions were tested for insecticidal activity against *A. pisum*, and, where necessary, further purified by preparative TLC (Table 5).

Table 5. List of solvents used for chromatographic separations (column chromatography or preparative TLC) and total number of homogeneous collected fractions.

Fungal Source	Eluent Used for Column Chromatography and Preparative TLC *	Total Number of Homogeneous Fractions
Paecilomyces lilacinus	EtOAc/petroleum ether (90:10, v:v)	5
Penicillium griseofulvum	EtOAc/petroleum ether (90:10, v:v)	8
Pochonia chlamydosporia	CH_2Cl_2/MeOH (90:10, v:v); fraction No. 4 out of 5 was further purified by preparative TLC on silica gel with CH_2Cl_2/MeOH (90:10, v:v)	5
Beauveria bassiana	EtOAc/petroleum ether (90:10, v:v)	5
Metarhizium anisopliae	EtOAc/petroleum ether (90:10, v:v)	7
Talaromyces pinophilus	CH_2Cl_2/MeOH (98:2, v:v); Fraction No. 2 out of 7 was further purified by preparative TLC with CH_2Cl_2/MeOH (98:2, v:v)	7

* EtOAc: ethyl acetate; CH_2Cl_2: Dichloromethane; MeOH: Methanol.

General experimental procedures. The isolated molecules were solubilised in 700 μL of deuterated chloroform (99.8% $CDCl_3$ – Sigma-Aldrich, Darmstadt, Germany) and transferred into a stoppered NMR tube (5 mm, 7″, 507-HP-7, NORELL, Morganton, NC, USA) where remaining void volume was gently degassed by a N_2 flux. Proton and carbon solvent signals were used as reference to calibrate both ^1H and ^{13}C frequency axes. A 400 MHz Bruker Avance spectrometer (Bruker Co., Billerica, MA, USA), equipped with a 5 mm Bruker Broad Band Inverse probe (BBI), working at the ^1H and ^{13}C frequencies of 400.13 and 100.61 MHz, respectively, was used for the NMR measurements (at 25 ±1 °C).

Monodimensional ^1H and ^{13}C acquisitions were conducted as follows: proton spectra were acquired with 2 s of thermal equilibrium delay (number of scans = 64), a 90° pulse length 7.7 μs, 50 transients and 16 ppm (6410.2 Hz) as spectral widths, whereas proton-decoupled carbon acquisitions were executed by both inverse-gated and DEPT 135° pulse sequences, adopting 7 and 5 s of equilibrium delay, 12,500 and 2400 transients, respectively, and a spectral width of 250 ppm (25.152 KHz). A time domain of 32,768 points was adopted for all cited mono-dimensional experiments. Homo-nuclear ^1H-^1H COSY (COrrelation SpectroscopY), TOCSY (TOtal Correlation SpectroscopY), NOESY (Nuclear Overhauser Enhancement SpectroscopY), and hetero-nuclear ^1H–^{13}C HSQC (Hetero-nuclear Single-Quantum Correlation) and HMBC (Hetero-nuclear Multiple Bond Coherence) experiments (2D) were used for structural identification of metabolites. 2D homo- and heteronuclear spectra experiments were acquired with 48 and 80 scans, respectively, 16 dummy scans, a time domain of 2k points (F2) and 256 experiments (F1). TOCSY and NOESY experiments were conducted with a mixing time of 80 and 1000 ms, respectively, while HSQC and HMBC experiments were optimized for 145 Hz short and 6.5 Hz long range J_{CH} couplings, respectively. All executed 2D experiments were gradient enhanced, except for the TOCSY acquisition. A Qsine weighting function associated to a magnitude mode was used to process NOESY spectrum with the purpose to emphasize the weak cross-peaks and minimize the noise artefacts. The free induction decay (FID) of mono-dimensional spectra was multiplied by an exponential factor corresponding to 0.1 Hz, for ^1H and ^{13}C acquisitions, and to 1 Hz for DEPT 135° experiment. All above mentioned spectra were baseline corrected and processed by using Bruker Topspin Software (v.4.0.2).

LC-MS/MS Q-TOF analysis were done on an Agilent HP 1260 Infinity Series liquid chromatograph equipped with a DAD system (Agilent Technologies, Santa Clara, CA, USA) coupled to a Q-TOF mass spectrometer model G6540B (Agilent Technologies). Separations were performed on a Zorbax Eclips Plus C18 column, 4.6 × 100 mm, with 3.5 μm particles (Agilent Technologies). The analyses were done at a constant temperature of 37 °C and using a linear gradient system composed of A: 0.1% (*v*/*v*) formic acid in water, and B: 0.1% (*v*/*v*) formic acid in acetonitrile. The flow was 0.6 mL/min, 95% A graduating to 100% B in 12 min, 100% B 12-15 min, 95% A 15-17 and equilibrating 95% A 17–20 min. The UV spectra were collected by DAD every 0.4 s from 190 to 750 nm with a resolution of

2 nm. The MS system was equipped with a Dual Electrospray Ionization (ESI) source and operated with Agilent MassHunter Data Acquisition Software, rev. B.05.01 in the positive or negative mode. Mass spectra were recorded in the range *m/z* 100–1600 as centroid spectra, with 3 scans per second. Two reference mass compounds were used to perform the real-time lock mass correction, purine ($C_5H_4N_4$ at *m/z* 121.050873, 10 μmol/L) and hexakis (1H,1H, 3H-tetrafluoropentoxy)-phosphazene ($C_{18}H_{18}O_6N_3P_3F_{24}$ at *m/z* 922.009798, 2 μmol/L). The capillary was maintained at 4000 V, fragmentor voltage at 180 V, cone 1 (skimmer 1) at 45 V, Oct RFV at 750 V. Gas temperature was 350 °C during the run at 11 L/min, and the nebulizer was set at 45 psig. The injected sample volume was 5 μL.

MS/MS spectra were simultaneously recorded for confirmation purposes of new compounds, using the operating parameters described above, unless otherwise stated. The instrument was operated in the range *m/z* 100-1000, recording two spectra per second in targeted acquisition mode (targeted mass: 244.1197, Z = 1, RT 5.88 ±0.5 min). The sample collision energy was set to 20 V.

LC-MS data were evaluated using MassHunter Qualitative Analysis Software B.06.00 and compared to known compounds included in an in-house database. The database contains information of about 4000 known secondary metabolites isolated from more than 80 different fungal genera, and recorded according to their name, molecular formula, monoisotopic mass and producing organism. Positive identifications of fungal metabolites were reported if the compound was detected with a mass error below 10 ppm and with a sufficient score. Standards were used to confirm the chemical identifications.

UV spectra were recorded with a V-730 UV-Visible Spectrophotometer JASCO (Mary's Court, Easton, MD, USA). Column chromatography was performed using silica gel (Merck silica gel 60 GF_{254}; Merck, Darmstadt, Germany), and TLC with glass pre-coated silica gel GF_{254} plates (Merck Kieselgel 60 GF_{254}, 0.25 mm). The compounds were detected on TLC plates using UV light (254 or 366 nm) and/or by dipping the plates in a 5% (*v/v*) H_2SO_4 solution in ethanol followed by heating at 110 °C for 10 min [28,29].

Acetylation of chlamyphilone. Acetic anhydride (40 μL) was added to chlamyphilone (**1**, 2.5 mg), dissolved in dry pyridine (80 μL 2.5 mg), and the residue was purified by preparative TLC on silica gel (petroleum ether/aceton, 20:80, *v/v*) to yield the acetyl derivative **6** (1.4 mg, 49%) [29]. The reaction was monitored by LC-MS analysis (Figure S1).

In vivo insecticidal assay. Insecticidal activity of organic extracts, each column fraction and pure metabolites were tested against the pea aphid *Acyrthosiphon pisum*. Aphids were reared on *Vicia faba* var. *aguadulce* in a growth chamber (20 ± 1 °C, 70% RH, 18 h light/6 h dark photoperiod). Small populations were synchronized to obtain newborn nymphs every 24 h. Third-instar nymphs were dipped in the assay solution for 10 s and put on a paper towel to dry. Then, each insect was carefully transferred by a soft paintbrush on a leaf plug [30,31]. Twenty treated (or untreated—solvent control) 3rd-instar nymphs were placed on two circular (35 mm diameter) leaf plugs on 2% water agar layer in a Petri dish, in order to keep the leaf turgid. Plates were incubated at the climatic conditions described above and the number of dead aphids was assessed 24, 48, and 72 h after treatments. Extracts, fractions, and pure compounds were tested at 500, 400, 300, 200, 100, and 50 μg/mL. Each metabolite solution was also tested for phytotoxicity on *V. faba*. The aphid mortality has been calculated in each treatment as the mean value ± standard deviation, in three replicates for each time point (24, 48, and 72 h). The experiments were repeated at least twice.

Statistical analysis. Data analysis was performed with SPSS 11.0 software (Statistics for Windows Version 24.0, IBM Corp., Armonk, NY, USA), and statistical analysis was done using one-way analysis of variance (ANOVA). The Least Significant Difference (LSD) post hoc test with $p < 0.05$ was used to analyse the multiple comparisons. The mean values between different treatments at the same time point and the mean values between the same treatments at different time points were compared.

5. Conclusions

In this work six known metabolites with activity against the pea aphid *A. pisum* have been isolated from culture filtrates of selected fungal strains; for some of them this is the first report of insecticidal activity (Table S1). Moreover, a novel anti-aphid natural product, named chlamyphilone, has been isolated form the culture filtrate of *P. chlamydosporia* and fully characterized. The isolation of natural products with beneficial effects for plants may help to formulate novel secondary metabolites–based biopesticides, which represent promising alternatives to synthetic chemicals in agriculture. Further investigations should optimize extraction protocols and define environmental parameters that can affect the commercial formulations based on these natural compounds.

Supplementary Materials: The following are available online, Figures S1–S11 and S13-S14: NMR and MS spectra of compounds. Figure S12: Conformational model of the new compound named chlamyphilone. Table S1: Biological activity of the isolated compounds.

Author Contributions: F.V. and F.L. conceived the experiments and analyzed all the results. R.M. analyzed all the results and characterized the fungal strains. M.C.D. and M.G. conducted with F.L. the experiments on insects. M.L., P.C., F.L., and S.L.W. conducted the experiments with fungal strains. P.M., A.P. and F.V. characterized all the isolated metabolites. All authors reviewed the manuscript.

Funding: This work was supported by the following projects: MIUR–PON [grant number Linfa 03PE_00026_1], [grant number Marea 03PE_00106].

Conflicts of Interest: The authors declare no conflicts of interest.

References

1. Dayan, F.E.; Cantrell, C.L.; Duke, S.O. Natural products in crop protection. *Bioorg. Med. Chem.* **2009**, *17*, 4022–4034. [CrossRef] [PubMed]
2. Copping, L.G.; Duke, S.O. Natural products that have been used commercially as crop protection agents. *Pest Manag. Sci.* **2007**, *63*, 524–554. [CrossRef] [PubMed]
3. Saxena, S.; Pandey, A.K. Microbial metabolites as eco-friendly agrochemicals for the next millennium. *Appl. Microbiol. Biot.* **2001**, *55*, 395–403. [CrossRef]
4. Tanaka, Y.; Omura, S. Agroactive compounds of microbial origin. *Annu. Rev. Microbiol.* **1993**, *47*, 57–87. [CrossRef] [PubMed]
5. Berdy, J. Bioactive microbial metabolites. *J. Antibiot.* **2005**, *58*, 1–26. [CrossRef] [PubMed]
6. Vinale, F.; Sivasithamparam, K.; Ghisalberti, E.L.; Ruocco, M.; Woo, S.; Lorito, M. *Trichoderma* secondary metabolites that affect plant metabolism. *Nat. Prod. Commun.* **2012**, *7*, 1545–1550. [PubMed]
7. Mukherjee, P.K.; Horwitz, B.A.; Kenerley, C.M. Secondary metabolism in *Trichoderma*—A genomic perspective. *Microbiology* **2012**, *158*, 35–45. [CrossRef] [PubMed]
8. Vinale, F.; Nicoletti, R.; Borrelli, F.; Mangoni, A.; Parisi, O.A.; Marra, R.; Lombardi, N.; Lacatena, F.; Grauso, L.; Finizio, S.; et al. Co-culture of plant beneficial microbes as source of bioactive metabolites. *Sci. Rep.* **2017**, *7*, 14330. [CrossRef] [PubMed]
9. Digilio, M.C.; Mancini, E.; Voto, E.; De Feo, V. Insecticide activity of mediterranean essential oils. *J. Plant Interact.* **2008**, *3*, 17–23. [CrossRef]
10. Fukushima, K.; Arai, T.; Mori, Y.; Tsuboi, M.; Suzuki, M. Studies on peptide antibiotics, leucinostatins. *J. Antibiot.* **1983**, *36*, 1613–1630. [CrossRef] [PubMed]
11. Park, J.O.; Hargreaves, J.R.; McConville, E.J.; Stirling, G.R.; Ghisalberti, E.L.; Sivasithamparam, K. Production of leucinostatins and nematicidal activity of Australian isolates of *Paecilomyces lilacinus* (Thom) Samson. *Lett. Appl. Microbiol.* **2004**, *38*, 271–276. [CrossRef] [PubMed]
12. Oxford, A.E.; Raistrick, H.; Simonart, P. Studies in the biochemistry of micro-organisms: Griseofulvin, $C_{17}H_{17}O_6Cl$, a metabolic product of *Penicillium griseofulvum* Dierckx. *Biochem. J.* **1939**, *33*, 240. [CrossRef] [PubMed]
13. Banani, H.; Marcet-Houben, M.; Ballester, A.R.; Abbruscato, P.; González-Candelas, L.; Gabaldón, T.; Spadaro, D. Genome sequencing and secondary metabolism of the postharvest pathogen *Penicillium griseofulvum*. *BMC Genom.* **2016**, *17*, 19. [CrossRef] [PubMed]

14. Dowd, P.F. Toxicity of the fungal metabolite griseofulvin to *Helicoverpa zea* and *Spodoptera frugiperda*. *Entomol. Exp. Appl.* **1993**, *69*, 5–11. [CrossRef]
15. Kimura, Y.; Shimada, A.; Kusano, M.; Yoshii, K.; Morita, A.; Nishibe, M.; Fujiok, S.; Kawano, T. Myxostiolide, myxostiol, and clavatoic acid, plant growth regulators from the fungus *Myxotrichum stipitatum*. *J. Nat. Prod.* **2002**, *65*, 621–623. [CrossRef] [PubMed]
16. Osmanova, N.; Schultze, W.; Ayoub, N. Azaphilones: A class of fungal metabolites with diverse biological activities. *Phytochem. Rev.* **2010**, *9*, 315–342. [CrossRef]
17. Gao, J.; Yang, S.; Qin, J. Azaphilones: Chemistry and Biology. *Chem. Rev.* **2013**, *113*, 4755–4811. [CrossRef] [PubMed]
18. Akihisa, T.; Tokuda, H.; Yasukawa, K.; Ukiya, M.; Kiyota, A.; Sakamoto, N.; Suzuki, T.; Tanabe, N.; Nishino, H. Azaphilones, furanoisophthalides, and amino acids from the extracts of *Monascus pilosus*-fermented rice (red-mold rice) and their chemopreventive effects. *J. Agric. Food Chem.* **2005**, *53*, 562–565. [CrossRef]
19. Wang, Y.L.; Li, L.F.; Li, D.X.; Wang, B.; Zhang, K.; Niu, X. Yellow pigment aurovertins mediate interactions between the pathogenic fungus *Pochonia chlamydosporia* and its nematode host. *J. Agric. Food Chem.* **2015**, *63*, 6577–6587. [CrossRef]
20. Logrieco, A.; Moretti, A.; Castella, G.; Kostecki, M.; Golinski, P.; Ritieni, A.; Chelkowski, J. Beauvericin production by *Fusarium* Species. *Appl. Environ. Microbiol.* **1998**, *64*, 3084–3088.
21. Daniel, J.F.; Silva, A.A.; Nakagawa, D.H.; de Medeiros, L.S.; Carvalho, M.G.; Tavarres, L.J.; Abreud, L.M.; Filhob, E.R. Larvicidal activity of *Beauveria bassiana* extracts against *Aedes aegypti* and identification of Beauvericins. *J. Braz. Chem. Soc.* **2017**, *28*, 1003–1013. [CrossRef]
22. Soledade, M.; Pedras, C.; Zaharia, L.I.; Ward, D.E. The destruxins: Synthesis, biosynthesis, biotransformation, and biological activity. *Phytochemistry* **2002**, *59*, 579–596.
23. Zhang, H.; Hu, W.; Xiao, M.; Ou, S.; Hu, Q. Destruxin A induces and binds HSPs in *Bombyx mori* Bm12 cells. *J. Agric. Food Chem.* **2017**, *65*, 9849–9853. [CrossRef] [PubMed]
24. Vinale, F.; Nicoletti, R.; Lacatena, F.; Marra, R.; Sacco, A.; Lombardi, N.; d'Errico, G.; Digilio, M.C.; Lorito, M.; Woo, S.L. Secondary metabolites from the endophytic fungus *Talaromyces pinophilus*. *Nat. Prod. Res.* **2017**, *31*, 1778–1785. [CrossRef] [PubMed]
25. Salvatore, M.M.; Della Greca, M.; Nicoletti, R.; Salvatore, F.; Vinale, F.; Naviglio, D.; Andolfi, A. Talarodiolide, a new 12-membered macrodiolide, and GC/MS Investigation of culture filtrate and mycelial extracts of *Talaromyces pinophilus*. *Molecules* **2018**, *23*, 950. [CrossRef] [PubMed]
26. Manter, D.K.; Vivanco, J.M. Use of the ITS primers, ITS1F and ITS4, to characterize fungal abundance and diversity in mixed-template samples by qPCR and length heterogeneity analysis. *J. Microbiol. Methods* **2007**, *71*, 7–14. [CrossRef] [PubMed]
27. Glass, N.L.; Donaldson, G.C. Development of Primer Sets Designed for Use with the PCR To Amplify Conserved Genes from Filamentous Ascomycetes. *Appl. Environ. Microbiol.* **1995**, *61*, 1323–1330.
28. Vinale, F.; Strakowska, J.; Mazzei, P.; Piccolo, A.; Marra, R.; Lombardi, N.; Manganiello, G.; Pascale, A.; Woo, S.L.; Lorito, M. Cremenolide, a new antifungal, 10-member lactone from *Trichoderma cremeum* with plant growth promotion activity. *Nat. Prod. Res.* **2016**, *30*, 2575–2581. [CrossRef]
29. Vinale, F.; Arjona Girona, I.; Nigro, M.; Mazzei, P.; Piccolo, A.; Ruocco, M.; Woo, S.; Ruano Rosa, D.; López Herrera, C.; Lorito, M. Cerinolactone, a hydroxy-lactone derivative from *Trichoderma cerinum*. *J. Nat. Prod.* **2011**, *75*, 103–106. [CrossRef]
30. De Feo, V.; Mancini, E.; Voto, E.; Curini, M.; Digilio, M.C. Bioassay-oriented isolation of an insecticide from *Ailanthus altissima*. *J. Plant Interact.* **2009**, *4*, 119–123. [CrossRef]
31. Chandrasena, D.; Difonzo, C.; Byrne, A. An Aphid-Dip Bioassay to Evaluate Susceptibility of Soybean Aphid (Hemiptera: Aphididae) to Pyrethroid, Organophosphate, and Neonicotinoid Insecticides. *J. Econ. Entomol.* **2011**, *104*, 1357–1363. [CrossRef] [PubMed]

Sample Availability: Samples of the compounds **1-9** are available from the authors.

molecules
MDPI

Article

The Issue of Misidentification of Kojic Acid with Flufuran in *Aspergillus flavus*

Marina DellaGreca [1], Gaetano De Tommaso [1], Maria Michela Salvatore [1], Rosario Nicoletti [2,3], Andrea Becchimanzi [3], Mauro Iuliano [1] and Anna Andolfi [1,*]

1 Department of Chemical Sciences, University of Naples 'Federico II', 80126 Naples, Italy; dellagre@unina.it (M.D.); gaetano.detommaso@unina.it (G.D.T.); mariamichela.salvatore@unina.it (M.M.S.); mauro.iuliano@unina.it (M.I.)
2 Council for Agricultural Research and Economics, Research Centre for Olive, Citrus and Tree Fruit, 81100 Caserta, Italy; rosario.nicoletti@crea.gov.it
3 Department of Agriculture, University of Naples 'Federico II', 80055 Portici, Italy; andrea.becchimanzi@unina.it
* Correspondence: andolfi@unina.it; Tel.: +39-081-2539179

Academic Editors: Francesco Vinale and Maria Luisa Balestrieri
Received: 11 April 2019; Accepted: 30 April 2019; Published: 2 May 2019

Abstract: In the course of investigations on the complex phenomenon of bee decline, *Aspergillus flavus* was isolated from the haemocoel of worker bees. Observations on the metabolomic profile of this strain showed kojic acid to be the dominant product in cultures on Czapek-Dox broth. However, an accurate review of papers documenting secondary metabolite production in *A. flavus* also showed that an isomer of kojic acid, identified as 5-(hydroxymethyl)-furan-3-carboxylic acid and named flufuran is reported from this species. The spectroscopic data of kojic acid were almost identical to those reported in the literature for flufuran. This motivated a comparative study of commercial kojic acid and 5-(hydroxymethyl)-furan-3-carboxylic acid, highlighting some differences, for example in the ^{13}C-NMR and UV spectra for the two compounds, indicating that misidentification of the kojic acid as 5-(hydroxymethyl)-furan-3-carboxylic acid has occurred in the past.

Keywords: kojic acid; flufuran; 5-(hydroxymethyl)-furan-3-carboxylic acid; *Aspergillus flavus*; bees; immunomodulation

---

1. Introduction

Fungi have evolved the capability to produce a great number of secondary metabolites involved in the improvement of their ecological fitness, and many of them play important biological roles as virulence factors, chemical defense agents, developmental regulators, insect attractants, and chemical signals for communication with other organisms. On these properties is founded the pharmacological exploitation of many products as antibiotic, antiviral, antitumor, antihypercholesterolemic, and immunosuppressant agents [1–5]. In this respect, fungi are prime targets of a vigorous investigational activity, based on the employment of the most advanced analytical and structure elucidation techniques [6,7].

A wide array of fungal secondary metabolites has been ascribed to the ascomycete genus *Aspergillus*, well-known for its ubiquity and cosmopolitan distribution [7,8]. The species *Aspergillus flavus* is well-known as a foodstuff contaminant and a mycotoxin producer, and in this respect its metabolomic profile has been quite well characterized [9,10]. However, this fungus has been also reported in association with plants and animals in many different environments [11,12]. Particularly, it has been directly isolated from bee (*Apis mellifera*) individuals in different developmental stages and health conditions [13]. Although reported as the agent of stonebrood disease, further elucidation is required if

A. flavus basically behaves like an entomopathogen, or if its relationships with bees are more complex and eventually involve modulation of the immunitary response to other pathogens and parasites of these insects. Generally, the fungus is considered an opportunistic pathogen of immunocompromised individuals, gaining access to the host through ingestion, or taking advantage from the interaction of bees with other pathogens and parasites which negatively affect host immunocompetence and cuticle integrity [13–16].

During investigations on the honeybee colony collapse, a multifactorial syndrome mainly related to the compresence of immunosuppressive viruses and *Varroa destructor*, a parasitic mite which is also known as a possible vector [16–18], strains of *A. flavus* were repeatedly isolated from the haemocoel of worker bees.

The present paper reports the identification of kojic acid (KA) as the main metabolite obtained from this source. Furthermore, revision of the structure of flufuran, previously characterized as 5-(hydroxymethyl)furan-3-carboxylic acid from *A. flavus* and other fungal species is proposed, based on a comparison of spectroscopic data of commercially available compounds and some derivatives.

2. Results

2.1. Comparison of NMR Data Obtained from KA (1) and 5-(Hydroxymethyl)furan-3-carboxylic Acid (2)

A white solid was obtained from the extraction with ethyl acetate (EtOAc) of the culture filtrate of strain AB1EET of *A. flavus*, which consisted in a main metabolite, as deduced by its ^{1}H and ^{13}C-NMR spectra. NMR data collected for the extracted compound showed significant similarities with proton and carbon chemical shifts reported in literature for both KA (**1**) [19] and 5-(hydroxymethyl)furan-3-carboxylic acid (**2**) [20], a compound characterized from *A. flavus* and other fungal species (Table 1). A full analysis of the previous reports revealed discrepancy between the carbon chemical shifts of furan compounds [21] and those assigned for 5-(hydroxymethyl)furan-3-carboxylic acid. In order to clarify this issue, commercially available KA and 5-(hydroxymethyl)furan-3-carboxylic acid were submitted to spectroscopic and potentiometric investigations.

Table 1. Previous reports dealing with flufuran identification.

Source	Spectroscopic Data	Activity	Ref.
Aspergillus flavus	^{1}H and ^{13}C-NMR, MS, IR, UV	Antifungal	[22]
A. flavus	—	Cytotoxic against human cancer cell lines	[23]
A. flavus		Bacteriostatic (*Escherichia coli*)	[24] [1]
A. flavus [2]	^{1}H and ^{13}C-NMR	Antibacterial, antioxidant	[25]
A. flavus	MS	—	[26]
Aspergillus oryzae	—	—	[27]
A. oryzae	[1]	Antioxidant, antimicrobial	[28]
Fusarium sp. [3]	^{1}H and ^{13}C-NMR	Antimicrobial	[29]
Penicillium dipodomyicola	^{1}H and ^{13}C-NMR	—	[30]
Penicillium polonicum	—	Monoamine oxidase inhibitor	[31]
Pestalotiopsis sp.	^{1}H and ^{13}C-NMR	Cytotoxic against HepG2 cells	[32]
Polyporus ciliatus	P.f., MS, ^{1}H and ^{13}C-NMR, MS, IR, UV	—	[20]

[1] This paper is in Chinese and could not be examined in detail. [2] 7-*O*-Acetyl derivative was also isolated. [3] Methyl ester was also isolated.

The comparison of ^{1}H-NMR spectra, recorded in CD_3OD, showed no significant differences between 5-(hydroxymethyl)furan-3-carboxylic acid and KA (Figures 1 and 2). Notwithstanding, the ^{1}H-NMR spectra recorded in DMSOd_6 for **2** revealed the coupling between the proton of the hydroxyl group in C-7 and protons of the methylene group which resonate as triplet and doublet, respectively, at δ 5.31 and 4.40 (J = 5.8 Hz) and the presence of the signal of proton of the carboxylic group at δ 12.60 (Figure 3).

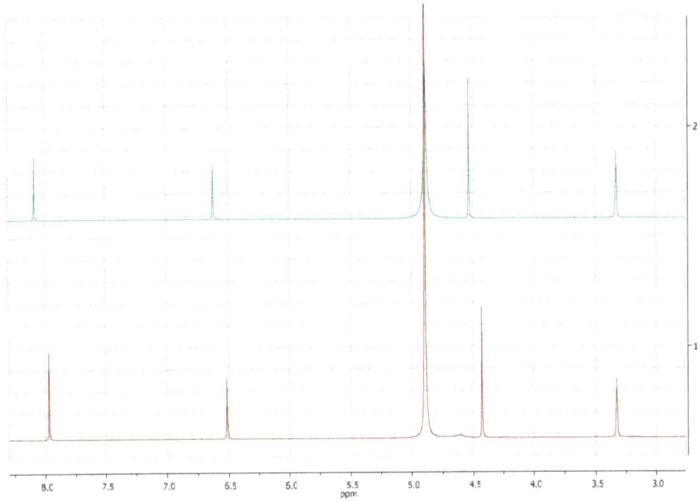

Figure 1. ^{1}H-NMR spectra of KA (**1**, red) and 5-(hydroxymetyhyl)-furan 3-carboxylic acid (**2**, green) recorded at 400 MHz in CD_3OD.

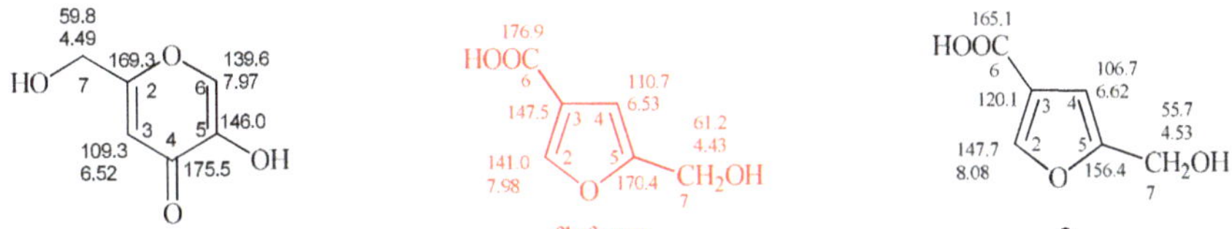

1 flufuran **2**

Figure 2. Proton and carbon chemical shifts of KA (**1**) and 5-(hydroxymethyl)furan-3-carboxylic acid (**2**). Data reported in red were erroneously assigned to flufuran. The spectra were recorded in CD_3OD.

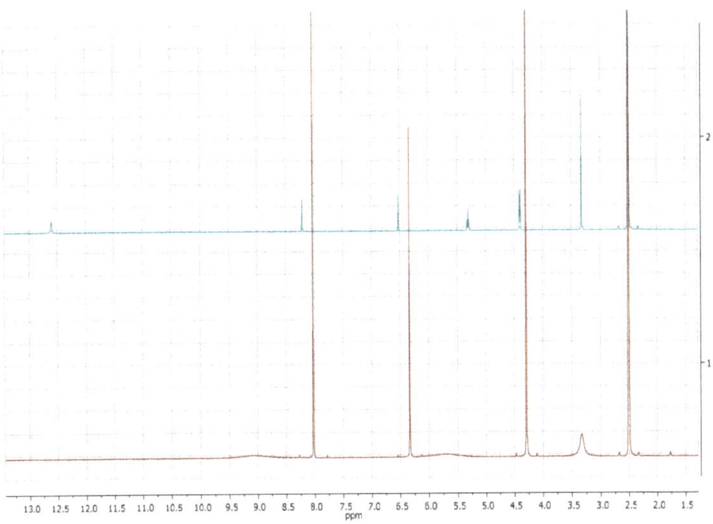

Figure 3. ^{1}H-NMR spectra of KA (**1**, red) and 5-(hydroxymetyhyl)-furan 3-carboxylic acid (**2**, green) recorded at 400 MHz in DMSOd_6.

The main differences between **1** and **2** were observed, as expected, in ^{13}C-NMR spectra. The C-3, C-5 and C-6 carbons of **2** resonate at δ 120.1, 156.4 e 165.1 (Figure 2 and Figure S1), respectively, assigned by 2D-NMR spectra (Figures S2 and S3). However, the ^{13}C-NMR spectrum of flufuran showed chemical shift values identical to KA (Figure 2), indicating a misinterpretation of the flufuran structure.

2.2. Comparison of Potentiometric, UV, and MS Spectrophotometric Measurements

The acid-base behavior of KA and 5-(hydroxymethyl)furan-3-carboxylic acid was determined by potentiometric and spectrophotometric methods. The calculated protonation constants cologarithm values are 7.68 ± 0.05 and 4.03 ± 0.05, for compound **1** and **2** respectively. Potentiometric data are reported in Figure 4.

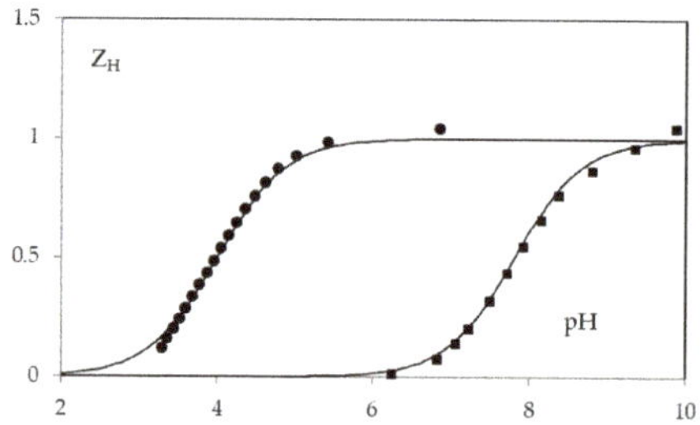

Figure 4. Z_H(pH) function in 0.1 M $NaClO_4$ for KA (squares) and 5-(hydroxymethyl)furan-3-carboxylic acid (circles). The solid lines are calculated with protonation constants cologarithm value 7.68 (on squared) and 4.03 (on circles).

In particular KA constant is very similar to that in 0.1 M KCl (7.7) [33]. While no value on the protonation of 5-(hydroxymethyl)furan-3-carboxylic acid had been previously reported.

UV spectra of the pure compounds, recorded in water in the interval 200–400 nm, show λ_{max} at 216 and 270 nm for KA, and at 240 nm for **2**. Similar values were observed when the spectra were recorded in methanol.

In agreement with acid-base behavior of both compounds, measurements conducted in pH function confirmed that KA spectra are highly influenced by pH [34]. In fact, absorbance value at 315 nm increases from 3 to 9 pH range, instead the peak at 270 nm decreases in the same interval. Conversely, UV spectra of compound **2** present minimal variation in function of pH (Figure 5). Finally with pH increasing, KA shows a isosbestic point at 285 nm, while 5-(hydroxymethyl)furan-3-carboxylic acid presents one at 240 nm.

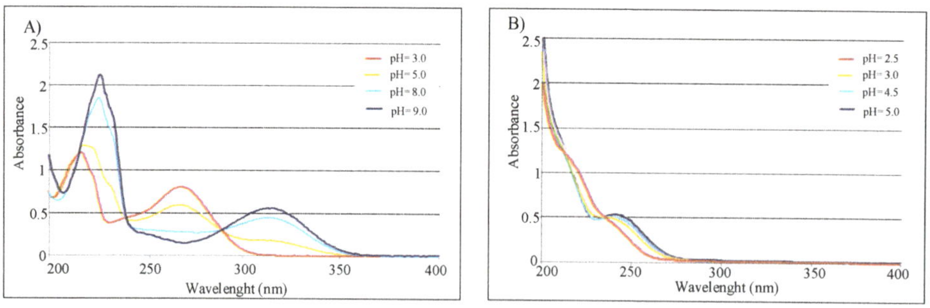

Figure 5. (**A**) UV spectra of 2.0×10^{-3} M in 0.1 M $NaClO_4$ for KA. (**B**) UV spectra of 2.0×10^{-3} M in 0.1 M $NaClO_4$ for 5-(hydroxymethyl)furan-3-carboxylic acid (optical path 0.2 cm).

The fragmentation pattern observed in MALDI TOF/MS spectra evidenced relevant analogies between the two compounds, as to be expected from considering the nature of the respective functional groups. In fact, from peaks corresponding to the protonated ions $[M+H]^+$ at *m/z* 143 of **1** and **2**, fragments deriving from the loss of OH, CHO, and COOH groups were observed in both cases.

2.3. *Comparison of KA and 5-(Hydroxymethyl)furan-3-carboxylic Acid Derivatives*

Considering reports concerning the presumptive isolation of the 7-*O*-acetyl derivative of flufuran [25] and of its methyl ester [29], in this work the differences between derivatives of **1** and **2** obtained from common acetylation and methylation reaction were also evaluated.

The proton spectrum of 5-(acetoxymethyl)furan 3-carboxylic acid (**4**, Figure S4), produced after acetylation of **2**, showed the down shift of Δδ 0.55 of the signal assigned to the methylene group C-7 which resonates as singlet at δ 5.08, and the presence of a further singlet at δ 2.07 assigned to the CH_3 of the acetyl group (Figure 6 and Figure S4). As discussed in relation to **1** and **2** the differences between the ^1H-NMR spectra of **4** and 7-*O*-acetylflufuran [25] are not so marked compared to the ones of ^{13}C-NMR spectra (Figure 6 and Figure S5).

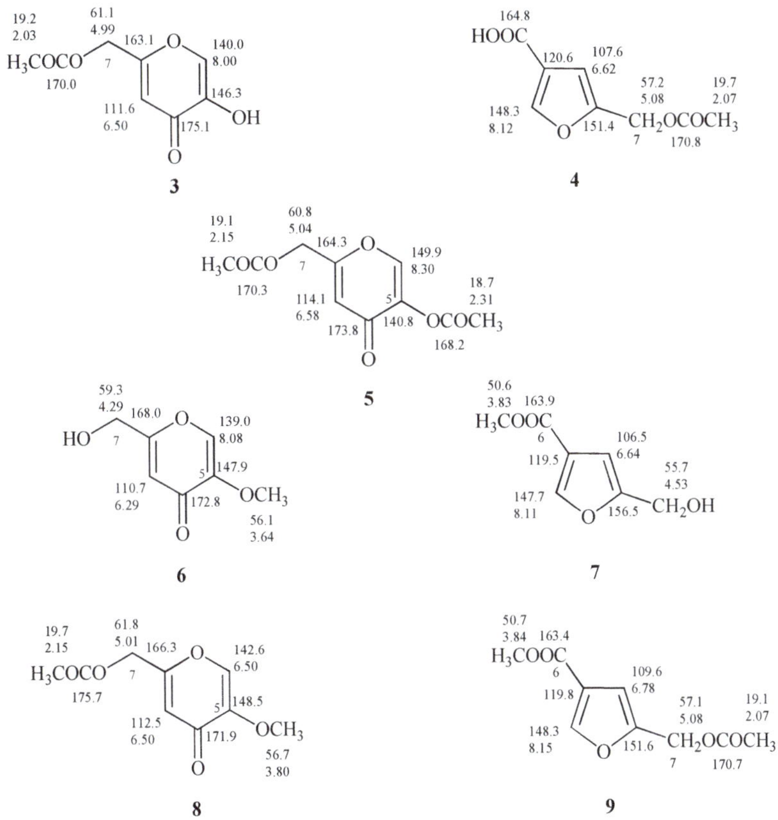

Figure 6. Proton and carbon chemical shifts of 7-*O*-acetylKA, 5,7-*O*,*O′*-diacetylKA, 5-*O*-methylKA, 5-*O*-methyl-7-*O′*-acetylKA (**3**, **5**, **6**, and **8**); 5-(acetoxymethyl)furan-3-carboxylic acid, methyl 5-(hydroxymethyl)furan-3-carboxylate, methyl 5-(acetoxymethyl)furan-3-carboxylate (**4**, **7**, and **9**).

From the data reported on the diacetylated flufuran [22], it is possible to exclude its formation from **2**; most likely the described compound is 5,7-*O*,*O′*-diacetylKA (**5**, Figure 6), derived from **1**.

The treatment of **1** and **2** with diazomethane led to the production of 5-*O*-methylKA (**6**) and methyl 5-(hydroxymethyl)furan-3-carboxylate (**7**, Figure 6) as deduced from ^1H and ^{13}C-NMR spectra.

In fact, NMR spectra of **7** (Figures S6 and S7) showed further singlets, respectively, at δ 3.83 and at δ 50.6 attributed to methoxy group. As observed for the other compounds under examination, the comparison of ^{13}C-NMR spectra of methyl derivatives of **1** and **2** revealed significant differences (**6** and **7**, Figure 6).

Finally, the preparation of 5-*O*-methyl-7-*O*′-acetylKA (**8**) and methyl 5-(acetoxymethyl)furan-3-carboxylate (**9**) and the interpretation of their one-dimensional NMR spectra (Figures S8 and S9) allowed to confirm the derivative structures, and to establish that the derivative i.e., methyl 5-(acetoxymethyl)furan-3-carboxylate) reported by Evidente et al. [22] is undoubtedly **8**.

3. Discussion

The furan compound 5-(hydroxymethyl)furan-3-carboxylic acid was identified for the first time as a natural product in cultures of *Polyporus ciliatus* [20], and afterwards reported from other fungi (Table 1). Although furan derivatives are not an extensive class of fungal compounds [35], over the past few years the literature concerning these natural products has been enriched by new reports [36–40]. According to their biosynthetic origin, the structure of natural furan derivatives shows the presence of furan rings substituted in position 2 and 5 [35]. Flufuran represented an exception to such a general model.

In the present paper, we produce evidence that in most of the previous reports the product identified as flufuran was instead KA. In addition to being known as a typical secondary metabolite of both *A. flavus* and *A. oryzae*, KA has been reported from several *Penicillium* spp. [41]. Other fungal species are claimed as producers of some derivatives [42], indicating that biosynthetic ability for this compound might be more widespread. To the best of our knowledge, our amendments to previous findings represents the first evidence of its production in important fungal genera, such as *Fusarium* and *Pestalotiopsis*, which deserve further confirmation also with reference to correct taxonomic identification of producing strains. In view of the present revision, the bioactivities erroneously assigned to flufuran should be instead referred to KA, integrating the known profile of biological properties of this γ-pyrone compound [43]. Actually, these data show that KA and 5-(hydroxymethyl)furan-3-carboxylic acid are susceptible of misidentification. Particularly, some instrumental techniques commonly used for the identification of secondary metabolites in fungal extracts (i.e., ^{1}H-NMR and ESI MS data) are not able to distinguish between these two compounds. However, huge differences can be observed between **1** and **2** in the UV spectra obtained with different solvents and pH. The acid-base properties are also valid to differentiate between **1** and **2**. Finally, observations concerning acetyl and methyl derivatives also confirmed the misidentification of KA with flufuran.

4. Materials and Methods

4.1. General Experimental Procedures

^{1}H and ^{13}C-NMR spectra were recorded at 400 and at 100 MHz, respectively, in CD_3OD unless otherwise noted, on a Bruker spectrometer (AscendTM400) (Bremen, Germany); the same solvent was used as internal standard. Potentiometric titrations were performed in an air-bath thermostat kept at (25.00 ± 0.05) °C. A programmable computer-controlled Data Acquisition Unit 34970A, (Agilent Tecnologies Inc., Santa Clara, CA, USA) was used to perform the potentiometric measurements. The glass electrodes were Metrohm (Herisau, Switzerland) of 60102-100 type, and Ag/AgCl electrode was utilized as reference. The EMF values were measured with a precision of ± 0.01 mV using a Keithley 642 type Digital Electrometer (Tektronix Inc., Beaveron, OR, USA). UV–VIS spectra were recorded by model Cary 5000 Spectrophotometer by Varian (Palo Alto, CA, USA) from 200 to 600 nm (optical path 0.2 cm) at 25.0 °C, under a constant flow of nitrogen.

MALDI-TOF/MS spectra were acquired by a 4700 Proteomics Analyzer (Applied Biosystems, Framingham, MA, USA). Compounds were detected in reflector mode using α-cyano-4-hydroxycinnamic acid (CHCA) as matrix.

Analytical and preparative TLC were performed on silica gel plates (Kieselgel 60, F_{254}, 0.25 mm) (Merck, Darmstadt, Germany). The spots were visualized by exposure to UV radiation (253 nm), or by spraying with 10% H_2SO_4 in MeOH followed by heating at 110 °C for 10 min. 5-(hydroxymethyl)furan-3-carboxylic acid and KA were purchased from Enamine Ltd. (Kyiv, Ukraine) and Alfa Aesar (Karlsruhe, Germany), respectively.

4.2. *Isolation of* A. flavus *from Bees*

Worker bee individuals were collected from experimental apiaries at the Department of Agriculture, University of Naples Federico II (Portici, Italy) in October–November 2018. Freshly emerged bees (<48 h) were anaesthetized by chilling on ice for 20 min, and surface-disinfected by dipping in 70% ethanol for 30 s. The intersegmental membrane between head and thorax was cut using a sterile scalpel. Ten microliters of the liquid matter exuding from haemocoel were aseptically collected using a micropipette, and spotted on potato-dextrose agar (PDA, Oxoid) amended with 50 mg/L chloramphenicol. Plates were incubated at 32 °C in the dark. Hyphal tips from the emerging fungal colonies were transferred to fresh PDA plates to obtain pure cultures for morphological identification and storage. The strains obtained were readily identified as belonging to the *Aspergillus* section *Flavi*, based on the formation of colonies producing a dense felt of yellow-green rough conidiophores with radiate heads, and subglobose echinulate conidia. An orange reverse could be observed in cultures on a specific medium (AFPA) [44]. Unequivocal ascription to the species *A. flavus* resulted after rDNA-ITS and calmodulin gene sequencing. To this purpose, total genomic DNA was extracted from fresh mycelium taken from pure culture of strain AB1EET using CTAB protocol [45]. According to reported methodology, primers ITS1F and ITS4 were used to amplify rDNA-ITS, while primers CF1M and CF4 were used to amplify the calmodulin gene [45,46]. The original DNA sequences obtained in this study have been deposited in GenBank under the codes MK611561 (ITS) and MK611938 (calmodulin). The calmodulin gene sequence proved to be more respondent for unequivocal species identification, yielding 100% homology with sequences from 60 strains of *A. flavus* available in GenBank.

4.3. *Production and Extraction of KA*

Strain AB1EET was cultured in Czapek-Dox broth (Oxoid) following a previously reported procedure [47]. The freeze-dried culture filtrates (750 mL) were dissolved in 100 mL, acidified with HCl 2 N at pH 3, and extracted three times with an equal volume of EtOAc. Organic phases were combined, dried with Na_2SO_4, and evaporated under reduced pressure originating a white solid residue (350.7 mg) identified as KA.

4.4. *Determination of Protonation Constants*

The evaluation of the protolysis constants was conducted through spectrophotometric and potentiometric titration, at 25 °C in 0.1 M $NaClO_4$ as ionic medium [48]. By experimental data the average number (Z_H) of protons released per molecule was assessed through the equation:

$$Z_H = ([H^+] - C_H - K_w/[H^+])/C_L \quad (1)$$

where $[H^+]$ is hydrogenionic concentration, C_H is the total acid concentration, C_L is the compound concentration and K_w is ionic product ($10^{-13.7}$, in 0.1 M $NaClO_4$). Experimental data were processed with Hyperquad softare [49].

4.5. *UV and MS Data of KA and 5-(Hydroxymethyl)furan-3-carboxylic Acid*

Kojic acid (**1**). UV λ_{max} nm (log ε): (H_2O) 216 (4.12), 270 (3.93); (MeOH) 225 (3.99), 270 (3.99); (pH 3.0) 215 (3.47), 270 (3.30); (pH 5.0) 216 (3.51), 268 (3.17), 318 (2.63); (pH 8.0) 227 (3.66), 315 (3.5); (pH 9.0) 227 (3.72), 315 (3.15). MALDI TOF/MS: *m/z* 143 $[M+H]^+$, 125 $[M-OH]^+$, 113 $[M-CHO]^+$, 97 $[M-COOH]^+$, 69 $[M-CH_2OH-COO]^+$.

5-(Hydroxymethyl)furan-3-carboxylic acid (**2**). UV λ_{max} nm (log ε): (H_2O) 240 (3.39); (MeOH) 240 (3.38); (pH 2.5) 243 (2.97); (pH 3.0) 243 (3.07); (pH 4.0) 243 (3.07); (pH 5.0) 243 (3.12). MALDI TOF/MS: *m/z* 143 $[M+H]^+$, 125 $[M\text{-}OH]^+$, 113 $[M\text{-}CHO]^+$, 97 $[M\text{-}COOH]^+$.

4.6. Sample Methylation

Fifteen milligrams of samples [KA, 5-(hydroxymethyl)furan-3-carboxylic acid, 5-(acetoxymethyl)furan-3-carboxylic acid (**1**, **2** and **4**)] were dissolved in MeOH (1.5 mL); an ethereal solution of CH_2N_2 was slowly added until a yellow color became persistent. The reaction mixtures were stirred at room temperature for 4 h. The solvent was evaporated under a N_2 stream at room temperature. Residues of each reaction were analyzed by TLC on silica gel; **6**, was evidenced at Rf 0.37 by eluting with EtOAc-MeOH (9:1), while Rf 0.54 and 0.82 corresponded to **7** and **9** respectively, as eluted with $CHCl_3$-i-PrOH (92:8).

4.7. Sample Acetylation

Ten mg of samples [KA, 5-(hydroxymethyl)furan-3-carboxylic acid, 5-*O*-methylkojic acid (**1**, **2**, **6**)], dissolved in pyridine (30 μL), were converted into the corresponding acetyl derivatives (**5**, **4**, **7**) by acetylation with Ac_2O (30 μL) at room temperature overnight. The reaction was stopped by addition of MeOH, and the azeotrope formed by addition of benzene was evaporated in a N_2 stream at 40 °C. Residues of each reaction were analyzed by TLC on silica gel; **5** was evidenced at Rf 0.44 by eluting with $CHCl_3$-i-PrOH (95:5), while Rf 0.54 and 0.82 corresponded to **7** and **9** respectively, as eluted with $CHCl_3$-i-PrOH (92:8).

Supplementary Materials: The following are available online, Figures S1–S3. NMR spectra of KA (**1**) and 5-(hydroxymethyl)furan-3-carboxylic acid (**2**); Figures S4–S9. ^1H and ^{13}C-NMR spectra of acetyl and methyl derivatives of **2**.

Author Contributions: M.D., R.N., and A.A. conceived and organized the manuscript, and wrote the text; R.N. and A.B. isolated and cultivated the fungal strain; M.M.S. and A.A. extracted the culture filtrate and prepared sample derivatives; M.D. and A.A. performed the NMR analysis; G.D.T. and M.I. performed potentiometric and spectrophotometric measurements; M.D., M.M.S, R.N., and A.A edited and reviewed the manuscript.

Funding: This research was funded by Finanziamento delle Attività Base della Ricerca (FFABR) 2017 of Ministero dell'Istruzione, dell'Università e della Ricerca (MIUR, Italy).

Conflicts of Interest: The authors declare no conflict of interest.

References

1. Chambergo, F.S.; Valencia, E.Y. Fungal biodiversity to biotechnology. *Appl. Microbiol. Biotechnol.* **2016**, *100*, 2567–2577. [CrossRef] [PubMed]
2. Meyer, V.; Andersen, M.R.; Brakhage, A.A.; Braus, G.H.; Caddick, M.X.; Cairns, T.C.; de Vries, R.P.; Haarmann, T.; Hansen, K.; Hertz-Fowler, C.; et al. Current challenges of research on filamentous fungi in relation to human welfare and a sustainable bio-economy: A white paper. *Fungal Biol. Biotechnol.* **2016**, *3*, 6.
3. Nicoletti, R.; Salvatore, M.M.; Andolfi, A. Secondary metabolites of mangrove-associated strains of *Talaromyces*. *Mar. Drugs* **2018**, *16*, 12. [CrossRef] [PubMed]
4. Félix, C.; Salvatore, M.M.; DellaGreca, M.; Meneses, R.; Duarte, A.S.; Salvatore, F.; Naviglio, D.; Gallo, M.; Jorrino-Novo, J.V.; Alves, A.; et al. Production of toxic metabolites by two strains of *Lasiodiplodia theobromae*, isolated from a coconut tree and a human patient. *Mycologia* **2018**, *110*, 642–653. [CrossRef] [PubMed]
5. Marra, R.; Nicoletti, R.; Pagano, E.; DellaGreca, M.; Salvatore, M.M.; Borrelli, F.; Lombardi, N.; Vinale, F.; Woo, S.L.; Andolfi, A. Inhibitory effect of trichodermanone C, a sorbicillinoid produced by *Trichoderma citrinoviride* associated to the green alga *Cladophora* sp., on nitrite production in LPS-stimulated macrophages. *Nat. Prod. Res.* **2018**. [CrossRef]
6. Keller, N.P.; Turner, G.; Bennett, J.W. Fungal secondary metabolism—from biochemistry to genomics. *Nat. Rev. Microbiol.* **2005**, *3*, 937. [CrossRef] [PubMed]

7. Salvatore, M.M.; Nicoletti, R.; Salvatore, F.; Naviglio, D.; Andolfi, A. GC–MS approaches for the screening of metabolites produced by marine-derived *Aspergillus*. *Mar. Chem.* **2018**, *206*, 19–33. [CrossRef]
8. Samson, R.A.; Visagie, C.M.; Houbraken, J.; Hong, S.B.; Hubka, V.; Klaassen, C.H.W.; Perrone, G.; Seifert, K.A.; Susca, A.; Tanney, J.B.; et al. Taxonomy, identification and nomenclature of the genus *Aspergillus*. *Stud. Mycol.* **2014**, *78*, 141–173. [CrossRef]
9. Rank, C.; Klejnstrup, M.L.; Petersen, L.M.; Kildgaard, S.; Frisvad, J.C.; Held Gotfredsen, C.; Ostenfeld Larsen, T. Comparative chemistry of *Aspergillus oryzae* (RIB40) and *A. flavus* (NRRL 3357). *Metabolites* **2012**, *2*, 39–56. [CrossRef]
10. Cary, J.W.; Gilbert, M.K.; Lebar, M.D.; Majumdar, R.; Calvo, A.M. *Aspergillus flavus* secondary metabolites: More than just aflatoxins. *Food Safety* **2018**, *6*, 7–32. [CrossRef]
11. Nicoletti, R.; Fiorentino, A. Plant bioactive metabolites and drugs produced by endophytic fungi of Spermatophyta. *Agriculture* **2015**, *5*, 918–970. [CrossRef]
12. Ramírez-Camejo, L.A.; Zuluaga-Montero, A.; Lázaro-Escudero, M.; Hernández-Kendall, V.; Bayman, P. Phylogeography of the cosmopolitan fungus *Aspergillus flavus*: Is everything everywhere? *Fungal Biol.* **2012**, *116*, 452–463. [CrossRef] [PubMed]
13. Foley, K.; Fazio, G.; Jensen, A.B.; Hughes, W.O. The distribution of *Aspergillus* spp. opportunistic parasites in hives and their pathogenicity to honey bees. *Veter. Microbiol.* **2014**, *169*, 203–210. [CrossRef] [PubMed]
14. Vojvodic, S.; Jensen, A.B.; James, R.R.; Boomsma, J.J.; Eilenberg, J. Temperature dependent virulence of obligate and facultative fungal pathogens of honeybee brood. *Veter. Microbiol.* **2011**, *149*, 200–205. [CrossRef]
15. Ilyasov, R.; Gaifullina, L.; Saltykova, E.; Poskryakov, A.; Nikolenko, A. Review of the expression of antimicrobial peptide defensin in honey bees *Apis mellifera* L. *J. Apicultural Sci.* **2012**, *56*, 115–124. [CrossRef]
16. Di Prisco, G.; Annoscia, D.; Margiotta, M.; Ferrara, R.; Varricchio, P.; Zanni, V.; Caprio, E.; Nazzi, F.; Pennacchio, F. A mutualistic symbiosis between a parasitic mite and a pathogenic virus undermines honey bee immunity and health. *Proc. Natl. Acad. Sci.* **2016**, *113*, 3203–3208. [CrossRef] [PubMed]
17. Benoit, J.B.; Yoder, J.A.; Sammataro, D.; Zettler, L.W. Mycoflora and fungal vector capacity of the parasitic mite *Varroa destructor* (Mesostigmata: Varroidae) in honey bee (Hymenoptera: Apidae) colonies. *Int. J. Acarology* **2004**, *30*, 103–106. [CrossRef]
18. Yang, X.; Cox-Foster, D.L. Impact of an ectoparasite on the immunity and pathology of an invertebrate: Evidence for host immunosuppression and viral amplification. *Proc. Natl. Acad. Sci.* **2005**, *102*, 7470–7475. [CrossRef]
19. Li, X.; Jeong, J.H.; Lee, K.T.; Rho, J.R.; Choi, H.D.; Kang, J.S.; Son, B.W. γ-Pyrone derivatives, kojic acid methyl ethers from a marine-derived fungus *Altenaria* sp. *Arch. Pharm. Res.* **2003**, *26*, 532–534. [CrossRef]
20. Cabrera, G.M.; Roberti, M.J.; Wright, J.E.; Seldes, A.M. Cryptoporic and isocryptoporic acids from the fungal cultures of *Polyporus arcularius* and *P. ciliatus*. *Phytochemistry* **2002**, *61*, 189–193. [CrossRef]
21. Alvarez-Ibarra, C.; Quiroga-Feijóo, M.L.; Toledano, E. An analysis of substituent effects on ^{1}H and ^{13}C-NMR parameters of substituted furans. Linear free energy relationships and PM3 semiempirical calculations. *J. Chem. Soc., Perkin Trans. 2* **1998**, *3*, 679–690. [CrossRef]
22. Evidente, A.; Cristinzio, G.; Punzo, B.; Andolfi, A.; Testa, A.; Melck, D. Flufuran, an antifungal 3,5-disubstituted furan produced by *Aspergillus flavus* Link. *Chem. Biodiver.* **2009**, *6*, 328–334. [CrossRef] [PubMed]
23. Balde, E.S.; Andolfi, A.; Bruyère, C.; Cimmino, A.; Lamoral-Theys, D.; Vurro, M.; Damme, M.V.; Altomare, C.; Mathieu, V.; Kiss, R.; et al. Investigations of fungal secondary metabolites with potential anticancer activity. *J. Nat. Prod.* **2010**, *73*, 969–971. [CrossRef] [PubMed]
24. Ma, C.; Ma, Y.M.; Li, T.; Wang, J. Structure and activity of secondary metabolites of endophytic fungus S19 strain in *Cephalotaxus fortune*. *Guizhou Nongye Kexue* **2014**, *42*, 152–156.
25. Ma, Y.M.; Ma, C.C.; Li, T.; Wang, J. A new furan derivative from an endophytic *Aspergillus flavus* of *Cephalotaxus fortunei*. *Nat. Prod. Res.* **2016**, *30*, 79–84. [CrossRef] [PubMed]
26. Saldan, N.C.; Almeida, R.T.; Avíncola, A.; Porto, C.; Galuch, M.B.; Magon, T.F.; Pilau, E.J.; Svidzinski, T.I.E.; Oliveira, C.C. Development of an analytical method for identification of *Aspergillus flavus* based on chemical markers using HPLC-MS. *Food Chem.* **2018**, *241*, 113–121. [CrossRef]
27. Lee, M.; Cho, J.Y.; Lee, Y.G.; Lee, H.J.; Lim, S.I.; Lee, S.Y.; Nam, Y.D.; Moon, J.H. Furan, phenolic, and heptelidic acid derivatives produced by *Aspergillus oryzae*. *Food Sci. Biotechnol.* **2016**, *25*, 1259–1264. [CrossRef]

28. Yang, X.; Wang, P.; Ma, Y.; Jia, Q. Metabolites of *Aspergillus oryzae*, an endophytic fungus associated with *Lycium ruthenicum* Murr. *Tianran Chanwu Yanjiu Yu Kaifa* **2015**, *27*, 1554–1557.
29. Zhang, H.C.; Ma, Y.M.; Liu, R. Antimicrobial materials derived from the endophytic fungus *Fusarium* sp. of *Eucommia ulmoides*. *Advan. Mater. Res.* **2012**, *531*, 346–349. [CrossRef]
30. He, S.; Yan, X.; Wang, T. Marine *Penicillium dipodomyicola* for manufacture of flufuran. 2014, CN 104031845 A 20140910. CN 104031845 A 20140910, 2014.
31. Elsunni, M.A.; Yang, Z.-D. Secondary metabolites of the endophytic fungi *Penicillium polonicum* and their monoamine oxidase inhibitory activity. *Chem. Nat. Comp.* **2018**, *54*, 1018–1019. [CrossRef]
32. Zhou, J.; Li, G.; Deng, Q.; Zheng, D.; Yang, X.; Xu, J. Cytotoxic constituents from the mangrove endophytic *Pestalotiopsis* sp. induce G0/G1 cell cycle arrest and apoptosis in human cancer cells. *Nat. Prod. Res.* **2018**, *32*, 2968–2972. [CrossRef]
33. Nurchi, V.M.; Crisponi, G.; Lachowicz, J.I.; Murgia, S.; Pivetta, T.; Remelli, M.; Rescigno, A.; Niclós-Gutìerrez, J.; González-Pérez, J.M.; Domìnguez-Martìn, A.; et al. Iron (III) and aluminum (III) complexes with hydroxypyrone ligands aimed to design kojic acid derivatives with new perspectives. *J. Inorg. Biochem.* **2010**, *104*, 560–569. [CrossRef] [PubMed]
34. Paterson, R.R.M.; Kemmelmeier, C. Neutral, alkaline and difference ultraviolet spectra of secondary metabolites from *Penicillium* and other fungi, and comparisons to published maxima from gradient high-performance liquid chromatography with diode-array detection. *J. Chromatogr. A* **1990**, *511*, 195–221. [CrossRef]
35. Klaiklay, S.; Rukachaisirikul, V.; Phongpaichit, S.; Buatong, J.; Preedanon, S.; Sakayaroj, J. Flavodonfuran: A new difuranylmethane derivative from the mangrove endophytic fungus *Flavodon flavus* PSU-MA201. *Nat. Prod. Res.* **2013**, *27*, 1722–1726. [CrossRef] [PubMed]
36. Kobori, H.; Sekiya, A.; Yasuda, N.; Noguchi, K.; Suzuki, T.; Choi, J.H.; Hirai, H.; Kawagishi, H. Armillariols A to C from the culture broth of *Armillaria* sp. *Tetrahedron Lett.* **2013**, *54*, 5481–5483. [CrossRef]
37. Ding, L.J.; Gu, B.B.; Jiao, W.H.; Yuan, W.; Li, Y.X.; Tang, W.Z.; Yu, H.B.; Liao, X.J.; Han, B.N.; Li, Z.Y.; et al. New furan and cyclopentenone derivatives from the sponge-associated fungus *Hypocrea koningii* PF04. *Mar. Drugs* **2015**, *13*, 5579–5592. [CrossRef]
38. Chen, L.L.; Wang, P.; Chen, H.Q.; Guo, Z.K.; Wang, H.; Dai, H.F.; Mei, W.L. New furan derivatives from a mangrove-derived endophytic fungus *Coriolopsis* sp. J5. *Molecules* **2017**, *22*, 261. [CrossRef] [PubMed]
39. Uchoa, P.K.S.; Pimenta, A.T.; Braz-Filho, R.; de Oliveira, M.D.C.F.; Saraiva, N.N.; Rodrigues, B.S.; Pfenning, L.H.; Abreu, L.M.; Wilke, D.V.; Florêncio, K.G.D.; et al. New cytotoxic furan from the marine sediment-derived fungi *Aspergillus niger*. *Nat. Prod. Res.* **2017**, *31*, 2599–2603. [CrossRef] [PubMed]
40. Wu, X.; Pang, X.J.; Xu, L.L.; Zhao, T.; Long, X.Y.; Zhang, Q.Y.; Qin, H.L.; Yang, D.F.; Yang, X.L. Two new alkylated furan derivatives with antifungal and antibacterial activities from the plant endophytic fungus *Emericella* sp. XL029. *Nat. Prod. Res.* **2018**, *32*, 2625–2631. [CrossRef] [PubMed]
41. Bentley, R. From miso, sake and shoyu to cosmetics: A century of science for kojic acid. *Nat. Prod. Rep.* **2006**, *23*, 1046–1062. [CrossRef] [PubMed]
42. Dembitsky, V.M.; Kilimnik, A. Anti-melanoma agents derived from fungal species. *Mathews, J. Pharm. Sci.* **2016**, *1*, 002.
43. Mohamad, R.; Mohamed, M.S.; Suhaili, N.; Salleh, M.M.; Ariff, A.B. Kojic acid: Applications and development of fermentation process for production. *Biotechnol. Mol. Biol. Rev.* **2010**, *5*, 24–37.
44. Pitt, J.I.; Hocking, A.D.; Glenn, D.R. An improved medium for the detection of *Aspergillus flavus* and *A. parasiticus*. *J. Appl. Bacteriol.* **1983**, *54*, 109–114. [CrossRef] [PubMed]
45. Cubero, O.F.; Crespo, A.; Fatehi, J.; Bridge, P.D. DNA extraction and PCR amplification method suitable for fresh, herbarium-stored, lichenized, and other fungi. *Plant Syst. Evol.* **1999**, *216*, 243–249. [CrossRef]
46. Peterson, S.W. Phylogenetic analysis of *Aspergillus* species using DNA sequences from four loci. *Mycologia* **2008**, *100*, 205–226. [CrossRef] [PubMed]
47. Nicoletti, R.; Buommino, E.; De Filippis, A.; Lopez-Gresa, M.P.; Manzo, E.; Carella, A.; Petrazzuolo, M.; Tufano, M.A. Bioprospecting for antagonistic *Penicillium* strains as a resource of new antitumor compounds. *World J. Microbiol. Biotechnol.* **2008**, *24*, 189–195. [CrossRef]

48. Meloun, M.; Havel, J.; Högfeldt, E. *Computation of Solution Equilibria: A Guide to Methods in Potentiometry, Extraction and Spectrophotometry*; Ellis Horwood Ltd.: Chichester, UK, 1988.
49. Gans, P.; Sabatini, A.; Vacca, A. Investigation of equilibria in solution. Determination of equilibrium constants with the HYPERQUAD suite of programs. *Talanta* **1996**, *43*, 1739–1753. [CrossRef]

Sample Availability: Samples of the compounds **1-9** are available from the authors.

molecules

MDPI

Article

Chemical Analysis of *Lepidium meyenii* (Maca) and Its Effects on Redox Status and on Reproductive Biology in Stallions †

Simona Tafuri [1,‡], Natascia Cocchia [1,*,‡], Domenico Carotenuto [2], Anastasia Vassetti [3], Alessia Staropoli [3,4], Vincenzo Mastellone [1], Vincenzo Peretti [1], Francesca Ciotola [1], Sara Albarella [1], Chiara Del Prete [1], Veronica Palumbo [1], Luigi Esposito [1], Francesco Vinale [3,4] and Francesca Ciani [1]

1 Department of Veterinary Medicine and Animal Production, University of Naples Federico II, 80137 Naples, Italy; stafuri@unina.it (S.T.); vincenzo.mastellone@unina.it (V.M.); vincenzo.peretti@unina.it (V.P.); francesca.ciotola@unina.it (F.C.); sara.albarella@unina.it (S.A.); chiara.delprete@unina.it (C.D.P.); veronica.palumbo@unina.it (V.P.); luigespo@unina.it (L.E.); ciani@unina.it (F.C.)

2 UNMSM, Universidad Nacional Mayor San Marcos, Lima 11-0058, Peru; domenicar@alice.it

3 Institute for Sustainable Plant Protection, National Research Council, 80055 Portici (Na), Italy; anastasia.vassetti@ipsp.cnr.it (A.V.); alessia.staropoli@unina.it (A.S.); francesco.vinale@ipsp.cnr.it (F.V.)

4 Department of Agricultural Sciences, University of Naples Federico II, 80055 Naples, Italy

* Correspondence: ncocchia@unina.it; Tel.: +39-81-253-6017; Fax: +39-81-253-6042

† Running title: Maca: Chemical Analysis and Biological Activity.

‡ The authors contributed equally to this work.

Academic Editor: Francesca Giampieri

Received: 15 March 2019; Accepted: 21 May 2019; Published: 23 May 2019

Abstract: The present study was conducted to assess the chemical composition of Yellow Maca (*Lepidium meyenii*) and its biological activity on stallions following oral administration of hypocotyl powder. Maca was subjected to methanolic extraction and the chemical analysis was carried out by LC-MS-QTOF (liquid chromatography-mass spectrometry). Our results showed that Maca contains some effective antioxidants, a high percentage of glucosinolates, and other important components with a high antioxidant capacity. To evaluate the plant biological activity in stallions fed with Maca powder for 60 days, the redox status and some reproductive parameters were investigated. Blood and semen samples were collected at 0, 30, 60, and 90 days from the beginning of this study. Blood samples showed a decrease of the reactive oxygen metabolites, evaluated by d-ROMs test, and an increase of the antioxidant barrier in terms of biological antioxidant potential (BAP test), powerful oxidant capacity (OXY-Adsorbent test), and thiols evaluation (-SHp test). Furthermore, semen samples showed a positive trend during Maca administration in the following parameters: ejaculate volumes and sperm concentrations, total and progressive motility, and acrosome integrity.

Keywords: *Lepidium meyenii* (Maca); stallion semen; oxidative stress; antioxidants; chemical analysis

1. Introduction

Maca (*Lepidium meyenii* Walp), belonging to the Brassicaceae family, was discovered more than 2000 years ago in the Andes highlands of Peru, where it grows exclusively between 3500 and 4500 m above sea level [1]. The hypocotyls have been widely used as a nutritional supplement and in folk medicine to increase fertility and sexual function [2]. The dried hypocotyls of Maca are rich in high nutritional value elements, such as carbohydrates, proteins, lipids, essential amino acids, and free fatty acids. Furthermore, Maca contains several secondary metabolites, such as macamides, macaridine, alkaloids, and glucosinolates [3]. The most abundant glucosinolates detected in Maca are aromatic glucosinolates, namely benzyl glucosinolate(glucotropaeolin) and *m*-methoxybenzyl glucosinolate (glucolimnanthin). Maca has been mainly classified in three ecotypes according to the color of the hypocotyls: red, yellow, and black. These showed different biological activity depending on the type of cultivation, processing, and extraction, and on the concentration of different bioactive metabolites [3,4]. The main chemical components of Maca were considered to be responsible, among other things, for its aphrodisiac, immunostimulant, and energizing properties [5,6].

Previous studies have demonstrated the effects of Maca on spermatogenesis, sperm count, and sperm motility [1,7]. Furthermore, in a rat model Maca restored the spermatogenesis decrease when altitude increases [8]. Additionally, diet supplementation of Maca showed a beneficial effect on semen quality of hypofertile and fertile stallions [9], as well as on the quantity of fresh semen and the quality of semen after storage at 5 °C up to 72 h [10].

The use of artificial insemination (AI) has greatly increased in recent decades. AI has several advantages compared to natural service, including higher pregnancy rates and reduced transportation costs of a mare to the stallion, with, consequently, lower stress and prevention of a stallion overuse. Semen used for AI should be preserved from damage occurring due to oxidative stress [10]. A strategy to prevent oxidative damage to spermatozoa may be found in food supplementation with antioxidants. The presence of compounds, such as macamides and glucosinolates, which are hydrolisated to isothiocyanates by myrosinase, allows Maca to scavenge free radicals and protect cells from oxidative stress (OS). Reactive oxygen species (ROS) are produced in different sites in mammalian bodies [11], and were found to accumulate during the normal testicular spermatogenesis and steroidogenesis [12]. Physiological low levels of ROS play an important role in processes such as capacitation, hyperactivation, acrosome reaction, and sperm-oocyte fusion, thus ensuring appropriate fertilization, whereas high levels of ROS may determine sperm pathologies, such as ATP depletion and loss of sperm motility and viability [13]. ROS activity is of major concern for sperm quality both in vivo and during in vitro incubation, as well as during its cool storage [14]. The cellular antioxidant defense systems are able to control the deleterious effects of ROS, but if an imbalance between oxidant and antioxidant systems determines ROS accumulation, OS may occur [15–17].

In the present study the chemical composition of Yellow Maca (*Lepidium meyenii*) was investigated by using LC-MS- QTOF. Moreover, the biological effects on stallions orally fed during the breeding season with Maca hypocotyl powder was evaluated in terms of antioxidant activity and semen quantity and quality.

2. Materials and Methods

2.1. Plant Material

Yellow Maca hypocotyls used in this study were harvested in the district of Junin, in the Andean Highlands of Peru (4100 m above the sea level). The taxonomic identification of the plant was performed by Professor Domenico Carotenuto (Figure 1).

Figure 1. Typical roots of Yellow Maca from Junin district, Peru.

Maca powder was obtained from hypocotyls exposed for two months at extreme temperature cycles, strong light conditions, and atmospheric pressure typical of the high altitude environment, using the traditional open-field drying process. After drying, hypocotiles were selected, washed, ground to obtain a flour with a particle size of 0.8 mm, and stored until used.

2.2. Extraction and Chemical Analysis of Maca

A total of 3 g of dried sample was extracted with 30 mL of methanol (MeOH) for 30 min under ultrasonic treatment and centrifuged. The solution was stored at 4 °C and filtered through a 0.22 μm membrane syringe filter before injection into the LC system. All analyses were performed on an Agilent high performance liquid chromatograph (HPLC) 1260 Infinity Series (Agilent Technologies, Santa Clara, CA, USA) equipped with a DAD (Diode Array Detector) system (Agilent Technologies) and coupled to a quadrupole-time of flight (Q-TOF) mass spectrometer model G6540B (Agilent Technologies) with a Dual ESI source (Agilent Technologies). Separations were performed on a Zorbax Eclipse Plus C-18 column (4.6 × 100 mm, with 3.5 μm particles, Agilent Technologies), held at a constant temperature of 30 °C. For elution, mobile phase consisted of A: 0.1% (*v/v*) formic acid (FA) in water (H_2O) and B: 0.1% formic acid (FA) in acetonitrile (ACN). Gradient elution was done at a flow rate of 0.3 mL/min, as follows: 0–1% B at 0–5 min, 1–30% B at 5–20 min, 30–50% B at 20–30 min, 50–70% B at 30–40 min, 70–95% B at 40–60 min, 95% B at 60–75 min, and 95–100% B at 75–80 min (as equilibration time). The injection volume was 5 μL. UV spectra were collected by DAD, setting the detection wavelength at 220 and 280 nm. MS and MS/MS parameters were set using the Agilent MassHunter Data Acquisition Software, rev. B.05.01 (Agilent Technologies). The system operated in positive and negative ion mode and for both cases MS spectra were recorded in *m/z* 50–1000 range as centroid spectra, with a speed of 1.5 spectra/s. The capillary was maintained at 4000 V, fragmentor voltage at 180 V, cone 1 (skimmer 1) at 45 V, Oct RFV at 750 V. Gas flow rate was set at 11 L/min, at 350 °C, and the nebulizer was set at 45 psig. A standard solution was infused by using an Isocratic pump (1260 Infinity Series, Agilent Technologies) in order to perform the real-time lock mass correction. The solution consisted of two reference mass compounds: purine ($C_5H_4N_4$ at *m/z* 121.050873, 10 μmol/L) and hexakis (1H,1H, 3H-tetrafluoropentoxy)-phosphazene ($C_{18}H_{18}O_6N_3P_3F_{24}$ at *m/z* 922.009798, 2 μmol/L) [18–20]. Flow rate was set at 0.06 mL/min, while the detection window and the minimum height were set at 1000 ppm and 10,000 counts, respectively, for reference mass correction. MS/MS spectra were recorded in *m/z* 50–500 mass range, with a speed of 2 spectra/s. Collision energy was set at 20 eV.

A recovery study was done according to Zhou et al. [18]; recovery percentage was calculated for three standard compounds, namely benzyl glucosinolate, *m*-methoxybenzyl glucosinolate, and *N*-benzylhexadecanamide, resulting in 96.51%, 97.62%, and 98.15%, respectively. Raw data were evaluated using Mass Hunter Qualitative Analysis Software, rev B.06.00 (Agilent Technologies), and compound identification was carried out using an in-house database (containing about 2000 different plant metabolites), and by comparison with standard compounds or with data present in the literature [18,21–23]. Among the detected molecules, only those with a mass error below 10 ppm and a sufficient score (>80% similarity) were reported.

2.3. Animals and Experimental Design

In the present study, active breeding healthy stallions weighing 400 to 600 kg were included. The study was performed in accordance with the code of ethics (Legislative Decree 26–04/03/2014) and the protocol and procedures were approved by the Ethics Committee of Department of Veterinary Medicine and Animal Production at the University of Naples Federico II, Italy (Protocol Number 0003909).

The study was conducted during the breeding season (lasting from February to July) starting from the beginning of April to the end of July. Animals of both control (C; $n = 5$) and treated (M; $n = 5$) groups received an identical diet based on hay and concentrate pellets. In M the concentrate was supplemented daily with 20 g of Yellow Maca powder, resulting in an average dosage of 4 g Maca/100 kg body weight (minimum: 3 g Maca/100 kg; maximum: 5 g Maca/100 kg) for 60 days. Blood and semen specimens were collected at the beginning of the experiment (D0) and every month (D30, D60, and D90, respectively) for a total of 4 samplings for each stallion. The dosage was based on studies performed on humans [1,24] and animals [8,10,25,26]. Blood analysis was used to monitor oxidative stress. Semen was collected from stallions of treated group (M) and both quantitative (ejaculate volume, concentration, and total sperm count) and qualitative (total and progressive motility and acrosome integrity) determinations were performed on 4 animals according to previous studies carried out in our laboratory [10,27,28].

2.4. Systemic Oxidative Stress Status

All kits for the evaluation of OS status were purchased from Diacron International, Grosseto, Italy. The pro-oxidative status was evaluated by measuring hydroperoxides in the serum using the d-ROMs test; the measurements of antioxidant capacity in serum samples were performed by BAP-test, OXY-Adsorbent test, and thiols by -SHp test. The ratio between the values of d-ROMs and OXY (×100) (Oxidative Stress index—OSi) is an arbitrary value, used as an index of plasma redox status; high values indicate a higher concentration of oxidized molecules than non-enzymatic antioxidants [29].

The d-ROMs test is a photometric test that allows the oxidant capacity of plasma to be determined, mainly related to the level of hydroperoxides (a subclass of reactive oxygen metabolites, ROM); it measures the oxidant ability of a serum sample towards a particular substance used as an indicator (chromogen, e.g., *N,N*-diethylparaphenylendiamin), producing a pink-colored derivative that is photometrically quantified at 505 nm [30]. The intensity of the developed color is directly proportional to the concentration of ROMs, according to the Lambert-Beer's law, and is expressed as Carratelli Units (1 CARR U = 0.08 mg hydrogen peroxide/dl).

In the Biological Antioxidant Potential (BAP) test the antioxidant power of the serum is evaluated by measuring the capacity of the sample to reduce a solution of ferric chloride from Fe^{3+} to Fe^{2+}. This test measures the directly-active fraction of the anti-oxidant barrier, encompassing both exogenous antioxidants (vitamin C, vitamin E, carotenoids, bioflavonoids) and endogenous ones (bilirubin, uric acid, cholesterol, protein). Some of these substances exhibit a scavenger-type activity, i.e., they neutralize free radicals by interacting directly with them. The values of the BAP tests are expressed as μM of a sample.

The OXY-Adsorbent test assesses the ability of the plasma to counteract the oxidation capacity of a hypochlorous acid (HClO) solution, which is both a powerful and physiological oxidant and is able to mimic situations occurring in vivo. Bilirubin, uric acid, vitamins C and E, albumin, and in general, macromolecular complexes that act as a shock absorber against free radicals, such as glycoproteins, help curb the oxidizing action of HClO. The values of the OXY-Adsorbent tests are expressed as μmol HClO/L of a sample.

The -SHp test measures thiols, organic compounds having a sulfhydryl group (-SH), that represent a significant qualitative component of the plasma antioxidant barrier against free radicals. Thiols also include cysteine and lipoic acid.

2.5. Statistical Analysis

The results are expressed as mean ± standard deviation (SD). Levels of OS in blood samples in C and M groups were compared using the Student *t*-test, as the data were normally distributed. A p value less than 0.05 was considered to be statistically significant. Due to the reduced number of semen samples obtained exclusively from animals fed with Maca (M), the Student *t*-test was performed on quantitative parameters of stallion semen between the samples collected at D30, D60, and D90, while samples collected at D0 were considered as controls.

3. Results

3.1. Chemical Analysis of Maca

LC-MS analysis (Figures S1 and S2) allowed to detect 88 different compounds, of which 80 were detected in ESI positive ion mode and 8 in ESI negative ion mode. Raw data were compared both with the in-house database and with data present in the literature [18,21–23]. From this analysis, 13 compounds were identified in positive mode via database search and 5 compounds using literature data (Table 1). Only *N*-benzylhexadecanamide (Compound No. 16) was identified by comparison with a commercial standard (Table 1). In negative mode 2 compounds were identified by comparing data with both existing literature and standard compounds (benzyl glucosinolate and *m*-methoxybenzyl glucosinolate, Table 2).

In Table 1 molecules belonging to alkaloids and alkamides are reported. Compounds 1,3-dibenzyl-2, 4, 5-trimethylimidazilium (called Lepidine B or Macaline B) belong to the former class (No. 1, 3–8 in Table 1). Alkamides are natural compounds found in many plants formed by fatty acids and different amide groups. Polyunsaturated fatty acids that are specific for Maca are called macaenes and the LC-MS analysis allowed one of them (5-oxo-6*E*, 8*E* octadecadienoic acid; N. 9 in Table 1) to be identified. Identification also allowed 9 macamides (No. 2, 10–18 in Table 1) to be found, which are amide derivatives of fatty acids exclusively found in Maca.

The metabolites reported in Table 2, benzyl glucosinolate and *m*-methoxybenzyl glucosinolate, are two important glucosinolates of Maca that may be converted into isothiocyanates when the plant is damaged [31,32].

3.2. Oxidative Stress Measurements in Blood Samples

The values of blood count did not show any difference between C and M groups, but they confirmed the good state of health of the animals during the whole experimental period (data not shown). Table 3 reports the pro- and antioxidant data of blood samples. The mean values obtained with the d-ROMs test showed a constant decrease in M compared to C group, with $p < 0.1$ at D60, the time point corresponding to the end of Maca administration (Table 3). Interestingly, the evaluation of the antioxidant barrier showed higher mean values of the OXY-Adsorbent, and BAP tests in M compared to C at D60 ($p < 0.1$) and -SHp at D90 ($p < 0.1$) (Table 3). A significant increase of BAP was highlighted at D90 ($p < 0.05$) in M versus C (Table 3).

OSi showed a progressive decrease in treated animals (M), as compared to controls (C), exhibiting the lowest value at D60, which corresponded to the end of Maca administration. Conversely, 30 days after suspension of supplementation (D90) a slight increase of OSi was observed; however, this value was lower than the correspondent in C and M at D0 (Table 4).

3.3. Semen Analysis

The parameters for semen quantity are shown in Table 5. The ejaculate volume increased gradually from D0 to D60 ($p < 0.05$), and decreased at D90. The sperm concentration and total sperm count steadily increased from D0 to D90 ($p < 0.05$), even if a slight decrease was observed at D60 (Table 5).

Table 1. Identification of the main *Lepidium meyenii* (Maca) metabolites by LC-MS-DAD and -qTOF analyses operating in positive ion mode (ESI +).

No.	Name	RT (min)	ESI(+)/Expected (*m/z*)	ESI(+)/Measured (*m/z*)	Delta (ppm)	Main Fragment (*m/z*)	λmax (nm)
1 *	(1*R*,3*S*)-1-methyltetrahydro-beta-5,6-hydride-carboline-3-carboxylic acid	13.842	233.1284	233.1276	3.431585341	91.054	215.5
2 *	*N*-(3-hydroxy-benzyl)-2*Z*-fivecarbon acrylamide	18.693	180.1019	180.1019	0.027	162.0914	220.1
3 *	1-dibenzyl-2-propane-4,5-dimethylimidazilium	18.997	227.1542	227.1546	−1.76091835	91.0544	219.5
4 **	3-benzyl-1,2-dihydro-*N*-hydroxypyridine-4-carbaldehyde	22.571	216.1175	216.1173	0.916936972	91.0543	213.2
5 *	1,3-dibenzyl-2,4,5-trimethylimidazilium	24.574	291.1854	291.1851	1.030271435	91.054	214.1
6 *	1,3-dibenzyl-2-phenyl-4,5-dimethylimidazilium	27.749	353.2014	353.2013	0.283124586	185.1072	221.1
7 **	*N*-ethyl-tetradecene ester	28.187	274.274	274.2743	−1.09379671	256.2637	221.2
8 **	1,3-dibenzyl-2-pentyl-4,5-dimethylimidazilium	29.578	347.2481	347.2482	−0.28797854	277.1705	222.7
9 **	5-oxo-6*E*,8*E*-octadecadienoic acid	40.444	295.2267	295.2266	0.338722751	277.2165	223.9
10 **	*N*-benzyl-5-oxo-6*E*,8*E*-octadecadienamide	43.789	384.2824	384.2822	0.520450585	306.2421	224.3
11 *	*N*-benzyl-9-oxo-12*E*-octadecenamide	44.976	386.2981	386.2979	0.517734879	261.2212	223.8
12 *	*N*-(3-methoxybenzyl)-(9*Z*,12*Z*,15*Z*)-octadecatrienamide	48.505	398.298	398.2976	1.004273182	381.2785	223.5
13 *	*N*-benzyl-(9*Z*,12*Z*,15*Z*)-octadecatrienamide	48.863	368.2875	368.288	−1.357635	351.2678	223.9
14 *	*N*-(3-Methoxybenzyl)-(9*Z*,12*Z*)-octadecadienamide	51.403	400.3137	400.3133	0.999216364	365.2838	224.3
15 *	*N*-Benzyl-(9*Z*,12*Z*)-octadecadienamide	51.868	370.3031	370.3034	−0.81014715	232.1694	224.7
16 *×	*N*-benzylhexadecanamide	54.34	346.3031	346.3034	−0.86629314	221.2261	221.3
17 *	*N*-benzyl-9-*Z*-octadecenamide	55.169	372.3188	372.3194	−1.61152217	247.2423	225.9
18 *	*N*-benzyloctadecanamide	59.184	374.3345	374.3344	0.267140752	296.3172	225.5

Note: * compound identified by comparison with in-house database of plant secondary metabolites; ** compound identified by comparison with literature [18,21–23]; × compound identified by comparison with a reference standard.

Table 2. Identification of the main *Lepidium meyenii* (Maca) metabolites by LC-MS-DAD and qTOF analyses operating in negative ion mode (ESI⁻).

No.	Name	RT (min)	ESI(-)/ Expected (*m*/*z*)	ESI(-)/ Measured (*m*/*z*)	Delta (ppm)	Main Fragment (*m*/*z*)	λmax (nm)
1 ** ×	Benzyl glucosinolate	14.174	408.0423	408.0424	−0.24507263	274.9903 241.0027 166.0333	217.2
2 ** ×	*m*-methoxybenzyl glucosinolate	15.421	438.0534	438.054	−1.36969602	358.0956 259.0126	219.4

Note: ** compound identified by comparison with literature [18,21–23]; × compound identified by comparison with a reference standard.

Table 3. Mean values (±SD) of pro- and antioxidant data of blood samples calculated in C and M groups, evaluated during the experimental trial.

	d-ROMs (U CARR)		OXY-Ads (μmol HClO/mL)		BAP (μEq/L)		-SHp (μM/L)	
	C	M	C	M	C	M	C	M
D0	336 (±34.4)	316.8 (±33.5)	325 (±9.9)	335.8 (±12.13)	1759.3 (±58.8)	1844.4 (±88.2)	532.7 (±57.5)	568.2 (±32.5)
D30	342.6 (±37.3)	329.8 (±31)	355.2 (±26.3)	375.6 (±27.4)	1652.4 (±122.7)	1722.6 (±151.2)	602.4 (±70.0)	645 (±48.6)
D60	315.8 (±12.1)	299.6 (±14.3)	354 (±58.5)	402 (±33.8)	1746.4 (±193.2)	1902.5 (±138.5)	579.8 (±67.7)	706.2 (±95.7)
D90	349.8 (±28.7)	336.8 (±29.9)	364.8 (±43.2)	379.8 (±52.7)	1580.8 (±51.6)	1683 * (±65.6)	469.5 (±49.6)	554.6 (±56.3)

C = control group; M = Maca treated group; D0, D30, D60, D90, samples taken at 0, 30, 60, 90 days from the beginning of the study, respectively; * $p < 0.05$ vs. corresponding C (Student *t*-test).

Table 4. OSi (%) is an arbitrary calculated value that indicates the plasma redox status for C and M groups.

	OSi (%)	
	C	M
D0	103.4	94.3
D30	96.5	87.8
D60	89.2	74.5
D90	95.9	88.7

Table 5. Mean values (±SD) of some quantitative parameters of stallion semen fed with Maca, evaluated during the experimental trial.

	D0	D30	D60	D90
Ejaculate volume (mL)	38.38 ± 12.12	48.13 ± 17.50 *	60.88 ± 29.09 *	47.5 ± 23.63 *
Sperm concentration (×10^6/mL)	124.38 ± 43.05	220.00 ± 90.69 *	178.88 ± 86.54 *	279 ± 126.87 *
Total Sperm Count (×10^9)	3.68 ± 1.18	10.15 ± 6.22 *	8.20 ± 7.65 *	8.83 ± 4.36 *

Note: * $p < 0.05$ vs. D0 (Student *t*-test).

The parameters for semen quality (total and progressive motility, and acrosome integrity) of 4 stallions in the M group are shown in Figure 2. The total motility reached a maximum value approximately after 60 days of Maca administration, while decreasing 30-days after Maca suspension (D90, Figure 2A). Conversely, both progressive motility and acrosome integrity increased gradually up to D90 (Figure 2B,C, respectively).

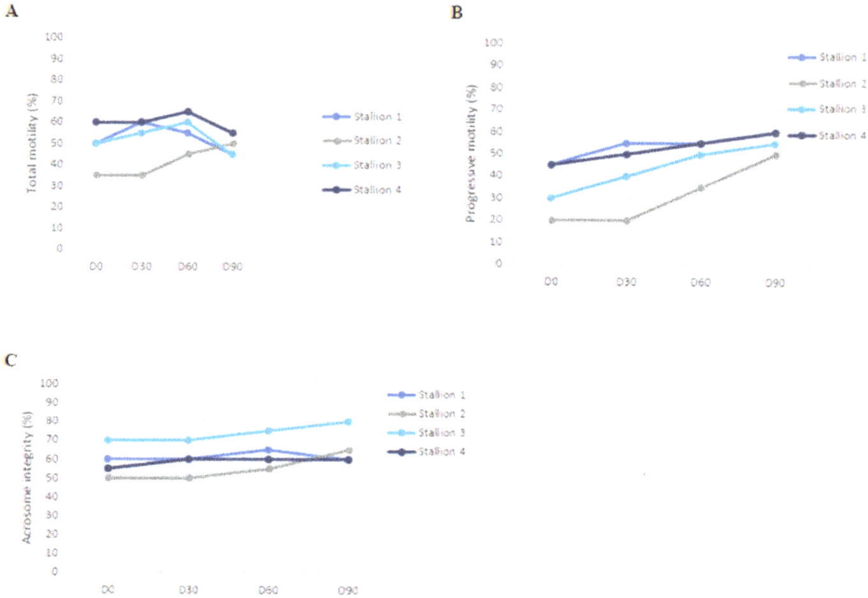

Figure 2. Trends of qualitative parameters of semen collected from four stallion fed with Maca during the experimental trial. (**A**) Total motility; (**B**) progressive motility; and (**C**) acrosome integrity.

4. Discussion

In this study *Lepidium meyenii* (Maca) was characterized from the chemical point of view, and its main bioactive compounds were identified. Moreover, the antioxidant capacity of Maca diet supplementation was evaluated by redox status of blood, and by qualitative and quantitative semen parameters in stallions.

In our study yellow Maca hypocotyls were found to contain several compounds with antioxidant activity, such as alkaloids and glucosinolates; benzyl glucosinolate and *m*-methoxybenzyl glucosinolate are commonly the most abundant molecules found in fresh and dry hypocotils of Maca [31]. According to the analysis, the extract contains a percentage of 24.37% for benzyl glucosinolate and 4.31% for *m*-methoxybenzyl glucosinolate. The level of glucosinolates and their derivatives in the dry product is considered a good indicator of the quality of the product and of the effectiveness of the drying process [33]. Moreover, plant glucosinolates may be converted by myrosinase in isothiocyanates, bioactive metabolites with protective properties. Myrosinase is also present in the gut microflora of mammals [30,31].

Macamides are specific alkamides of Maca that are known for their antioxidant effect. Their concentrations are generally low, but their presence can be used to measure and standardize Maca quality [34,35]. The total amount of alkaloids in the extract is 40.74% and the percentage of other unidentified compounds is 31.21%. The drying process may affect the amount of macamides. Esparza et al. [33] showed that the traditional open-field drying method led the accumulation of macamide precursors.

In the present study, the biological effect of Maca administration was evaluated by the systemic redox status. The antioxidant barrier evaluation provides indications on the type of defenses that the body can adopt to obtain the balance between pro- and antioxidant species. BAP, OXY-Adsorbent, and -SH tests measure the antioxidant capacity in blood from different points of view; thus, altogether give a complete picture of the antioxidant arrangement. In our study, a progressive increase of all antioxidants occurred during the first part of the experimental trial, with the maximum values obtained at D60, which corresponded to the suspension of diet supplementation with Maca. Subsequently, at D90 the values decreased, possibly as a consequence of the suspension of Maca supplementation, but they were significantly higher in M vs. C in BAP. Conversely, the detection of d-ROMs has shown a progressive decrease from D0 to D60 in M group, as compared to controls (C), while only at D90 an increase was highlighted.

OSi allows a global view of the degree of oxidative stress, because it relates to the pro- and antioxidant status, thus avoiding data misinterpretation. The lowest value of OSi was found in M group at D60; this may indicate that after 60 days of Maca administration the redox balance in blood samples was hanging towards the antioxidant barrier. It is well known that high values of OSi could reflect an imbalance between oxidant and antioxidant systems, with an over-accumulation of ROS that can affect membrane structure and consequently changes in membrane fluidity [13,36]. However, despite the decrease of the antioxidant markers and increase of the oxidants at D90 with respect to D60, these final values were always higher in M group compared to the correspondent controls, and for the BAP were also statistically significant ($p < 0.05$ Student *t*-test). A range of the normal values of the redox status markers has been established for humans [30], but not for horses. In the literature very few data are available currently [37]; thus far, our results may provide knowledge for the study of the blood redox status in horses and could be useful in establishing species-specific references.

Finally, the biological effects of Maca administration as diet supplementation to stallions were evaluated in the reproductive season from April to July. Semen samples were collected at the same time points as blood samples (D0, D30, D60, D90). Quantitative semen parameters showed a progressive and significant increase during the experimental trial. The increase of ejaculate volume is in agreement with Gonzales et al. [1] and Melnikovova et al. [38], who highlighted this effect in rats and in humans, respectively. Sperm concentrations and total sperm count showed the same trend as the ejaculate volume, and this confirms the observations in a previous study by our group [10]. These results may also present interesting economic advantages, as an increased volume of the ejaculate allows preparation of a greater number of pellets to be used for AI and a greater concentration allows a higher probability of fertilization.

Data reported in Figure 2 show the trend of qualitative semen parameters of stallions belonging to M group. It was not possible to perform a reliable statistical analysis of the results, due to the reduced number of animals and for a lack of a control group, but the beneficial trend observed in terms of total and progressive motility and acrosomal integrity during the treatment with Maca confirms our previous findings [10].

In conclusion, this study revealed that Yellow Maca composition, and particularly the presence of specific bioactive metabolites, such as glucosinolates and macamides, may be responsible for the plant antioxidant activity. Moreover, our data showed that Maca supplementation positively affected the redox status and some reproductive parameters in stallions.

Supplementary Materials: The following are available online, Figures S1 and S2.

Author Contributions: Conceptualization, S.T., F.C. (Francesca Ciani) and N.C.; methodology, S.T., F.C. (Francesca Ciani), N.C., F.V. and F.C. (Francesca Ciotola); validation, V.P. (Veronica Palumbo), C.D.P. and L.E.; formal analysis, V.P. (Vincenzo Peretti), V.M. and L.E.; investigation, S.T., N.C., A.V., A.S., V.M., V.P. (Vincenzo Peretti), F.C. (Francesca Ciotola), S.A., C.D.P., V.P. (Veronica Palumbo), D.C., L.E., F.V., F.C. (Francesca Ciani); resources, D.C., V.P. (Vincenzo Peretti), F.V., S.T., F.C. (Francesca Ciani), N.C.; data curation, A.V., S.A., F.V., S.T., F.C. (Francesca Ciani), C.D.P. and N.C.; writing-original draft preparation, S.T., F.C. (Francesca Ciani), N.C., A.V. and F.V.; writing-review and editing, S.T., F.C. (Francesca Ciani) , N.C., A.V., A.S. and F.V.; visualization, A.V., C.D.P., F.C. (Francesca Ciotola), F.C. (Francesca Ciani), S.T., N.C. and F.V.; supervision, F.C. (Francesca Ciani), S.T., N.C. and F.V.; project administration, F.C. (Francesca Ciani) and F.V.; funding acquisition, F.C. (Francesca Ciani), V.M., N.C., V.P. (Vincenzo Peretti) and F.V.

Funding: The research of F.V., A.S. and A.V. was funded by MIURPON [grant number Linfa 03PE_00026_1; grant number Marea 03PE_00106]; MIUR-GPS [grant number Sicura DM29156]; POR FESR CAMPANIA 2014/2020- O.S. 1.1 [grant number Bioagro 559].

Acknowledgments: This study was supported by a grant from the Department of Veterinary Medicine and Animal Productions (N. 37-15/6/2015), University of Naples Federico II, Italy.

Conflicts of Interest: None of the authors of this article has a financial or personal relationship with other people or organizations that could inappropriately influence or bias the content of the article.

References

1. Gonzales, G.F.; Ruiz, A.; Gonzales, C.; Villegas, L.; Cordova, A. Effect of *Lepidium meyenii* (Maca) roots on spermatogenesis of male rats. *Asian J. Androl.* **2001**, *3*, 231–233.
2. Canales, M.; Aguilar, J.; Prada, A.; Marcelo, A.; Huamán, C.; Carbajal, L. Nutritional evaluation of *Lepidium meyenii* (MACA) in albino mice and their descendants. *Arch. Latinoam. Nutr.* **2000**, *50*, 126–133.
3. Gonzales, G.F. Ethnobiology and ethnopharmacology of *Lepidium meyenii* (Maca), a plant from the Peruvian Highlands. *Evid. Based Compl. Alt. Med.* **2012**, *2012*, 1–10. [CrossRef]
4. Gonzales, G.F.; Miranda, S.; Nieto, J.; Fernandez, G.; Yucra, S.; Rubio, J.; Yi, P.; Gasco, M. Red maca (*Lepidium meyenii*) reduced prostate size in rats. *Reprod. Biol. Endrocrinol.* **2005**, *20*, 3–5.
5. Oshima, M.; Gu, Y.; Tsukada, S. Effects of *Lepidium meyenii* Walp and Jatropha macrantha on blood levels of estradiol-17 beta, progesterone, testosterone and the rate of embryo implantation in mice. *J. Vet. Med. Sci.* **2003**, *65*, 1145–1146. [CrossRef]
6. Valentova, K.; Buckiova, D.; Kren, V.; Peknicova, J.; Ulrichova, J.; Simanek, V. The in vitro biological activity of *Lepidium meyenii* extracts. *Cell Biol. Toxicol.* **2006**, *22*, 91–99. [CrossRef] [PubMed]
7. Gonzales, G.F.; Cordova, A.; Gonzales, C.; Chung, A.; Vega, K.; Villena, A. *Lepidium meyenii* (Maca) improved semen parameters in adult men. *Asian J. Androl.* **2001**, *3*, 301–303.
8. Gonzales, G.F.; Gasco, M.; Córdova, A.; Chung, A.; Rubio, J.; Villegas, L. Effect of *Lepidium meyenii* (Maca) on spermatogenesis in male rats acutely exposed to high altitude (4340 m). *J. Endocrinol.* **2004**, *180*, 87–95. [CrossRef]
9. Del Prete, C.; Cocchia, N.; Ciani, F.; Tafuri, S.; Carotenuto, D.; Napoleone, G.; Pasolini, M.P. Influence of a diet supplementation with *Lepidium meyenii* on quality of cooled equine semen. *Reprod. Domest. Anim.* **2015**, *50*, 30. [CrossRef]
10. Del Prete, C.; Tafuri, S.; Ciani, F.; Pasolini, M.P.; Ciotola, F.; Albarella, S.; Carotenuto, D.; Peretti, V.; Cocchia, N. Influences of dietary supplementation with *Lepidium meyenii* (Maca) on stallion sperm production and on preservation of sperm quality during storage at 5 °C. *Andrology* **2018**, *6*, 351–361. [CrossRef] [PubMed]
11. Ciani, F.; Cocchia, N.; d'Angelo, D.; Tafuri, S. Influence of ROS on Ovarian Functions. In *New Discoveries in Embryology*; Wu, B., Ed.; Intechopen: London, UK, 2015; pp. 41–73. [CrossRef]
12. Mathur, P.P.; D'Cruz, S.C. The effect of environmental contaminants on testicular function. *Asian J. Androl.* **2011**, *13*, 585–591. [CrossRef]
13. Tafuri, S.; Ciani, F.; Iorio, E.L.; Esposito, L.; Cocchia, N. Reactive oxygen species (ROS) and male fertility. In *New Discoveries in Embryology*; Wu, B., Ed.; Intechopen: London, UK, 2015; pp. 19–40. [CrossRef]
14. Agarwal, A.; Said, T.M. Role of sperm chromatin abnormalities and DNA damage in male infertility. *Hum. Reprod. Update* **2003**, *9*, 331–345. [CrossRef]
15. Sies, H.; Mehlhorn, R. Mutagenicity of nitroxide-free radicals. *Arch. Biochem. Biophys.* **1986**, *251*, 393–396. [CrossRef]
16. Costantino, M.; Giuberti, G.; Caraglia, A.; Caraglia, M.; Lombardi, A.; Misso, G.; Abbruzzese, A.; Ciani, F.; Lampa, E. Possible antioxidant role of SPA therapy with chlorine-sulphur-bicarbonate mineral water. *Amino Acids* **2009**, *36*, 161–165. [CrossRef]
17. Del Prete, C.; Ciani, F.; Tafuri, S.; Pasolini, M.P.; Valle, G.D.; Palumbo, V.; Abbondante, L.; Calamo, A.; Barbato, V.; Gualtieri, R.; et al. Effect of superoxide dismutase, catalase, and glutathione peroxidase supplementation in the extender on chilled semen of fertile and hypofertile dogs. *J. Vet. Sci.* **2018**, *19*, 667–675. [CrossRef] [PubMed]
18. Zhou, Y.; Li, P.; Brantner, A.; Wang, H.; Shu, X.; Yang, J.; Si, N.; Han, L.; Zhao, H.; Bian, B. Chemical profiling analysis of Maca using UHPLC-ESI-Orbitrap MS coupled with UHPLC-ESI-QqQ MS and the neuroprotective study on its active ingredients. *Sci. Rep.* **2017**, *7*, 44660. [CrossRef] [PubMed]

19. Vinale, F.; Strakowska, J.; Mazzei, P.; Piccolo, A.; Marra, R.; Lombardi, N.; Manganiello, G.; Pascale, A.; Woo, S.L.; Lorito, M. Cremenolide, a new antifungal, 10-member lactone from Trichoderma cremeum with plant growth promotion activity. *Nat. Prod. Res.* **2016**, *31*, 2207. [CrossRef] [PubMed]
20. Vinale, F.; Nicoletti, R.; Lacatena, F.; Marra, R.; Sacco, A.; Lombardi, N.; d'Errico, G.; Digilio, M.C.; Woo, S.L. Secondary metabolites from the endophytic fungus *Talaromyces pinophilus*. *Nat. Prod. Res.* **2017**, *31*, 1778–1785. [CrossRef]
21. Bennet, R.N.; Mellon, F.A.; Kroon, P.A. Screening crucifer seeds as sources of specific intact glucosinolates using ion-pair High-Performance Liquid Chromatography Negative Ion Electrospray Mass Spectrometry. *J. Agric. Food Chem.* **2004**, *52*, 428–438. [CrossRef] [PubMed]
22. Feng, J.; Dou, J.; Wu, Z.; Yin, D.; Wu, W. Controlled release of biological control agents for preventing aflatoxin contamination from starch-alginate beads. *Molecules* **2019**, *24*, 1858. [CrossRef] [PubMed]
23. Fahey, J.W.; Zalcmann, A.T.; Talalay, P. The chemical diversity and distribution of glucosinolates and isothiocyanates among plants. *Phytochemistry* **2001**, *56*, 5–51. [CrossRef]
24. Gonzales, G.F.; Còrdova, A.; Vega, K.; Chung, A.; Villena, A.; Gònez, A.; Castillo, S. Effect of *Lepidium meyenii* (MACA) on sexual desire and its absent relationship with serum testosterone levels in adult healthy men. *Andrologia* **2002**, *34*, 367–372. [CrossRef] [PubMed]
25. Zheng, B.L.; He, K.; Kim, C.H.; Rogers, L.; Shao, Y.; Huang, Z.Y.; Lu, Y.; Yan, S.J.; Qien, L.C.; Zheng, Q.Y. Effect of a lipidic extract from *Lepidium meyenii* on sexual behavior in mice and rats. *Urology* **2000**, *55*, 598–602. [CrossRef]
26. Cicero, A.F.G.; Bandieri, E.; Arletti, R. *Lepidium meyenii* Walp. improves sexual behaviour in male rats independently from its action on spontaneous locomotor activity. *J. Ethnopharmacol.* **2001**, *75*, 225–229. [CrossRef]
27. Cocchia, N.; Ciani, F.; El-Rass, R.; Russo, M.; Borzacchiello, G.; Esposito, V.; Montagnaro, S.; Avallone, L.; Tortora, G.; Lorizio, R. Cryopreservation of feline epididymal spermatozoa from dead and alive animals and its use in assisted reproduction. *Zygote* **2010**, *18*, 1–8. [CrossRef] [PubMed]
28. Cocchia, N.; Corteggio, A.; Altamura, G.; Tafuri, S.; Rea, S.; Rosapane, I.; Sica, A.; Landolfi, F.; Ciani, F. The effects of superoxide dismutase addition to the transport medium on cumulus–oocyte complex apoptosis and IVF outcome in cats (*Felis catus*). *Reprod. Biol.* **2015**, *15*, 56–64. [CrossRef]
29. Tafuri, S.; Marullo, A.; Ciani, F.; Della Morte, R.; Montagnaro, S.; Fiorito, F.; De Martino, L. Reactive oxygen metabolites in alpha-herpesvirus-seropositive Mediterranean buffaloes (*Bubalus bubalis*): A preliminary study. *Pol. J. Vet. Sci.* **2018**, *21*, 639–642. [PubMed]
30. Alberti, A.; Bolognini, L.; Macciantelli, D.; Carratelli, M. The radical cation of *N,N*-diethylpara-phenylendiamine: A possible indicator of oxidative stress in biological samples. *Res. Chem. Intermediat.* **2000**, *26*, 253–267. [CrossRef]
31. Muhammad, I.; Zhao, J.; Khan, I.A. Maca (*Lepidium meyenii*). In *Encyclopedia of Dietary Supplement*, 2nd ed.; Coates, P., Blackman, M.R., Cragg, G., Levine, M., Moss, J., White, J., Eds.; Markel Dekker Inc.: New York, NY, USA, 2010; Supplement 1, pp. 522–531. [CrossRef]
32. Piacente, S.; Carbone, V.; Plaza, A.; Zampelli, A.; Pizza, C. Investigation of the tuber constituents of Maca (*Lepidium meyenii* Walp.). *J. Agric. Food Chem.* **2002**, *50*, 5621–5625. [CrossRef]
33. Esparza, E.; Hadzich, A.; Kofer, W.; Mithofer, A.; Cosio, E.G. Bioactive maca (*Lepidium meyenii*) alkamides are a result of traditional Andean post harvest drying practices. *Phythochemistry* **2015**, *116*, 138–148. [CrossRef]
34. Zhao, J.; Muhammad, I.; Dunbar, C.; Mustafa, J.; Khan, I.A. New alkamides from Maca (*Lepidium meyenii*). *J. Agric. Food Chem.* **2005**, *53*, 690–693. [CrossRef]
35. Chain, F.E.; Grau, A.; Martins, J.C.; Catalan, C.A.N. Macamides from wild "Maca", *Lepidium meyenii* Walpers (Brassicaceae). *Phytochem. Lett.* **2014**, *8*, 145–148. [CrossRef]
36. Martınez-Soto, J.C.; Landeras, J.; Gadea, J. Spermatozoa and seminal plasma fatty acids as predictors of cryopreservation success. *Andrology* **2013**, *1*, 365–375. [CrossRef]
37. Fazio, F.; Casella, S.; Giannetto, C.; Caola, G.; Piccione, G. Serum homocysteine and oxidative stress evaluation during exercise in horse. *Pol. J. Vet. Sci.* **2009**, *12*, 169–174.

38. Melnikovova, I.; Fait, T.; Kolarova, M.; Fernandez, E.C.; Milella, L. Effect of Lepidium meyenii Walp. on Semen Parameters and Serum Hormone Levels in Healthy Adult Men: A Double-Blind, Randomized, Placebo-Controlled Pilot Study. *Evid. Based Complement. Alternat. Med.* **2015**, 324369.

Sample Availability: Not available.

molecules

MDPI

Article

Exploring the Phytochemical Landscape of the Early-Diverging Flowering Plant *Amborella trichopoda* Baill.

Sheng Wu [1,2], Alexander E. Wilson [3,4], Lijing Chang [1,2] and Li Tian [1,2,3,*]

1 Shanghai Key Laboratory of Plant Functional Genomics and Resources, Shanghai Chenshan Botanical Garden, Shanghai 201602, China; wusheng@sioc.ac.cn (S.W.); changlijing@csnbgsh.cn (L.C.)
2 Shanghai Chenshan Plant Science Research Center, Chinese Academy of Sciences, Shanghai 201602, China
3 Department of Plant Sciences, University of California, Davis, Davis, CA 95616, USA; aewilson@ucdavis.edu
4 Department of Chemistry, Northern Michigan University, Marquette, MI 49855, USA
* Correspondence: ltian@ucdavis.edu; Tel.: +1-530-752-0940

Academic Editors: Francesco Vinale and Maria Luisa Balestrieri
Received: 26 July 2019; Accepted: 21 October 2019; Published: 23 October 2019

Abstract: Although the evolutionary significance of the early-diverging flowering plant *Amborella* (*Amborella trichopoda* Baill.) is widely recognized, its metabolic landscape, particularly specialized metabolites, is currently underexplored. In this work, we analyzed the metabolomes of *Amborella* tissues using liquid chromatography high-resolution electrospray ionization mass spectrometry (LC-HR-ESI-MS). By matching the mass spectra of *Amborella* metabolites with those of authentic phytochemical standards in the publicly accessible libraries, 63, 39, and 21 compounds were tentatively identified in leaves, stems, and roots, respectively. Free amino acids, organic acids, simple sugars, cofactors, as well as abundant glycosylated and/or methylated phenolic specialized metabolites were observed in *Amborella* leaves. Diverse metabolites were also detected in stems and roots, including those that were not identified in leaves. To understand the biosynthesis of specialized metabolites with glycosyl and methyl modifications, families of small molecule UDP-dependent glycosyltransferases (UGTs) and *O*-methyltransferases (OMTs) were identified in the *Amborella* genome and the InterPro database based on conserved functional domains. Of the 17 phylogenetic groups of plant UGTs (A–Q) defined to date, *Amborella* UGTs are absent from groups B, N, and P, but they are highly abundant in group L. Among the 25 *Amborella* OMTs, 7 cluster with caffeoyl-coenzyme A (CCoA) OMTs involved in lignin and phenolic metabolism, whereas 18 form a clade with plant OMTs that methylate hydroxycinnamic acids, flavonoids, or alkaloids. Overall, this first report of metabolomes and candidate metabolic genes in *Amborella* provides a starting point to a better understanding of specialized metabolites and biosynthetic enzymes in this basal lineage of flowering plants.

Keywords: *Amborella trichopoda*; metabolome; specialized metabolites; *O*-methyltransferase; OMT; UDP-dependent glycosyltransferase; UGT

1. Introduction

Amborella (*Amborella trichopoda* Baill.), a short shrub native to the tropical rainforests of New Caledonia, is the only living species in the Amborellales, the earliest diverging order of flowering plants (angiosperms) [1]. Despite its widely recognized importance in understanding flowering plant phylogeny and land plant evolution, little is known about the metabolomes, in particular specialized metabolites (secondary metabolites), of *Amborella*. To date, only three specialized metabolites, procyanidin, kaempferol-3-*O*-glucoside, and kaempferol-3-*O*-rutinoside, have been tentatively identified from *Amborella* leaves according to the R_f values of the compounds measured using paper and column chromatography [2].

Plant specialized metabolites participate in beneficial and defensive interactions between plants and the environment [3–5]. In addition, their bioactivities in humans have been exploited as a source of nutraceuticals (e.g., phytonutrients) and pharmaceuticals (e.g., anticancer agents) [6]. Plant specialized metabolites can be classified into three major groups: terpenoids (isoprenoids), phenolics, and alkaloids. Terpenoids are divided into mono- (C_{10}), sesqui- (C_{15}), di- (C_{20}), sester- (C_{25}), tri- (C_{30}), tetra- (C_{40}), and poly- (C_n) terpenoids based on the number of C_5 isoprene units in the carbon skeleton [7]. A majority of plant phenolics are produced initially by the general phenylpropanoid pathway and subsequently branch into groups of flavonoids, isoflavonoids, anthocyanins, and proanthocyanidins [8]. Unlike the hydrophobic nature of most terpenoid and phenolic aglycones, alkaloids are water-soluble, alkaline molecules that contain heterocyclic (true alkaloids) or exocyclic (proto-alkaloids) nitrogen atoms. Alkaloids can be categorized based on either the biosynthetic precursors (e.g., phenylalanine, tyrosine, lysine, arginine, etc.) or the N-containing heterocyclic/exocyclic ring structures (e.g., monoterpenoid indole alkaloids, benzylisoquinoline alkaloids, tropane alkaloids, pyrrolizidine alkaloids, etc.) [9].

There are estimated over 200,000 specialized metabolites produced by plants [10]. The rich diversity of plant specialized metabolites is conferred by enzymes that modify the core compound structures (i.e., modification enzymes). Glycosylation and *O*-methylation, catalyzed by UDP-dependent glycosyltransferases (UGTs) and *O*-methyltransferases (OMTs), respectively, represent common modifications of plant specialized metabolites [11]. Conserved functional domains have been identified in plant UGT and OMT proteins. Plant small molecule UGTs contain a 44 amino acid plant secondary product glycosyltransferase (PSPG) motif for interacting with the sugar donor [12]. *S*-adenosyl-L-methionine (SAM)-dependent OMTs contain a GxG/GxGxG motif for binding the adenosyl part of SAM and an acid reside for hydrogen bonding with the ribose part of SAM. Plant small-molecule OMTs can be divided into three groups. Class I OMTs require divalent ions (e.g., Mg^{2+}) for activity and include caffeoyl-coenzyme A OMTs (CCoA OMTs). Class II OMTs do not require divalent ions for catalysis and encompass OMTs that methylate substrates with diverse structures. A third group, the SABATH (salicylic acid carboxyl methyltransferase, benzoic acid carboxyl methyltransferase, and theobromine synthase) OMTs, catalyzes the formation of volatile esters and does not share significant sequence homology with the classes I and II OMTs [13].

To better understand the evolutionary roles of phytochemicals in plant–environment interactions, we performed metabolite profiling in an early diverging lineage of flowering plants *Amborella*. The metabolomes of *Amborella* leaves, stems, and roots were analyzed using liquid chromatography high-resolution electrospray ionization mass spectrometry (LC-HR-ESI-MS), which revealed, for the first time, the accumulation of diverse primary and specialized metabolites in these tissues. In addition, the fully sequenced *Amborella* genome allowed us to explore candidate biosynthetic enzymes that give rise to these structurally diverse phytochemicals, particularly those with glycosyl and methyl modifications. Families of small molecule UGTs and OMTs were identified in *Amborella,* and their phylogenetic association with other plant UGTs and OMTs were examined. These metabolic and molecular data from *Amborella* provide a reference point for interrogating the evolution, function, and environmental interactions of phytochemicals and their biosynthetic genes.

2. Results

2.1. *Abundant Phenolic Compounds with Glycosylation and Methylation Modifications in Amborella Leaves*

To examine the metabolomes of Amborella, methanolic extracts of leaf, stem, and root tissues were analyzed using LC-HR-ESI-MS (Figure 1). The Amborella metabolites were tentatively identified by matching their MS and MS/MS spectra with those of authentic phytochemical standards in the publicly available mass spectral libraries, followed by manual inspection. After data processing, 63, 39, and 21 compounds were tentatively identified in leaves, stems, and roots of *Amborella*, respectively, with peak areas ranging from 5.4×10^6 to 2.7×10^{10} (Tables S1–S3).

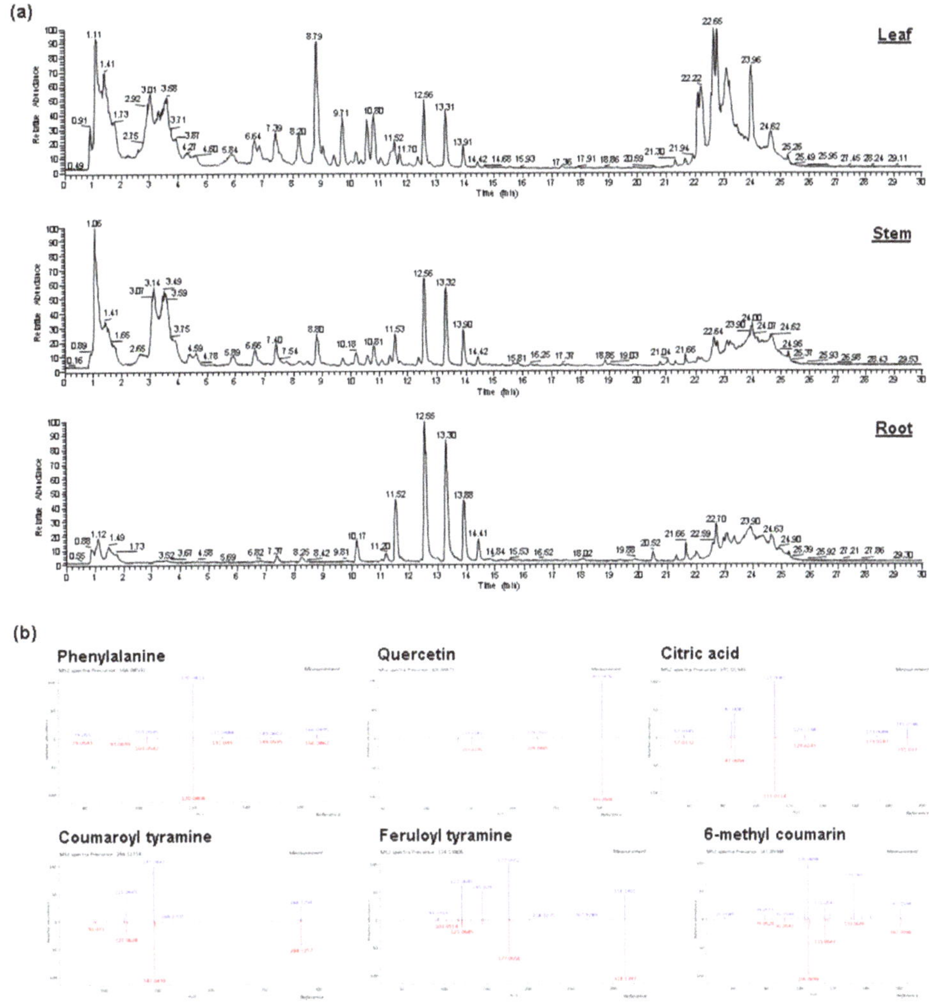

Figure 1. (**a**) Total ion chromatograms (TICs) of metabolites extracted from the leaf, stem, and root tissues of *Amborella trichopoda*. (**b**) Representative MS/MS spectral comparisons of the *A. trichopoda* metabolites (measurement) with the corresponding phytochemical standards in the publicly available mass spectral libraries (reference).

In *Amborella* leaves, the detectable free amino acids include isoleucine, arginine, aspartic acid, glutamine, and glutamic acid as well as the three aromatic amino acids, phenylalanine, tyrosine, and tryptophan, and their acetylated derivatives acetylphenylalanine and acetyltryptophan (Table S1). A group of organic acids including gluconate, citrate, malic acid, citramalate, 2-isopropylmalic acid, glutaric acid, 2-hydroxyisocaproic acid, and azelaic acid was found. Several cofactors, such as oxidized glutathione, flavin adenine dinucleotide, and riboflavin, as well as the nucleobase adenine and its nucleoside derivative adenosine, were also detected. In addition, *Amborella* leaves contain simple sugars (sucrose and raffinose), nucleotide sugars (UDP-glucose and UDP-xylose), and phosphate sugars (mannose 6-phosphate) (Table S1).

Phenylalanine serves as the biosynthetic precursor of phenylpropanoids. In addition to phenylalanine (in the free amino acid form), metabolites in the general phenylpropanoid pathway and their derivatives were found in high abundance in *Amborella* leaves, including coumaric acid, methyl cinnamate, 4-hydroxy-3-methoxycinnamate, 3,5-dimethoxycinnamic acid, 3-phenyllactic acid, feruloyl lactate, and feruloyl hexoside (Figure 2; Table S1). Products of the general phenylpropanoid pathway led to the biosynthesis of several groups of phenolic compounds detected in *Amborella* leaves, including coumarins (6-methylcoumarin, 6,8-dimethyl-4-hydroxycoumarin), isoflavonoids (genistein), anthocyanins (leucocyanidin, cyanidin 3-glucoside, delphinidin 3-rutinoside), proanthocyanidins (procyanidin B1), and flavonoids (Table S1).

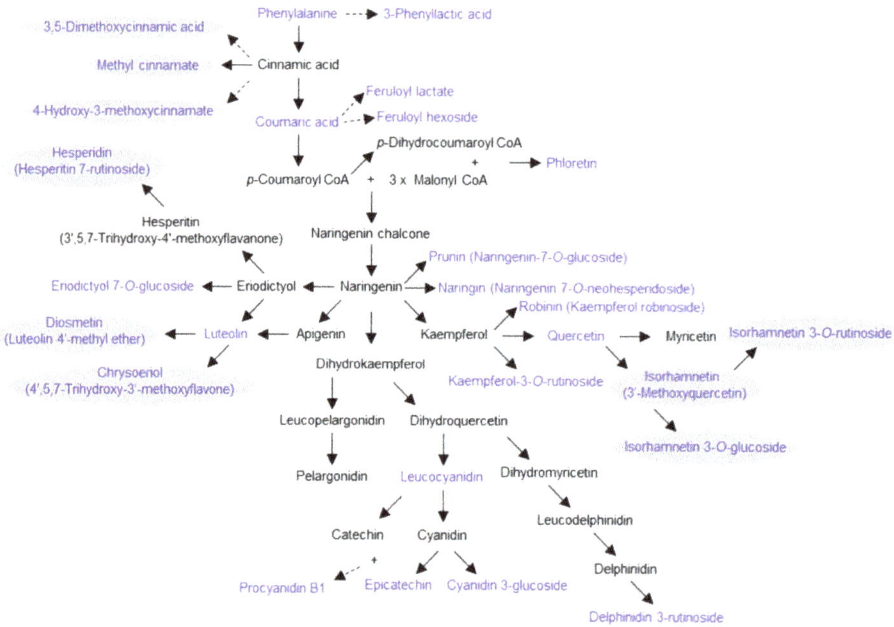

Figure 2. A simplified scheme of the general phenylpropanoid, flavonoid, anthocyanin, and proanthocyanidin biosynthetic pathways in *Amborella trichopoda*. These aglycones lead to the formation of diverse glycosidic derivatives (Tables S1–S3). Metabolites present in *Amborella* leaves are shown in blue, and methylated compounds are shaded in gray. Dotted arrows denote multiple enzymatic steps.

A wide range of flavonoid aglycones and glycosides are present in *Amborella* leaves (Figure 2; Table S1). In particular, there are 6 flavonol aglycones and glycosides, robinin (kaempferol robinoside), kaempferol-3-*O*-rutinoside, quercetin, isorhamnetin (3′-methoxyquercetin), isorhamnetin 3-*O*-glucoside, and isorhamnetin 3-*O*-rutinoside. In addition, chalcones (phloretin (dihydronaringenin)), glycosides of flavanones (prunin (naringenin-7-*O*-glucoside), naringin (naringenin 7-*O*-neohesperidoside), eriodictyol-7-*O*-glucoside, hesperidin (hesperitin-7-rutinoside)), flavones and flavone glucosides (luteolin, diosmetin (luteolin 4′-methyl ether), chrysoeriol (4′,5,7-trihydroxy-3′-methoxyflavone)), and flavan-3-ols (epicatechin) were also found (Table S1). Along with glycosylation, methylation is another modification commonly observed for many phenolic compounds in *Amborella* leaves (Figure 2; Table S1).

2.2. *Diverse Metabolites in Amborella Stems and Roots*

Amborella stems contain the free amino acids lysine, 5-oxo-proline, and 2-aminoadipic acid, in addition to arginine, aspartic acid, glutamic acid, glutamine, tyrosine, phenylalanine, and acetyltryptophan that are also identified in leaves (Table S2). Three organic acids, malic acid, saccharic acid, and 2-hydroxyisocaproic acid, are present in stems. Besides sucrose, raffinose, mannose 6-phosphate, and UDP-glucose that are detectable in both leaves and stems, stems also accumulate trehalose (Table S2).

Like leaves, phenolic compounds are the most abundant specialized metabolites in stems (Table S2). These include coumaric acid, feruloyl hexoside, 6-methylcoumarin, kaempferol 3-*O*-glucoside, kaempferol 3-*O*-rutinoside, quercetin 7-*O*-rhamnoside, rutin (quercetin rutinoside), isorhamnetin (3′-methoxyquercetin), isorhamnetin 3-*O*-glucoside, isorhamnetin 3-*O*-rutinoside, naringenin, naringin (naringenin 7-*O*-neohesperidoside), prunin (naringenin-7-*O*-glucoside), and epicatechin. Several phenolics were found in stems but not leaves, such as coumaroyl putrescin, coumaroyl hexoside, sinapyl aldehyde, feruloyl tyramine, and catechin. A cyclohexenone glucoside, roseoside, was also detected in stems (Table S2).

Only a few metabolites were identified in *Amborella* roots, of which arginine, glutamic acid, citric acid, glutaric acid, malic acid, azaleic acid, adenine, UDP-glucose, trehalose, 3-phenyllactic acid, 3,5-dimethoxycinnamic acid, naringenin, and feruloyl tyramine are in common with leaves and/or stems (Tables S1–S3). However, several metabolites were found only in roots, but not leaves and stems, including citraconic acid, galactarate, 4-hydroxy-proline, 2-deoxy-D-glucose, pantothenic acid, coumarol tyramine, 2-propanamidoacetic acid, and heptadecanoic acid (Table S3).

2.3. *Large Families of UDP-Dependent Glycosyltransferases (UGTs) and O-Methyltransferases (OMTs) in the Amborella Genome*

To explore the UGTs and OMTs that generate the diverse glycosylated and/or methylated specialized metabolites in *Amborella* tissues, the fully sequenced *Amborella* genome [14] and the InterPro database for protein sequence analysis and classification were searched using functional domains common to plant small molecule UGTs or OMTs. There are 87 putative UGTs in *Amborella* that contain the classic PSPG motif conserved in plant UGTs and are over 340 aa in length (the minimum size of a functionally characterized plant UGT) (Figure S1). To understand the evolutionary relationship among the UGTs, a neighbor-joining tree was built with the *Amborella* UGTs and selected UGTs representing different plant UGT phylogenetic groups (Figure 3). Of the 17 UGT phylogenetic groups delineated in plants to date (A–Q), *Amborella* UGTs occupy 14 groups (A, C–M, O, and Q), but they are absent in groups B, N, and P (Figure 3). While groups C, D, H, I, J, M, and O contain 2 to 5 *Amborella* UGTs per group, only one *Amborella* UGT is present each in groups F, K, and Q. On the other hand, groups G, A, and E are relatively abundant with *Amborella* UGTs, containing 7, 9, and 13 UGTs, respectively. Mostly notably, 25% of the *Amborella* UGTs (22 out of 87) belong to group L (Figure 3). There are also 9 *Amborella* UGTs in the outgroup clade of plant UGTs that glycosylate sterols and lipids [15]. Interestingly, the *Amborella* UGT W1PRF4 is not associated with any of the currently defined plant UGT clades (Figure 3).

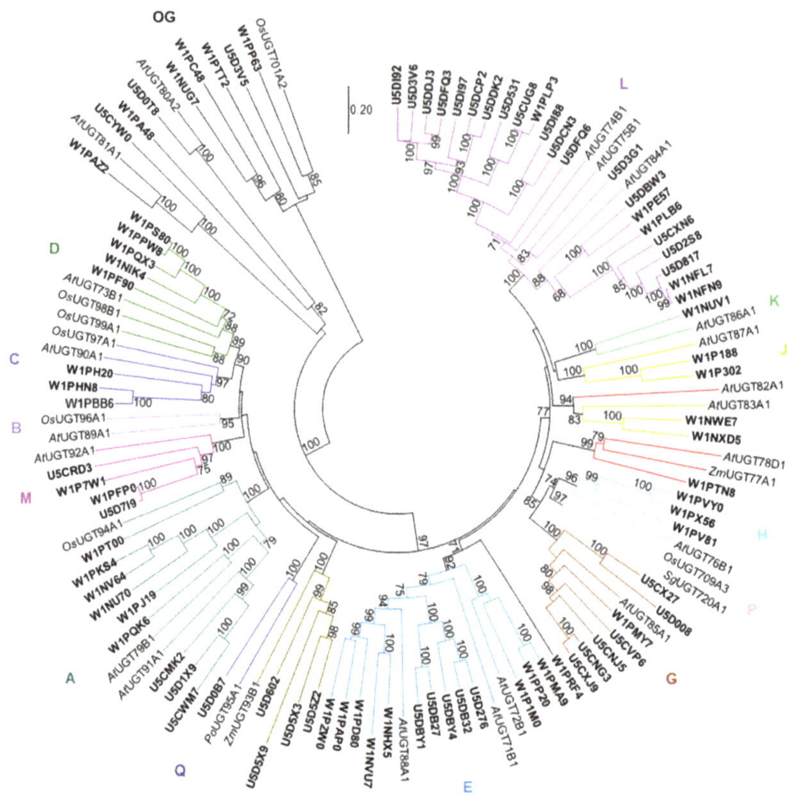

Figure 3. A neighbor-joining tree of *Amborella trichopoda* UGTs and selected UGTs representing different plant UGT phylogenetic groups. The *Amborella* UGTs are highlighted in bold. Bootstrap values greater than 60 are shown next to the branches. OG, outgroup.

Twenty-five *Amborella* proteins (51 to 395 aa, average size 216 ± 111 aa) were predicted to contain the *O*-methyltransferase, class I-like SAM-dependent *O*-methyltransferase, O-methyltransferase COMT-type, or plant methyltransferase dimerization domain (Figure 4). The protein sequences of *Amborella* OMTs and selected functionally characterized plant small-molecule OMTs were aligned (Figure S2). Because some of the pairwise distances could not be estimated from the multiple sequence alignment, a character-based method, the maximum likelihood method, was used for building the OMT phylogeny instead of neighbor-joining, which requires a distance matrix (Figure 4). Eighteen *Amborella* OMTs (including a tight cluster of 13 OMTs) were associated with plant OMTs functioning towards hydroxycinnamic acids, flavonoids, or alkaloids (Figure 4). Interestingly, within the same clade, W1NLC7 (358 aa) grouped closely (bootstrap value 72) with two monocot OMTs, *Zm*COMT (maize) and *Ta*OMT2 (wheat), which utilize caffeic acid as substrate. Seven *Amborella* OMTs clustered with plant CCoA OMTs, including U5CZ55, U5D229, U5CZN4, U5D0E8, and U5D520 that fall in the same group as *Sl*CCoAOMT and *Mc*CCoAOMT, U5CX90 that is associated with other plant CCoA OMTs, and W1PM29 that is more distant from the other *Amborella* CCoA OMTs (Figure 4). It should be noted that the SABATH OMTs from *Arabidopsis thaliana*, *Antirrhinum majus*, and *Clarkia breweri* were used as the outgroup for the phylogenetic analysis. Thirteen *Amborella* proteins contain the SAM-dependent carboxyl methyltransferase domain (present in SABATH OMTs) but were not included in the phylogenetic analysis.

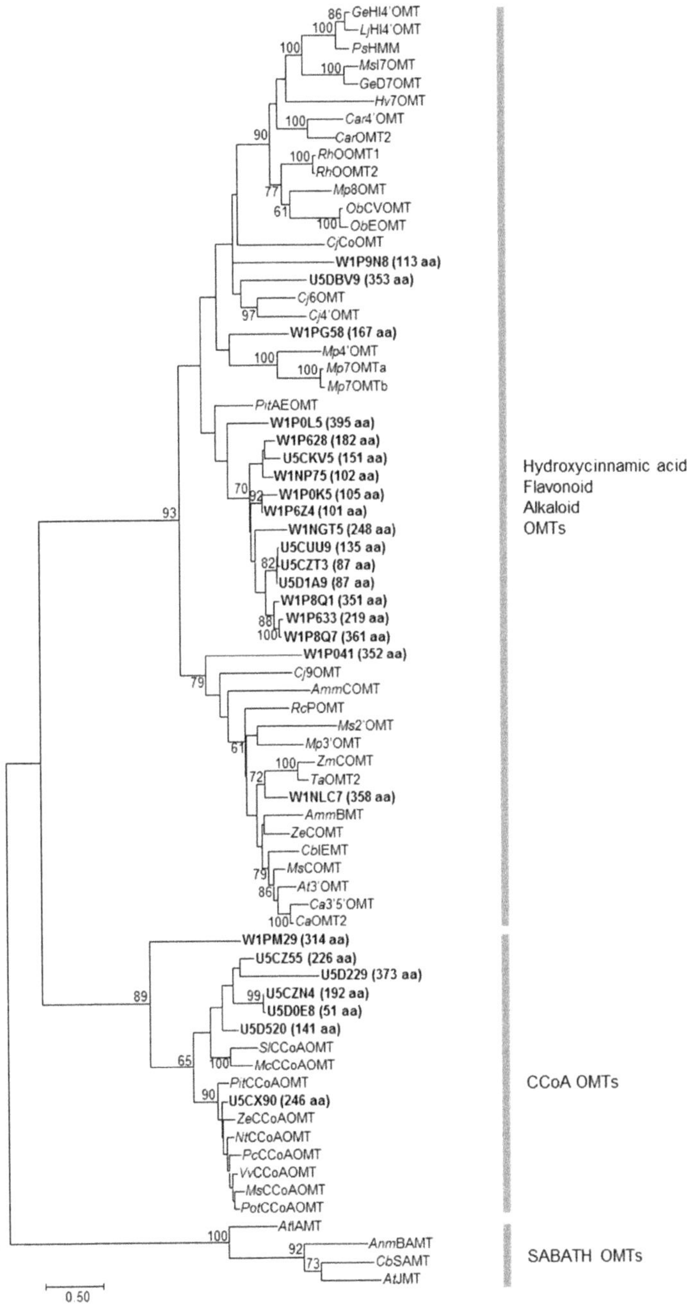

Figure 4. A maximum likelihood tree of *Amborella trichopoda O*-methyltransferases (OMTs) and selected functionally characterized plant small-molecule OMTs. The size of the predicted *Amborella* OMT is shown next to the identifier (highlighted in bold). Bootstrap values greater than 60 are shown next to the branches.

3. Discussion

Overall, this nontargeted, metabolite-profiling study revealed the presence of diverse groups of phytochemicals in *Amborella* tissues. Putative metabolite identification was accomplished by querying the *Amborella* metabolites against authentic standards in multiple mass spectral libraries followed by manual inspection of the matched spectra. The accumulation of phenolic specialized metabolites in *Amborella* tissues suggests a major role of these compounds in *Amborella* interactions with the environment. On the other hand, the low abundance of terpenoids and alkaloids in these tissues could be due to either the lack of biosynthetic genes and enzymes or inducible biosynthesis that only occurs under stress conditions.

An interesting observation was that feruloyl tyramine and coumaroyl tyramine were identified in stems and/or roots, but not leaves (Tables S1–S3). Hydroxycinnamoyl tyramines are reportedly phytoalexins with increased production in response to wounding [16] and inoculation of pathogens [17] in other plant species. This poses the question of whether feruloyl tyramine and coumaroyl tyramine constitute a form of chemical defense against abiotic and biotic stresses in *Amborella* stems and roots. If this is the case, it remains to be answered whether these compounds in *Amborella* have coevolved with pathogens in the environment.

Notably, various glycosylated and methylated phenolics are produced by *Amborella* tissues, particularly leaves (Tables S1–S3). The diverse glycosylated flavonoids, anthocyanins, and proanthocyanidins present in *Amborella* leaves may protect the plants from UV radiation, as have been demonstrated in other plants [18]. Although the role of methylated flavonoids in foliar tissues has not been well characterized, methylated isoflavonoids were shown to act as phytoalexins in different plants [19], suggesting that methylated flavonoids in *Amborella* may also be involved in defense against biotic stress. In addition, the phenylpropanoid pathway derivative methylcinnamate acts as a signaling molecule in plant–insect interactions [20]. Methylcinnamate and its associated compounds, 3,5-dimethoxycinnamic acid and 4-hydroxy-3-methoxycinnamate, were identified in *Amborella* leaves, suggesting that they could be mediators of *Amborella* and insect relations.

Intrigued by the occurrence of multiple glycosylated and methylated specialized metabolites, the fully sequenced *Amborella* genome was explored to identify candidate genes encoding modification enzymes of these compounds [14] (Figures 3 and 4). Phylogenetic analysis showed that a large number of *Amborella* UGTs (22 out of 87) belonged to group L (Figure 3). Retention of duplicated genes after whole-genome duplication (WGD) may have led to the large group L UGTs in *Amborella*. On the other hand, recent gene duplications after the divergence of *Amborella* from other flowering plants may have also contributed to the expansion of group L UGTs, as suggested by W1NFN9 and W1NFL7 that share 97.1% identity and 97.5% similarity (Figure 3). Group L UGTs from different plants have shown to form glycosides and glucose esters of a wide range of compounds, such as flavonoids, isoflavonoids, benzoic acid derivatives, cinnamic acid derivatives, lignans, hydroxy coumarins, phenylethanoids, hydroquinones, diterpenes, triterpenes, glucosinolates, epoxy sesquiterpenoids, auxins, and xenobiotics [21–27]. Functional characterization of the group L UGTs in *Amborella* will help understand whether they are responsible for generating the diverse glycosylated specialized metabolites reported here. *Amborella* UGTs are absent in groups B, N, and P (Figure 3). Although the activity of group N UGTs has not been elucidated, group B UGTs are active towards flavonoids, benzoic acid derivatives, and xenobiotics [23,25], whereas group P UGTs glycosylate monoterpenes and triterpenes [26,28]. The lack of groups B and P UGTs in *Amborella* suggests that glycosylation of flavonoids and terpenoids relies on UGTs in other phylogenetic groups (e.g., group L).

Seven *Amborella* OMTs are clustered with plant CCoA OMTs (bootstrap value 89) and may be involved in monolignol biosynthesis (Figure 4). Interestingly, of the seven *Amborella* CCoA OMTs, five group with two CCoA OMTs from the Caryophyllales, *Sl*CCoAOMT and *Mc*CCoAOMT, which exhibited activities towards caffeoyl esters and a broad range of flavonoid substrates [29]. U5CX90 is located in the branch of CCoA OMTs from various plant species for lignin biosynthesis. On the other hand, W1PM29 is more distantly related to the other six *Amborella* CCoA OMTs within the same clade (Figure 4). Thirteen *Amborella* OMTs are clustered together and group with plant OMTs that use hydroxycinnamic acids, flavonoids or alkaloids as substrates. Notably, another OMT of this clade, W1NLC7, is strongly associated (bootstrap value 72) with two monocot COMTs, *Zm*COMT and *Ta*OMT2, which are able to methylate caffeic acid (Figure 4). The recombinant *Ta*OMT2 protein also carried out sequential methylations of the flavone tricetin [30]. These phylogenetic associations of *Amborella* OMTs with plant OMTs of various activities instigate an exciting next step of characterization of their biochemical properties.

Overall, *Amborella* tissues are rich in phenolic specialized metabolites, and its genome is abundant in enzymes that modify the core structure of compounds. In the future, the role of these specialized metabolites may be investigated within the context of *Amborella* interacting with the environment. Functional characterization of the candidate UGTs and OMTs in *Amborella* will allow for comparative analysis with UGTs and OMTs from other plant lineages for convergence or divergence in enzyme evolution. Glycosylation and methylation have shown to improve the bioavailability and bioactivity of the core molecules (e.g., flavonoids) [31,32]. Elucidating the catalytic features of *Amborella* enzymes capable of producing specialized metabolites with unique structures (e.g., regiospecific) will enable valuable pharmaceutical biotechnology applications.

4. Materials and Methods

4.1. Plant Materials

An *Amborella* plant was obtained from UC Santa Cruz in 2014 by the UC Davis Botanical Conservatory; this plant was propagated vegetatively from an *Amborella* plant that was originally collected from New Caledonia by Dr. Ray Collett (*Amborella* is endemic to New Caledonia). A voucher specimen was deposited at the UC Davis herbarium (No. b.2014.123). Leaf and stem tissues were collected from the *Amborella* plant growing at the UC Davis Botanical Conservatory in April 2017. Cuttings were previously made from this plant in August 2016 for vegetative propagation. Root tissues were obtained from one of the rooted cuttings in April 2017. The plant tissues were harvested using a razor blade and immediately frozen in liquid nitrogen. The leaf, stem, and root tissues were taken and analyzed in triplicate.

4.2. Metabolite Analysis

Amborella leaves, stems, and roots were ground into fine powder in liquid nitrogen using a mortar and pestle and then freeze-dried. Fifty milligrams of the lyophilized tissue was extracted using 1 mL of 70% methanol with sonication, followed by centrifugation at 13,000 rpm for 10 min. The supernatant was filtered through a syringe filter (MilliporeSigma, Burlington, MA, USA) and subjected to LC-HR-ESI-MS analysis on an ultra-performance liquid chromatography (UPLC) (Waters, Milford, MA, USA) coupled to a Q Exactive mass spectrometer (Thermo Scientific, Waltham, MA, USA) as previously described [33]. Briefly, metabolite separation was conducted using a BEH C_{18} column (Acquity UPLC®, 100 mm × 2.1 mm, particle size 1.7 µm; Waters, Milford, MA, USA) at a column temperature of 30 °C and with gradient elution between solvents (A) 0.1% formic acid in water and (B) acetonitrile. The injection volume was 5 µL, and the total run time was 30 min. The gradient was as follows: 0–2 min, 3% B; 2–19 min, 3%–35% B; 19–22 min, 35%–90% B; 22–24 min, 90% B; 24–24.1 min, 90%–3% B; and 24.1–30 min, 3% B. The flow rate was maintained at 0.25 mL min^{-1}. The UPLC chromatograms were monitored at 254, 280, and 320 nm.

Mass spectra were obtained in the positive and negative ion modes over the mass range of m/z 120–1500 at an ion spray voltage of 4 and 3 kV, respectively. For both types of analysis, the capillary temperature was kept at 320 °C and source temperature at 200 °C. Sheath gas and auxiliary gas were nitrogen at a flow rate of 35 arbitrary units (arb) and 8 arb, respectively. The stepped normalized collision energy (NCE) for MS/MS analysis was at 15% and 40%.

4.3. Metabolite Identification

The raw data obtained from the LC-HR-ESI-MS analysis were analyzed using MS-DIAL version 3.20 [34]. Multiple reference mass spectral libraries were used for querying the unknown metabolites, including MassBank [35], RIKEN tandem mass spectral database for phytochemicals (ReSpect) (http://spectra.psc.riken.jp/), the Global Natural Product Social Molecular Networking system (GNPS) (https://gnps.ucsd.edu/ProteoSAFe/libraries.jsp), the Critical Assessment of Small Molecule Identification (CASMI) 2016 library [36], the Fiehn lab hydrophilic interaction liquid chromatography (HILIC) MS/MS library (available at http://prime.psc.riken.jp/Metabolomics_Software/MS-DIAL/index.html), the Bruker MetaboBASE plant library (https://www.bruker.com/products/mass-spectrometry-and-separations/ms-oftware/metabolomics-spectral-libraries/overview.html), RIKEN PlaSMA (http://plasma.riken.jp/), and Karolinska Institute and Gunma (GIAR) zic-HILIC deconvoluted MS2 spectra in data independent acquisition [37]. Tentative compound identification was based on the weighted similarity score of accurate mass, isotope ratio, and MS/MS spectra with a cut-off value set at 80. The accurate mass tolerance for MS was set at 0.01 Da and MS/MS at 0.05 Da. The tentatively identified metabolites were further examined by careful manual inspection of the matching MS/MS spectra between the *Amborella* metabolites and the authentic standards.

4.4. Phylogenetic Analysis

The *Amborella* UGT sequences were obtained from the Ensembl Plants database (https://plants.ensembl.org/info/website/ftp/index.html). There are 87 *Amborella* proteins that contain the conserved PSPG motif and are longer than 340 amino acids, which is the shortest length reported for a functionally characterized plant UGT. The accession numbers are: U5CMK2 (467), U5CNG3 (472), U5CNJ5 (477), U5CRD3 (382), U5CUG8 (447), U5CVP6 (486), U5CWM7 (431), U5CX27 (363), U5CXJ9 (482), U5CXN6 (489), U5CYW0 (453), U5D008 (374), U5D0B7 (562), U5D0T8 (551), U5D1 × 9 (469), U5D276 (470), U5D2S8 (473), U5D3G1 (450), U5D3V5 (434), U5D3V6 (463), U5D531 (408), U5D5 × 3 (411), U5D5 × 9 (438), U5D5Z2 (488), U5D602 (452), U5D7I9 (490), U5D817 (481), U5DB27 (427), U5DB32 (469), U5DBW3 (470), U5DBY1 (482), U5DBY4 (478), U5DCN3 (473), U5DCP2 (463), U5DDJ3 (467), U5DDK2 (467), U5DFQ3 (467), U5DFQ6 (467), U5DI88 (510), U5DI92 (419), U5DI97 (469), W1NFL7 (481), W1NFN9 (481), W1NHX5 (466), W1NIK4 (481), W1NU70 (500), W1NUG7 (539), W1NUV1 (607), W1NV64 (484), W1NVU7 (388), W1NWE7 (342), W1NXD5 (462), W1P188 (462), W1P1M0 (474), W1P302 (465), W1P7W1 (486), W1PA48 (541), W1PAP0 (479), W1PAZ2 (518), W1PBB6 (485), W1PC48 (514), W1PD80 (476), W1PE57 (469), W1PF90 (490), W1PFP0 (486), W1PH20 (465), W1PHN8 (448), W1PJ19 (438), W1PKS4 (481), W1PLB6 (487), W1PLP3 (363), W1PMA9 (471), W1PMY7 (390), W1PP20 (470), W1PP63 (543), W1PPW8 (500), W1PQK6 (475), W1PQX3 (486), W1PRF4 (436), W1PS80 (500), W1PT00 (452), W1PTN8 (451), W1PTT2 (531), W1PV81 (398), W1PVY0 (447), W1PX56 (479), and W1PZW0 (469). The size of the protein is indicated in parenthesis next to the gene identifier.

The *Amborella* OMT sequences were retrieved from the InterPro database for protein sequence analysis and classification (http://www.ebi.ac.uk/interpro/) [38]. Seventeen *Amborella* proteins were found for both domain IPR016461 (*O*-methyltransferase COMT-type) and domain IPR001077 (*O*-methyltransferase) searches, including U5CKV5 (151), U5CUU9 (135), U5CZT3 (87), U5D1A9 (87), U5DBV9 (353), W1NGT5 (248), W1NLC7 (358), W1NP75 (102), W1P041 (352), W1P0K5 (105), W1P0L5 (395), W1P628 (182), W1P633 (219), W1P6Z4 (101), W1P8Q1 (351), W1P8Q7 (361), and W1PG58 (167). An additional protein, W1P9N8 (113), possesses the plant methyltransferase dimerization domain (IPR012967). Furthermore, the class I-like SAM-dependent *O*-methyltransferase (domain IPR002935) family contains seven *Amborella* proteins, including U5CX90 (246), U5CZ55 (226), U5CZN4 (192), U5D0E8 (51), U5D229 (373), U5D520 (141), and W1PM29 (314). A search of SAM-dependent carboxyl methyltransferase (domain IPR005299; putative SABATH OMTs) identified 13 proteins, including U5D0V3 (357), U5DA47 (359), U5DCV1 (358), W1NFR3 (363), W1NGL4 (349), W1NHF8 (358), W1NQ15 (406), W1P226 (347), W1PFG6 (116), W1PG85 (109), W1PGE4 (367), W1PNB8 (387), and W1PZL5 (318). The size of the protein is indicated in parenthesis next to the gene identifier.

For phylogenetic analysis of UGTs, the *Amborella* UGTs and selected UGTs representing different plant UGT phylogenetic groups were aligned using multiple sequence comparison by log-expectation (MUSCLE) [39]. A neighbor-joining tree was constructed in Molecular Evolutionary Genetics Analysis (MEGA7) with 1000 bootstrap replicates [40]. The AGI (*Arabidopsis* sequences) and GenBank (other sequences) accession numbers of the selected plant UGTs are: *At*UGT71B1 (AT3G21750), *At*UGT72B1 (AT4G01070), *At*UGT73B1 (AT4G34138), *At*UGT74B1 (AT1G24100), *At*UGT75B1 (AT1G05560), *At*UGT76B1 (AT3G11340), *At*UGT78D1 (AT1G30530), *At*UGT79B1 (AT5G54060), *At*UGT80A2 (AT3G07020), *At*UGT81A1 (AT4G31780), *At*UGT82A1 (AT3G22250), *At*UGT83A1 (AT3G02100), *At*UGT84A1 (AT4G15480), *At*UGT85A1 (AT1G22400), *At*UGT86A1 (AT2G36970), *At*UGT87A1 (AT2G30150), *At*UGT88A1 (AT3G16520), *At*UGT89A1 (AT1G51210), *At*UGT90A1 (AT2G16890), *At*UGT91A1 (AT2G22590), *At*UGT92A1 (AT5G12890), *Os*UGT94A1 (BAC15998), *Os*UGT96A1 (BAD09654), *Os*UGT97A1 (BAD36519), *Os*UGT98B1 (BAB17059), *Os*UGT99A1 (ABF96059), *Os*UGT701A2 (CAE05712), *Os*UGT709A3 (BAC80059), *Po*UGT95A1 (ACB56927), *Zm*UGT77A1 (CAA31855), and *Zm*UGT93B1 (AAK53551). The *Sg*UGT720A1 sequence was obtained from the supplementary data file of [26].

For phylogenetic analysis of OMTs, the *Amborella* OMTs and selected functionally characterized plant small-molecule OMTs were aligned using MUSCLE. A maximum likelihood tree was constructed in MEGA7 with 100 bootstrap replicates. The GenBank accession numbers of the selected plant OMTs are: *Amm*BMT (AAR24096), *Amm*COMT (AAR24095), *Anm*BAMT (AF198492), *At*3′OMT (AAB96879), *At*IAMT (NP_200336), *At*JMT (AAG23343), *Ca*3′5′OMT (AAA80579), *Ca*OMT2 (AAA86982), *Car*4′OMT (AAR02419), *Car*OMT2 (AAM97497), *Cb*IEMT (AAC01533), *Cb*SAMT (AF133053), *Cj*4′OMT (BAB08005), *Cj*6OMT (BAB08004), *Cj*9OMT (BAA06192), *Cj*CoOMT (BAC22084), *Ge*D7OMT (BAC58012), *Ge*HI4′OMT (BAC58011), *Hv*7OMT (CAA54616), *Lj*HI4′OMT (BAC58013), *Mc*CCoAOMT (AAN61072), *Mp*7OMTa (AAR09598), *Mp*7OMTb (AAR09599), *Mp*8OMT (AAR09600), *Mp*3′OMT (AAR09601), *Mp*4′OMT (AAR09602), *Ms*2′OMT (AAB48059), *Ms*CCoAOMT (AAC28973), *Ms*COMT (AAB46623), *Ms*I7OMT (AAC49928), *Nt*CCoAOMT (AAC49913), *Ob*CVOMT (AAL30423), *Ob*EOMT (AAL30424), *Pc*CCoAOMT (AAA33851), *Pit*AEOMT (AAC49708), *Pit*CCoAOMT (AAD02050), *Pot*CCoAOMT (CAA12198), *Ps*HMM (AAC49856), *Rc*POMT (BAD18975), *Rh*OOMT1 (AAM23004), *Rh*OOMT2 (AAM23005), *Sl*CCoAOMT (AAB61680), *Ta*OMT2 (ABB03907), *Vv*CCoAOMT (CAA90969), *Ze*CCoAOMT (AAA59389), *Ze*COMT (AAA86718), and *Zm*COMT (AAB03364). (*Amm, Ammi majus; Anm, Antirrhinum majus; At, Arabidopsis thaliana; Ca, Chrysosplenium americanum; Car, Catharanthus roseus; Cb, Clarkia breweri; Cj, Coptis japonica; Ge, Glycyrrhiza echinata; Hv, Hordeum vulgare; Lj, Lotus japonicas; Mc, Mesembryanthemum crystallinum; Mp, Mentha x piperita; Ms, Medicago sativa; Nt, Nicotiana tobacum; Ob, Ocimum basilicum; Os, Oryza sativa; Pc, Petroselium crispum; Pit, Pinus taeda; Po, Pilosella officinarum; Pot, Populus trichocarpa; Rc, Rosa chinensis; Rh, Rosa hybrida; Sg, Siraitia grosvenorii; Sl, Stellaria longipes; Ta, Triticum aestivum; Vv, Vitis vinifera; Ze, Zinnia elegans; Zm, Zea mays.*)

Supplementary Materials: The following are available online. Table S1: High-resolution electrospray ionization mass spectrometry (HR-ESI-MS) data of metabolites tentatively identified in leaves of *Amborella trichopoda*. The MS/MS fragments that match those of authentic standards are highlighted in bold. Table S2: High-resolution electrospray ionization mass spectrometry (HR-ESI-MS) data of metabolites tentatively identified in stems of *Amborella trichopoda*. The MS/MS fragments that match those of authentic standards are highlighted in bold. Table S3: High-resolution electrospray ionization mass spectrometry (HR-ESI-MS) data of metabolites tentatively identified in roots of *Amborella trichopoda*. The MS/MS fragments that match those of authentic standards are highlighted in bold. Figure S1: Alignment of *Amborella trichopoda* UDP-dependent glycosyltransferases (UGTs) and selected UGTs representing different plant UGT phylogenetic groups. Figure S2: Alignment of *Amborella trichopoda O*-methyltransferases (OMTs) and selected functionally characterized plant small molecule OMTs.

Author Contributions: Conceptualization, A.E.W. and L.T.; investigation, S.W., A.E.W., L.C., and L.T.; writing—original draft preparation, S.W. and L.T.; writing—review and editing, S.W., A.E.W., L.C., and L.T.

Funding: The metabolite analysis described in this work was supported by the Science and Technology Commission of Shanghai Municipality under grant 14DZ2260400, and the Special Fund for Scientific Research of Shanghai Landscaping and City Appearance Administrative Bureau under grants G172403 and G182403. A.E.W. was supported by a UC Davis, Department of Plant Sciences Graduate Research Fellowship and the Henry A. Jastro Research Fellowship.

Acknowledgments: We thank Ernesto Sandoval, manager and curator at the UC Davis Botanical Conservatory, for assistance with collecting the *Amborella trichopoda* tissues.

Conflicts of Interest: The authors declare no conflicts of interest.

References

1. Drew, B.T.; Ruhfel, B.R.; Smith, S.A.; Moore, M.J.; Briggs, B.G.; Gitzendanner, M.A.; Soltis, P.S.; Soltis, D.E. Another look at the root of the angiosperms reveals a familiar tale. *Syst. Biol.* **2014**, *63*, 368–382. [CrossRef]
2. Young, D.A. Leaf flavonoids of *Amborella trichopoda*. *Biochem. Syst. Ecol.* **1982**, *10*, 21–22. [CrossRef]
3. Mithöfer, A.; Boland, W. Plant defense against herbivores: Chemical aspects. *Annu. Rev. Plant Biol.* **2012**, *63*, 431–450. [CrossRef] [PubMed]
4. Dudareva, N.; Klempien, A.; Muhlemann, J.K.; Kaplan, I. Biosynthesis, function and metabolic engineering of plant volatile organic compounds. *New Phytol.* **2013**, *198*, 16–32. [CrossRef] [PubMed]
5. Klee, H.J. Improving the flavor of fresh fruits: Genomics, biochemistry, and biotechnology. *New Phytol.* **2010**, *187*, 44–56. [CrossRef] [PubMed]
6. De Luca, V.; Salim, V.; Atsumi, S.M.; Yu, F. Mining the biodiversity of plants: A revolution in the making. *Science* **2012**, *336*, 1658–1661. [CrossRef] [PubMed]
7. Ogura, K.; Koyama, T. Enzymatic aspects of isoprenoid chain elongation. *Chem. Rev.* **1998**, *98*, 1263–1276. [CrossRef] [PubMed]
8. Vogt, T. Phenylpropanoid biosynthesis. *Mol. Plant* **2010**, *3*, 2–20. [CrossRef]
9. Ziegler, J.; Facchini, P.J. Alkaloid biosynthesis: Metabolism and trafficking. *Annu. Rev. Plant Biol.* **2008**, *59*, 735–769. [CrossRef]
10. Dixon, R.A.; Strack, D. Phytochemistry meets genome analysis, and beyond. *Phytochemistry* **2003**, *62*, 815–816. [CrossRef]
11. Tian, L.; Pang, Y.; Dixon, R. Biosynthesis and genetic engineering of proanthocyanidins and (iso) flavonoids. *Phytochem. Rev.* **2008**, *7*, 445–465. [CrossRef]
12. Hughes, J.; Hughes, M. Multiple secondary plant product UDP-glucose glucosyltransferase genes expressed in cassava (*Manihot esculenta* Crantz) cotyledons. *DNA Seq.* **1994**, *5*, 41–49. [CrossRef]
13. D'Auria, J.C.; Chen, F.; Pichersky, E. The SABATH family of MTS in *Arabidopsis thaliana* and other plant species. In *Recent Advances in Phytochemistry*; Romeo, J.T., Ed.; Elsevier: Amesterdam, The Netherlands, 2003; Volume 37, pp. 253–283.
14. Amborella Genome Project. The Amborella genome and the evolution of flowering plants. *Science* **2013**, *342*, 1241089. [CrossRef] [PubMed]
15. Paquette, S.; Møller, B.L.; Bak, S. On the origin of family 1 plant glycosyltransferases. *Phytochemistry* **2003**, *62*, 399–413. [CrossRef]
16. Kapteyn, J.; Qualley, A.V.; Xie, Z.; Fridman, E.; Dudareva, N.; Gang, D.R. Evolution of cinnamate/*p*-coumarate carboxyl methyltransferases and their role in the biosynthesis of methylcinnamate. *Plant cell* **2007**, *19*, 3212–3229. [CrossRef] [PubMed]
17. Agati, G.; Tattini, M. Multiple functional roles of flavonoids in photoprotection. *New Phytol.* **2010**, *186*, 786–793. [CrossRef] [PubMed]
18. Pearce, G.; Marchand, P.A.; Griswold, J.; Lewis, N.G.; Ryan, C.A. Accumulation of feruloyltyramine and *p*-coumaroyltyramine in tomato leaves in response to wounding. *Phytochemistry* **1998**, *47*, 659–664. [CrossRef]
19. Newman, M.-A.; von Roepenack-Lahaye, E.; Parr, A.; Daniels, M.J.; Dow, J.M. Induction of hydroxycinnamoyl-tyramine conjugates in pepper by *Xanthomonas campestris*, a plant defense response activated by *hrp* gene-dependent and *hrp* gene-independent mechanisms. *Mol. Plant Microbe Interact.* **2001**, *14*, 785–792. [CrossRef]
20. Liu, C.-J.; Deavours, B.E.; Richard, S.B.; Ferrer, J.-L.; Blount, J.W.; Huhman, D.; Dixon, R.A.; Noel, J.P. Structural basis for dual functionality of isoflavonoid *O*-methyltransferases in the evolution of plant defense responses. *Plant cell* **2006**, *18*, 3656–3669. [CrossRef]
21. Jackson, R.G.; Lim, E.-K.; Li, Y.; Kowalczyk, M.; Sandberg, G.; Hoggett, J.; Ashford, D.A.; Bowles, D.J. Identification and biochemical characterization of an Arabidopsis indole-3-acetic acid glucosyltransferase. *J. Biol. Chem.* **2001**, *276*, 4350–4356. [CrossRef]
22. Lim, E.; Li, Y.; Parr, A.; Jackson, R.; Ashford, D.; Bowles, D. Identification of glucosyltransferase genes involved in sinapate metabolism and lignin synthesis in Arabidopsis. *J. Biol. Chem.* **2001**, *276*, 4344–4349. [CrossRef]

23. Lim, E.; Doucet, C.; Li, Y.; Elias, L.; Worrall, D.; Spencer, S.; Ross, J.; Bowles, D. The activity of Arabidopsis glycosyltransferases toward salicylic acid, 4-hydroxybenzoic acid, and other benzoates. *J. Biol. Chem.* **2002**, *277*, 586–592. [CrossRef]
24. Lim, E.-K.; Baldauf, S.; Li, Y.; Elias, L.; Worrall, D.; Spencer, S.P.; Jackson, R.G.; Taguchi, G.; Ross, J.; Bowles, D.J. Evolution of substrate recognition across a multigene family of glycosyltransferases in Arabidopsis. *Glycobiology* **2003**, *13*, 139–145. [CrossRef] [PubMed]
25. Brazier-Hicks, M.; Gershater, M.; Dixon, D.; Edwards, R. Substrate specificity and safener inducibility of the plant UDP-glucose-dependent family 1 glycosyltransferase super-family. *Plant Biotechnol. J.* **2018**, *16*, 337–348. [CrossRef] [PubMed]
26. Itkin, M.; Davidovich-Rikanati, R.; Cohen, S.; Portnoy, V.; Doron-Faigenboim, A.; Oren, E.; Freilich, S.; Tzuri, G.; Baranes, N.; Shen, S.; et al. The biosynthetic pathway of the nonsugar, high-intensity sweetener mogroside V from *Siraitia grosvenorii*. *Proc. Natl. Acad. Sci. USA* **2016**, *113*, E7619–E7628. [CrossRef] [PubMed]
27. Grubb, C.D.; Zipp, B.J.; Kopycki, J.; Schubert, M.; Quint, M.; Lim, E.-K.; Bowles, D.J.; Pedras, M.S.C.; Abel, S. Comparative analysis of Arabidopsis UGT74 glucosyltransferases reveals a special role of UGT74C1 in glucosinolate biosynthesis. *Plant J.* **2014**, *79*, 92–105. [CrossRef] [PubMed]
28. Miettinen, K.; Dong, L.; Navrot, N.; Schneider, T.; Burlat, V.; Pollier, J.; Woittiez, L.; van der Krol, S.; Lugan, R.; Ilc, T.; et al. The seco-iridoid pathway from *Catharanthus roseus*. *Nat. Commun.* **2014**, *5*, 3606. [CrossRef]
29. Ibdah, M.; Zhang, X.-H.; Schmidt, J.; Vogt, T. A novel Mg^{2+}-dependent *O*-Methyltransferase in the phenylpropanoid metabolism of *Mesembryanthemum crystallinum*. *J. Biol. Chem.* **2003**, *278*, 43961–43972. [CrossRef]
30. Zhou, J.-M.; Gold, N.D.; Martin, V.J.J.; Wollenweber, E.; Ibrahim, R.K. Sequential *O*-methylation of tricetin by a single gene product in wheat. *Biochim. Biophys. Acta* **2006**, *1760*, 1115–1124. [CrossRef]
31. Wen, L.; Jiang, Y.; Yang, J.; Zhao, Y.; Tian, M.; Yang, B. Structure, bioactivity, and synthesis of methylated flavonoids. *Ann. N. Y. Acad. Sci.* **2017**, *1398*, 120–129. [CrossRef]
32. Bowles, D.; Isayenkova, J.; Lim, E.; Poppenberger, B. Glycosyltransferases: Managers of small molecules. *Curr. Opin. Plant Biol.* **2005**, *8*, 254–263. [CrossRef] [PubMed]
33. Wilson, A.E.; Wu, S.; Tian, L. *Pg*UGT95B2 preferentially metabolizes flavones/flavonols and has evolved independently from flavone/flavonol UGTs identified in *Arabidopsis thaliana*. *Phytochemistry* **2019**, *157*, 184–193. [CrossRef]
34. Tsugawa, H.; Cajka, T.; Kind, T.; Ma, Y.; Higgins, B.; Ikeda, K.; Kanazawa, M.; VanderGheynst, J.; Fiehn, O.; Arita, M. MS-DIAL: Data-independent MS/MS deconvolution for comprehensive metabolome analysis. *Nat. Methods* **2015**, *12*, 523–526. [CrossRef] [PubMed]
35. Horai, H.; Arita, M.; Kanaya, S.; Nihei, Y.; Ikeda, T.; Suwa, K.; Ojima, Y.; Tanaka, K.; Tanaka, S.; Aoshima, K.; et al. MassBank: A public repository for sharing mass spectral data for life sciences. *J. Mass. Spectrom.* **2010**, *45*, 703–714. [CrossRef] [PubMed]
36. Nikolić, D. CASMI 2016: A manual approach for dereplication of natural products using tandem mass spectrometry. *Phytochem. Lett.* **2017**, *21*, 292–296. [CrossRef]
37. Naz, S.; Gallart-Ayala, H.; Reinke, S.N.; Mathon, C.; Blankley, R.; Chaleckis, R.; Wheelock, C.E. Development of a liquid chromatography–high resolution mass spectrometry metabolomics method with high specificity for metabolite identification using all ion fragmentation acquisition. *Anal. Chem.* **2017**, *89*, 7933–7942. [CrossRef]
38. Mitchell, A.L.; Attwood, T.K.; Babbitt, P.C.; Blum, M.; Bork, P.; Bridge, A.; Brown, S.D.; Chang, H.-Y.; El-Gebali, S.; Fraser, M.I.; et al. InterPro in 2019: Improving coverage, classification and access to protein sequence annotations. *Nucl. Acids Res.* **2019**, *47*, D351–D360. [CrossRef]
39. Edgar, R. MUSCLE: Multiple sequence alignment with high accuracy and high throughput. *Nucl. Acids Res.* **2004**, *32*, 1792–1797. [CrossRef]
40. Kumar, S.; Stecher, G.; Tamura, K. MEGA7: Molecular evolutionary genetics analysis version 7.0 for bigger datasets. *Mol. Biol. Evol.* **2016**, *33*, 1870–1874. [CrossRef]

Sample Availability: Samples of some of the phenolic compounds are available from the authors.

molecules

Article

The Ancient Neapolitan Sweet Lime and the Calabrian Lemoncetta Locrese Belong to the Same Citrus Species

Domenico Cautela [1,*], Maria Luisa Balestrieri [2], Sara Savini [3], Anna Sannino [3], Giovanna Ferrari [4], Luigi Servillo [2], Luigi De Masi [5], Annalisa Pastore [6] and Domenico Castaldo [1,4,7]

1 Stazione Sperimentale per le Industrie delle Essenze e dei Derivati dagli Agrumi (SSEA), Azienda Speciale CCIAA di Reggio Calabria, via G. Tommasini 2, 89125 Reggio Calabria, Italy; dcastaldo@ssea.it
2 Dipartimento di Medicina di Precisione, Università degli Studi della Campania "Luigi Vanvitelli", Via L. De Crecchio 7, 80138 Napoli, Italy; marialuisa.balestrieri@unicampania.it (M.L.B.); luigi.servillo@unicampania.it (L.S.)
3 Stazione Sperimentale per le Industrie delle Conserve Alimentari (SSICA), Fondazione di Ricerca CCIAA di Parma, Viale Tanara 31/A, 43100 Parma, Italy; sara.savini.borsista@ssica.it (S.S.); anna.sannino@ssica.it (A.S.)
4 Dipartimento di Ingegneria Industriale e ProdALscarl, Università degli Studi di Salerno, Via Ponte Don Melillo 1, 84084 Fisciano, Italy; gferrari@unisa.it
5 National Research Council (CNR), Institute of Biosciences and BioResources (IBBR), Via Università 133, 80055 Portici, Italy; luigi.demasi@cnr.it
6 United Kingdom Dementia Research Institute Centre, Maurice Wohl Clinical Neuroscience Institute, Institute of Psychiatry, Psychology and Neuroscience, King's College London, 125 Coldharbour Lane, London SE5 9NU, UK; annalisa.pastore@crick.ac.uk
7 Ministero dello Sviluppo Economico (MiSE), Via Molise 2, 00187 Roma, Italy
* Correspondence: dcautela@ssea.it

Academic Editor: Celestino Santos-Buelga
Received: 21 November 2019; Accepted: 23 December 2019; Published: 27 December 2019

Abstract: "Neapolitan limmo" is an ancient and rare sweet Mediterranean lime, now almost extinct but used until a few decades ago for the production of a fragrant liqueur called the "four citrus fruits". The objective of this work was to compare, through the use of chemical (flavonoids, volatile organic compounds, and chiral compounds) and molecular (DNA fingerprint based on RAPD-PCR) markers, the residual population of Neapolitan limmo with other populations of sweet limes, identified in Calabria and known as "lemoncetta Locrese". We report for the first time specific botanical characteristics of the two fruits and unequivocally show that the ancient sweet Mediterranean limes Neapolitan limmo and lemoncetta Locrese are synonyms of the same Citrus species. Owing to the biodiversity conserved in their places of origin, it will now be possible to recover, enhance and implement the use of this ancient sweet lime for agro-industrial purposes.

Keywords: agrobiodiversity; lemoncetta Locrese; Mediterranean sweet lime; Neapolitan limmo; Neapolitan "four citrus fruits" liqueur; taxonomy

1. Introduction

"Limmo" or "limo" or also "limma", not to be confused with the more famous lime (*C. aurantifolia*), is an ancient Neapolitan citrus of the genus *Citrus*. The first traces of the presence of limmo in the Neapolitan province date back to the end of the seventeenth century [1]. It was described as the fruit of Lomia or Lumia, a species of sweet and sweet-smelling *Citrus* fruits similar to lemon but smaller [2]. Limmo has a strongly rounded shape of about 5–6 cm diameter (Figure 1a). According to TG/203/1 UPOV guidelines [3], it is morphologically characterized by a base with a depressed, slightly rounded distal part and a nipple of conical and umbonate shape, sunken at the base of ca. 1–2 cm (Figure 1b). It is similar to the "Sicilian lumia", recently described by Raimondo et al. [4].

Figure 1. Neapolitan limmo or lemoncetta Locrese. (**a**,**b**) Ripe fruits are sulfur-yellow in color and have a diameter of 4–5 cm. (**c**) Fruit section with 8–10 loggias. (**d**) Ripe fruits with greenish color. (**e**) Young leaves and flowers. (**f**) Leaves and small fruits. (**g**) Neapolitan limmo tree.

Limmo has a thin yellow skin (Figure 1c) in the ripe fruit, consisting of 8–10 loggias containing few seeds, with segments of color between yellow and green, of delicate flavor, not sour, aromatic and sometimes also with greenish notes in the peel of the ripe fruit (Figure 1d). The limmo flowers, compared to lemon, are on average smaller, of medium size, fragrant, with white petals and buds (Figure 1e,f). Small is also the limmo tree with leaves similar to those of lemon (Figure 1f,g). The cultivation of this sweet lime is now completely amateurish. We could count only a few plants in gardens of Naples and of the Neapolitan province.

Limmo belongs to the group of the Mediterranean sweet limes and lemons [5]. It appears nevertheless distinct from them for the color of the flower petals and for the low acidity of the fruit juice [4,6,7]. The acidity is instead high in lemons (*C. limon*) and medium high in most of the common limes (*C. limetta* Risso, subsect. *Limonoides*) [8], like the acidic "limonette de Marrakech" [9], and the Mediterranean sweet lime, *C. lumia* Risso. The latter is an acid-less variety of *C. limetta* Risso, (subsect. *Decumanoides*-sect. *Citrophorum* according to Tanaka) [8] which has a long history of cultivation in Italy as early as the seventeenth and eighteenth centuries [10,11]. The acid-less phenotype of *C. limetta* is due to the inability of producing anthocyanin pigments in leaves and flowers and proanthocyanidins in seeds [6,7] with low citric acid contents [12–14] and juice pH values also above 6 [5–7,15].

Neapolitan limmo stands out among other citrus fruits for its fragrant aroma. The flavor is not very sweet, rather watered down, almost totally devoid of acidity, not savory and, therefore, unappetizing; the latter characteristics were in ancient times systematically exploited in Naples and its surroundings as a defense from thieves. At the end of the nineteenth century, traditional and patrician gardens were surrounded by limmo trees to protect the property from the street urchins, who were Neapolitan boys accustomed by the adversity and poverty of that time to survive in the street thanks to small daily thefts like stealing seasonal fruits from city gardens. Limmo was also used in Neapolitan families to prepare the ancient liqueur "with the four citrus fruits", today almost disappeared. The liqueur was obtained by cold maceration in ethylic alcohol of the slightly unripe peels of oranges, mandarins, lemons, and Neapolitan limmo. This ancient liqueur was much more sought after than the more popular and widespread liqueur "limoncello", another Neapolitan liqueur obtained using IGP lemons (Protected Geographical Indication) from the Amalfi coast and Sorrento [16,17].

The use of limmo in natural medicine, together with that of other citrus fruits, was reported in the 1825 edition of Phamacopoeia by Antonio Ferrarini (pharmacist, Member of the Health Commission of Bologna City and surroundings, and Lecturer at the Faculty of Pharmacy) [18]. Along with other citrus fruits, limmo was used for the preparation of "aromatic cedar water" or "citron aromatic water". In traditional medicine, the limmo juice was once used as a remedy for cough mixed with prickly pear juice [19].

No studies on limmo are present in the literature. Early taxonomists hypothesized that lemons and limes are derivatives or hybrids of citrons. However, a definitive classification and origin of the species was not proposed. It is a shared opinion that cultivated limes, sweet limes, and lemons originate from interspecific hybridization of cedar (*C. medica* L.) in combination with sour orange (*C. aurantium* L.) while the *C. maxima* × *C. reticulata* hybrid gives rise to the *C. limettioide* subgroup Palestinian sweet lime and *C. meyeri* Meyer lemon [5,20].

Since in Naples limmo was also described as a "sweet bergamot", we searched for this sweet lime in Calabria as well, with the intention of verifying the presence of limmo. Calabria is a region of Southern Italy with extensive citrus fruit cultivations, especially in the Ionic area of the province of Reggio Calabria, the area of origin and production of bergamot (*Citrus Bergamia* Risso) [21,22].

The results of this survey showed the presence of a discreet population of sweet limes in the region East to Reggio Calabria, the Locri area. The local fruit is morphologically similar to Neapolitan limmo and locally known as "lemoncetta Locrese" or "pirettu Locrese". This fruit is of no agro-industrial use and thus of little economic importance. As for Neapolitan limmo, it was never characterized compositionally.

We thus decided to compare the populations of Neapolitan limmo with the Calabrian lemoncetta Locrese with chemical and genetic approaches. We measured chemical markers of citrus fruits, such as glycoside flavanones and determined the profile of volatile organic compounds (VOCs) in juice and peel and the enantiomeric distribution of volatile organic chiral compounds. We then analyzed the genetic diversity of the two populations by Random Amplified Polymorphic DNA (RAPD) analysis.

The results of this study aimed at characterizing and comparing the two apparently distinct fruits led us to conclude that they are compositionally and genetically indistinguishable within variations due to climatic and soil differences. These results will help to restore the use of the ancient Mediterranean limmo to produce the Neapolitan "four citrus fruits" liqueur and as a promising and precious resource for the essential oil industry.

2. Results and Discussion

2.1. Flavonoids, Organic Acids and Proximate Constituents

Flavanones and to a lesser extent flavonols are the predominant flavonoids in the genus *Citrus* [23]. The quali-quantitative distribution of these phenols is largely influenced by the specie and/or the variety [24–26]. Flavonoids are thus commonly used as chemotaxonomic markers and evaluate the quality and genuineness of citrus juices [24,27–30]. Since flavanones constitute virtually all of the total flavonoids present (e.g., 98% in grapefruit, 90% in lemons, and 96% in limes) [29], we focused on the major aglycone flavanones with their rutinose or neohesperidose glycosides as markers to differentiate Neapolitan limmo and lemoncetta Locrese from other citrus juices and between the two populations. The same approach was utilized by Mouly et al. [24] to effectively differentiate lemon and lime, varieties of grapefruits (white, pink, red, and green), and sweet oranges (Valencia, navel, blood, Thomson, and Malta). The flavonoid profile is also a method widely utilized to detect the possible mixture of different juices as for instance the addition of bergamot to lemon juice [30]. The range of variability of flavonoids and organic acids for citrus juices are reported in the code of practice of the International Federation of Fruit Juice Producers (IFFJP).

We thus compared by HPLC the population of flavonoids in Neapolitan limmo with those of lemoncetta Locrese (Figure 2 and Table 1). An identical flavonoid profile unites both analyzed populations (Figure 2). Neapolitan limmo had flavanone profiles more like lemoncetta Locrese. The insignificant difference ($p < 0.05$) observed in flavonoid contents are within the normal limits of environmental variability of these two juices.

Both Neapolitan limmo and lemoncetta Locrese are qualitatively characterized by a common presence of five different flavonoids: the three rutinosidic flavanones, hesperidin, eriocitrin, and narirutin and the two flavones *O*-glycosides, rutin and diosmin. The identification of these flavonoids is confirmed not only by the retention times but also by spectra analysis compared to their standards (data not shown). The samples do not contain flavanone *O*-neohesperidose and the non-bitter flavanone neoponcirin.

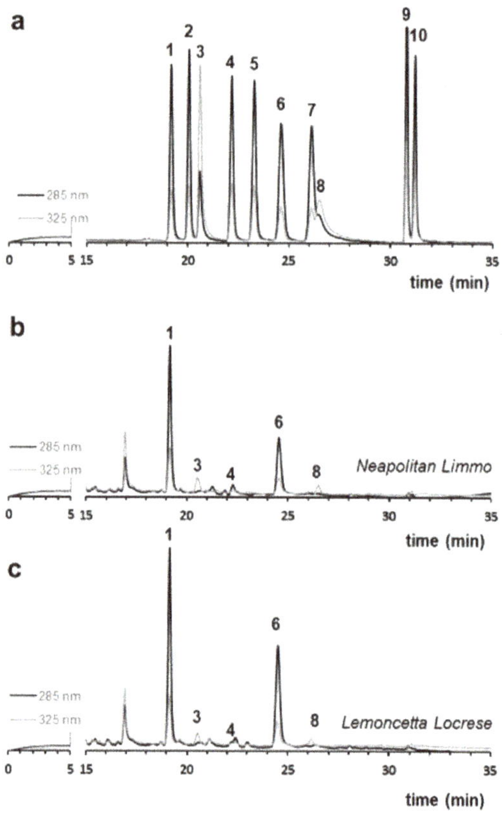

Figure 2. Liquid chromatography (LC) flavonoid profile of Neapolitan limmo and lemoncetta locrese. (**a**) Standard containing a mixture (50 mg/L) of eriocitrin (1), neo-eriocitrin (2), rutin (3), narirutin (4), naringin (5), hesperidin (6), neohesperidin (7), diosmin (8), poncirin (9), didymin (10). (**b**) LC chromatograms of Neapolitan limmo (**c**) LC chromatograms of lemoncetta Locrese. Flavanones were monitored at 285 nm (black line), flavones at 325 nm (gray line).

Of particular interest between the present flavonoids is eriocitrin, a flavonoid that is exclusively characteristic of lemon juice [27,29], and is almost absent in orange and grapefruit juice. Eriocitrin in limmo and lemoncetta strengthens the genetic closeness to lemon of these Mediterranean limes, both hybrids of citron (*C. medica* L.) being a sour orange × Citron cross [5].

The flavonoid composition of limmo and lemoncetta was also compared with that of two other limes, both previously characterized by Nogata et al. [23]. The first is classified under the *Limonoides* subset according to the Tanaka's system [8], with the common name Sweet lemon and the scientific name *C. limetta* or *C. limetta* Risso, with slightly acidulous pulp [9,12,13]. The second is classified in the subsection of *Decumanoides* (sect. *Citrophorum*) [8], with common name lumie and the scientific name *C. lumia* or *C. lumia* Risso, with sweet and non-acidic pulp [4,13]. This citrus is most diffused in Italy and in some southern regions of France [31]. A prevalent similarity emerges from the comparison between the flavonoid profiles of *C. limetta* and *C. lumia* and those of Neapolitan limmo or lemoncetta Locrese (Table 1): Neapolitan limmo, lemoncetta Locrese and *C. lumia* have a significant content of the flavanone *O*-rutinosides esperidine and eriocitrin. They also have a reduced content of the other flavanone rutinosides narirutin (Table 1) and of the flavones rutin and diosmin. Significative is the common absence of the flavanones neoponcirin, naringin, neohesperidin, neoeriocitrin e poncirin in limmo, lemoncetta and *C. lumia* but not in *C. limetta*.

Unfortunately, no other paper on the traditional Italian sweet lime varieties besides Nogata et al. [23] report data on flavonoids [12,13]. The data are anyway consistent with those recently found by Smeriglio et al. [32] on *C. Lumia* which reported a similar significant presence of hesperidin and eriocitrin.

Table 1. Flavonoid content (mg/Kg) in Neapolitan limmo and lemoncetta Locrese estimated by HPLC. (n.d. ≤ 0.5 mg/Kg).

Peak	Compounds	Neapolitan Limmo		Lemoncetta Locrese		*p*-Value
		Mean ± sd	Min–Max	Mean ± sd	Min–Max	
1	Eriocitrin (ERC)	166 ± 15	96–229	261 ± 64	176–327	0.003
2	Neo-Eriocitrin (NER)	n.d.		n.d.		
3	Rutin (RUT)	4 ± 1	3–5	9 ± 4	5–12	0.05
4	Narirutin (NRT)	1	trace-2	1	trace-2	
5	Naringin (NRG)	n.d.		n.d.		
6	Hesperidin (HSP)	130 ± 17	45–189	270 ± 80	178–322	0.001
7	Neohesperidin (NHP)	n.d.		n.d.		
8	Diosmin (DSM)	12 ± 3	10–17	24 ± 11	15–30	0.04
9	Poncirin (PON)	n.d.		n.d.		
10	Didimin (DDM)	n.d.		n.d.		

A further indication that the Neapolitan limmo can be with good reasons classified in the *Citrus lumie* group is also offered by two recent papers by two distinct research groups [6,7]. These authors demonstrated by independent methodological approaches that in acid-less varieties of citrus, exceptionally low fruit acidity is associated with absence of anthocyanin pigments in leaves and flowers and of proanthocyanidins in seeds and flowers without pigmentation or white, like those of Neapolitan limmo (Figure 1e).

Next, we extended our investigation to the quali-quantitative distribution of organic acids, the overall acidity, and the pH of the juice. These parameters can give useful indications on the nature of the lime type discriminating between acid ecotypes. Both limmo and lemoncetta have qualitatively a common acidic chromatographic profile characterized by the presence of five organic acids: malic, citric, quinic, tartaric, and fumaric acids (Figure 3). Similar are also the quantitative data (Table 2).

Malic acid is the dominant organic acid of these sweet Mediterranean limes with average values of 1.57 ± 0.03 g/L in Neapolitan limmo and slightly higher, 1.88 ± 0.02 g/L, in lemoncetta. It is probably this significant presence in the acidic profile that confers to the juices of these fruits (acid-less sweet tasting) that smooth tartness acidity given by malic acid. This taste is clearly different from the sensorial sour quality given by citric acid in juices when this is dominant [33].

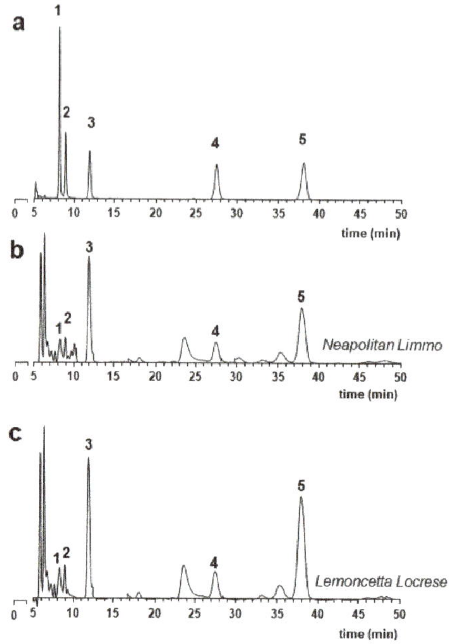

Figure 3. Organic acid profile of Neapolitan limmo and lemoncetta locrese. (**a**) Standard containing a mixture of tartaric acid (1) 0.5 mg/mL; quinic acid (2) 0.5 g/L; malic acid (3) 0.5 g/L; citric acid (4) 0.5 g/L; fumaric acid (5) 2.5 mg/L. (**b**) LC chromatograms of Neapolitan limmo (**c**) LC chromatograms of lemoncetta Locrese.

Both Neapolitan limmo and lemoncetta showed reduced contents of citric acid with average values of about 0.94 g/L in the group of Neapolitan limmo and even lower in lemoncetta Locrese (0.48 g/L), with an average pH > 5.7 and values of total acidity < 1.4 g/L (Table 2). This is consistent with the phenotypes of the sweet forms of *C. limetta* Risso—Mediterranean sweet lime—sweet Roman [12], Roman [13,34], Lima Dulce, or Dulce lime [14].

Quinic acid is the most expressed acidic compound after malic acid and citric acids. The average levels of malic acid are between 0.10 and 0.43 g/L with higher average values for lemoncetta Locrese compared to Neapolitan limmo (Table 2). Also, for this acid, the quantitatively expressed levels appear on average higher in the group of lemoncetta Locrese than in those of Neapolitan limmo. The average values are however completely comparable with each other. Finally, fumaric acid is much less expressed and generally does not exceed 0.01 g/L.

Table 2. Proximate constituents. pH, soluble solids (°Brix), titratable acidity (as citric monohydrate acid g/L), and organic acids (g/L) in Neapolitan limmo and lemoncetta Locrese.

Proximate Constituents	Neapolitan Limmo		Lemoncetta Locrese		*p*-Value
	Mean ± sd	Min–Max	Mean ± sd	Min–Max	
Total Soluble Solids	7.9 ± 0.5	7.6–8.5	8.4 ± 0.5	8.0–8.9	0.001
pH	5.8 ± 0.2	5.7–5.9	5.9 ± 0.2	5.8–6.0	0.991
Titratable Acidity	1.22 ± 0.2	0.98–1.32	0.98 ± 0.2	0.84–1.20	0.079
Tartaric Acid	0.15 ± 0.02	0.13–0.17	0.22 ± 0.04	0.17–0.26	0.071
Quinic Acid	0.15 ± 0.05	0.10–0.22	0.32 ± 0.10	0.19–0.43	0.045
Malic Acid	1.57 ± 0.10	1.45–1.70	1.88 ± 0.31	1.42–2.14	0.467
Citric Acid	0.94 ± 0.01	0.85–1.02	0.48 ± 0.25	0.13–0.70	0.003
Fumaric Acid	0.01 ± 0.01	0–0.01	0.01 ± 0.01	0–0.01	0.767

This analysis makes us conclude that limmo and lemoncetta are chemically similar although there is an appreciable quantitative difference in some substances likely due to climatic and soil composition and other environmental differences. The values of flavonoids and other metabolites are, for instance, different likely because of the different degree of activivity of phenylalanine ammonium lyase, the enzyme central to the production of the biosyntesis precursor of flavonoids cinnamic acid [35].

2.2. Chirospecific Analysis

Biological activity is often correlated with chiral properties. In citrus fruits, chiral compounds are widely used as indicators of adulteration or fraud of essential oils by addition of synthetic or natural compounds of different botanical origin. The GC profiles of volatile aromatic compounds of essential oils from Neapolitan limmo and lemoncetta Locrese (Figure 4a) were initially compared and analyzed by heart-cutting multidimensional GC [36] to estimate the enantiomeric distribution (ee%) of chiral β-pinene, sabinene, limonene, linalool and linalyl acetate (Figure 4b).

As for flavonoids and organic acids, an identical metabolic profile of volatile compounds was common to both citrus populations (Figure 4a). Forty-three volatile aromatic compounds were identified. In both populations, the more expressed were limonene (61.8 ≥ ± 14.4%) ≥ linalyl acetate (9.2 ± 0.5%) ≥ linalool (6.6 ± 0.4%) ≥ β-pinene (4.4 ± 2.9%) ≥ myrcene (1.3 ± 0.6%) ≥ sabinene (0.8 ± 0.4%) ≥ α-terpineol (0.7 ± 0.6%) ≥ α-pinene (0.5 ± 0.4%) ≥ geranial (0.4 ± 0.3%) ≥ neral (0.3 ± 0.1%) ≥ β-bisabolene (0.2 ± 0.1%) ≥ nerol (0.2 ± 0.1%) ≥ terpinene and citronellol ranged from 0.05 to 0.1%. Camphene, octanal, α-phellandrene, terpinolene, and terpinen-4-ol were less than 0.05%.

The data obtained by four heart-cut multidimensional GC are even more interesting; the enantiomers of β-pinene, sabinene, limonene, linalool, and linalyl acetate were all well-separated on a DiActButylsilyl γ-CDX chiral column (Figure 4b). The dominant enantiomeric form for limonene was (*R*)-(+). Both populations showed (*R*)-(+) for β-pinene and sabinene (*S*)-(−), and (*R*)-(−) for linalyl acetate and linalool (Table 3).

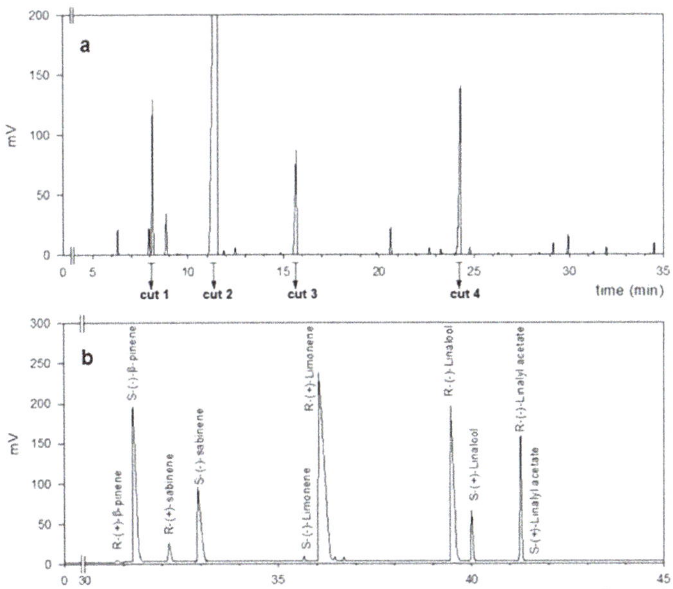

Figure 4. Analysis of the chirality of Neapolitan limmo and lemoncetta Locrese (**a**) GC chromatogram of Neapolitan limmo essential oil with the four heart-cuts. (**b**) Enantio-MDGC chromatogram of selected chiral compounds in Neapolitan limmo.

Table 3. Enantiomeric distribution of chiral compounds in Neapolitan limmo and lemoncetta Locrese essential oil.

Compound		Enantiomeric Ratio		Enantiomeric Excess, ee (%)	
		Neapolitan Limmo	Lemoncetta Locrese	Neapolitan limmo	Lemoncetta Locrese
β-pinene	*R*-(+)	0.5	0.5	99.2	99.1
β-pinene	*S*-(−)	99.5	99.5		
Sabinene	*R*-(+)	15.4	15.8	69.3	68.4
Sabinene	*S*-(−)	84.6	84.2		
Limonene	*S*-(−)	0.6	0.5	99.2	99.0
Limonene	*R*-(+)	99.4	99.5		
Linalool	*R*-(−)	8.3	83.6	66.9	67.2
Linalool	*S*-(+)	16.7	16.4		
Linalyl acetate	*R*(−)	98.8	98.7	97.6	97.6
Linalyl acetate	*S*-(+)	1.2	1.3		

2.3. Volatile Organic Compounds Analysis

Comparison of the total ion chromatograms of the aroma components collected in the juices of Neapolitan limmo and lemoncetta Locrese showed the presence of 8 terpenes, 5 monoterpenoid alcohols, and 3 sesquiterpene hydrocarbons in the juices. Both fruits presented the same volatile compounds (Figures 5 and 6).

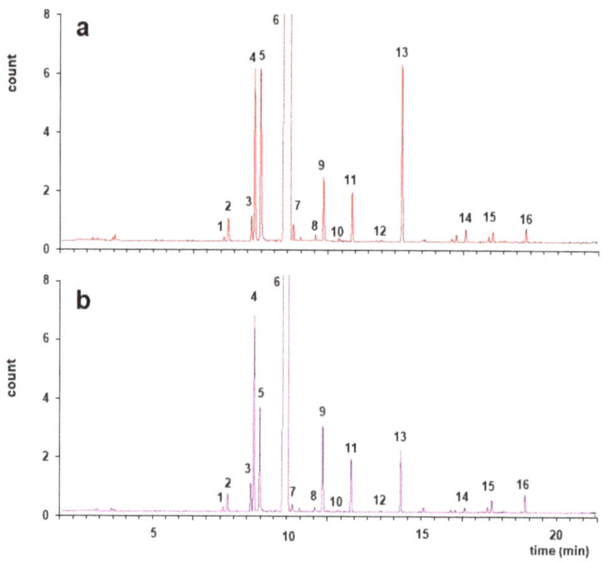

Figure 5. Total ion chromatograms of the juices collected by SPME/GC/MS. (**a**) Neapolitan limmo. (**b**) lemoncetta Locrese. IS: Internal Standard. The peak numbers correspond to the numbers in the first column of Table 4.

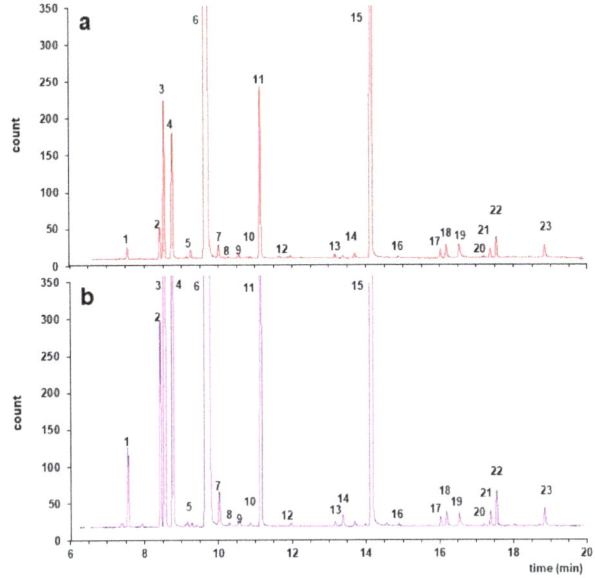

Figure 6. Total ion chromatograms of the peels collected by SPME/GC/MS. (**a**) Neapolitan limmo. (**b**) lemoncetta Locrese. Peak numbers correspond to the numbers in the first column of Table 5.

In the variety lemoncetta Locrese 12 volatile compounds were in concentrations significantly higher than in limmo (Table 4).

Table 4. Volatile compounds identified in the juices of the Neapolitan limmo and lemoncetta Locrese. Data were expressed as: Area volatile compound/Area Internal Standard (A/As.i.) and Area percent (%). Values correspond to average ± standard deviation. Values are significant at $p < 0.05$.

Peak	Volatile Compounds	Neapolitan Limmo		Lemoncetta Locrese		*p*-Value
		A/A s.i.	%	A/A s.i.	%	
		Terpenes				
1	α-phellandrene	0.005 ± 0.00	0.24 ± 0.00	0.007 ± 0.00	0.33 ± 0.00	0.009
2	α-pinene	0.004 ± 0.00	0.18 ± 0.00	0.004 ± 0.00	0.18 ± 0.00	0.06
3	β-phellandrene	0.006 ± 0.00	0.29 ± 0.00	0.009 ± 0.00	0.45 ± 0.00	0.009
4	β-pinene	0.03 ± 0.00	1.38 ± 0.00	0.039 ± 0.01	1.87 ± 0.01	0.009
5	β-myrcene	0.035 ± 0.00	1.62 ± 0.00	0.031 ± 0.01	1.49 ± 0.02	0.009
6	D-limonene	2.051 ± 0.01	93.82 ± 0.01	1.943 ± 0.02	93.39 ± 0.03	0.009
7	β-ocimene	0.002 ± 0.00	0.08 ± 0.00	0.002 ± 0.00	0.10 ± 0.00	0.009
8	Careen	0.001 ± 0.00	0.03 ± 0.00	0.001 ± 0.00	0.04 ± 0.00	0.009
		Monoterpenoid Alcohols				
9	Linalool	0.008 ± 0.00	0.35 ± 0.00	0.015 ± 0.00	0.70 ± 0.00	0.009
10	Eucalyptol	0.001 ± 0.00	0.06 ± 0.00	0.002 ± 0.00	0.07 ± 0.00	0.009
11	Isoborneol	0.003 ± 0.00	0.15 ± 0.00	0.004 ± 0.00	0.18 ± 0.00	0.009
12	Borneol	0.001 ± 0.00	0.05 ± 0.00	0.001 ± 0.00	0.06 ± 0.00	0.009
13	Bergamol	0.034 ± 0.01	1.54 ± 0.01	0.015 ± 0.01	0.70 ± 0.01	0.009
		Sesquiterpene Hydrocarbons				
14	Caryophyllene	0.001 ± 0.00	0.06 ± 0.00	0.002 ± 0.00	0.08 ± 0.00	0.009
15	trans-α-bergamotene	0.002 ± 0.00	0.09 ± 0.00	0.004 ± 0.00	0.19 ± 0.00	0.009
16	β-bisabolene	0.002 ± 0.00	0.11 ± 0.00	0.004 ± 0.00	0.18 ± 0.00	0.009

D-limonene was the main flavor compound found in both varieties, but its concentration was significantly higher in Neapolitan limmo compared to lemoncetta Locrese ($p < 0.05$). Other major components (except limonene) are β-pinene, β-myrcene, and bergamol (linalyl acetate) for both varieties. The last two compounds are significantly higher in Neapolitan limmo ($p < 0.05$). Moreover, other volatile aromas found at small percentage such as α-phellandrene, α-pinene, β-phellandrene, linalool, trans-α-bergamotene, and β-bisabolene were considered to be important compounds influencing the entire aroma [37]. Their concentrations were significantly higher in lemoncetta ($p < 0.05$). Only the volatile compound α-pinene was not significantly different in the two varieties ($p > 0.05$).

Twenty-three volatile compounds were identified in the peels: 10 terpenes, 8 monoterpenoid alcohols, and 5 sesquiterpene hydrocarbons (Table 5). Both varieties exhibited the same volatile compounds with different intensity (Figure 6). The peel of lemoncetta Locrese presented significantly higher areas than Neapolitan limmo. The main component of the peels is Limonene followed by bergamol, linalool, β-pinene, and β-myrcene with different intensities. Limonene value was not reported because is no longer linear and the areas were off the charts. Other relevant compounds are α-pinene, β-phellandrene, β-pinene, β-myrcene, linalool, and bergamol which are significantly higher in lemoncetta compared to Neapolitan limmo ($p < 0.05$). Only two volatile compounds (nerol acetate and geraniol acetate) are significantly higher in Neapolitan limmo than in lemoncetta Locrese ($p < 0.05$).

Table 5. Volatile compounds identified in the peels of Neapolitan limmo and lemoncetta Locrese. Data were expressed as average area of volatile compounds ± standard deviation. Values are significant at $p < 0.05$.

Peak	Volatile Compounds	Neapolitan Limmo (Area)	lemoncetta Locrese (Area)	*p*-Value
		Terpens		
1	α-phellandrene	$5.17 \times 10^6 \pm 5.00 \times 10^3$	$1.42 \times 10^7 \pm 5.00 \times 10^4$	0.009
2	α-Pinene	$3.87 \times 10^7 \pm 1.00 \times 10^5$	$2.82 \times 10^8 \pm 2.00 \times 10^6$	0.009
3	β-Phellandrene	$1.08 \times 10^8 \pm 1.00 \times 10^6$	$7.24 \times 10^8 \pm 1.50 \times 10^6$	0.009
4	β-Pinene	$5.54 \times 10^8 \pm 2.50 \times 10^6$	$4.34 \times 10^9 \pm 2.00 \times 10^7$	0.009
5	β-Myrcene	$4.39 \times 10^8 \pm 3.50 \times 10^6$	$1.66 \times 10^9 \pm 5.00 \times 10^6$	0.009
6	D-limonene	Off the chart	Off the chart	
7	β-ocimene	$4.23 \times 10^7 \pm 1.50 \times 10^5$	$1.27 \times 10^8 \pm 5.00 \times 10^5$	0.009
8	carene	$4.99 \times 10^6 \pm 3.50 \times 10^4$	$8.59 \times 10^6 \pm 4.00 \times 10^4$	0.009
9	cis-β-terpineol	$1.38 \times 10^7 \pm 2.50 \times 10^5$	$2.36 \times 10^7 \pm 3.00 \times 10^5$	0.009
10	terpinolene	$9.53 \times 10^6 \pm 1.50 \times 10^5$	$1.54 \times 10^7 \pm 2.00 \times 10^5$	0.009
		Monoterpenoid Alcohols		
11	linalool	$6.09 \times 10^8 \pm 3.00 \times 10^6$	$2.47 \times 10^9 \pm 2.00 \times 10^7$	0.009
12	α-terpineol	$1.16 \times 10^7 \pm 5.00 \times 10^4$	$2.05 \times 10^7 \pm 1.00 \times 10^5$	0.009
13	acetic acid octyl ester	$1.31 \times 10^7 \pm 5.00 \times 10^4$	$4.79 \times 10^7 \pm 4.00 \times 10^5$	0.009
14	trans geraniol	$2.28 \times 10^7 \pm 3.00 \times 10^5$	$2.78 \times 10^7 \pm 4.00 \times 10^5$	0.009
15	bergamol	$7.27 \times 10^9 \pm 3.00 \times 10^7$	$1.51 \times 10^{10} \pm 1.50 \times 10^8$	0.009
16	α-terpineol acetate	$3.29 \times 10^7 \pm 3.00 \times 10^5$	$3.44 \times 10^7 \pm 1.50 \times 10^5$	0.009
17	nerol acetate	$5.77 \times 10^7 \pm 3.00 \times 10^5$	$5.62 \times 10^7 \pm 1.50 \times 10^5$	0.009
18	geraniol acetate	$7.26 \times 10^7 \pm 3.00 \times 10^5$	$6.22 \times 10^7 \pm 5.00 \times 10^4$	0.009
		Sesquiterpene Hydrocarbons		
19	α-bergamotene	$7.94 \times 10^6 \pm 6.50 \times 10^4$	$1.12 \times 10^7 \pm 1.00 \times 10^5$	0.009
20	caryophyllene	$3.23 \times 10^7 \pm 1.50 \times 10^5$	$5.56 \times 10^7 \pm 2.50 \times 10^5$	0.009
21	trans-α-bergamotene	$7.44 \times 10^7 \pm 2.00 \times 10^5$	$1.28 \times 10^8 \pm 1.00 \times 10^6$	0.009
22	cis-α-bisabolene	$4.58 \times 10^6 \pm 3.50 \times 10^4$	$6.13 \times 10^6 \pm 1.00 \times 10^4$	0.009
23	β-bisabolene	$5.74 \times 10^7 \pm 1.50 \times 10^5$	$7.71 \times 10^7 \pm 1.50 \times 10^5$	0.009

The compounds found in these varieties were also reported in other Citrus varieties. Particularly, in the peels were found 10 additional volatile compounds: cis-β-terpineol, terpinolene, α-terpineol, acetic acid octyl ester, trans-geraniol, α-terpineol acetate, nerol acetate, geraniol acetate, α-bergamotene, and cis-α-bisabolene. The absence of these compounds in the juices is probably due to the juice squeezing. The extraction pressure conditions will determine different aroma components in juices and in peels.

Our results revealed that there are not qualitative differences between the two varieties. The aromatic profiles are identical and there are not specific volatile compounds that could be used to differentiate the varieties. The main differences are connected only to the intensity of the aromatic profile. The use of SPME-GC-MS thus resulted to be a valuable tool to analyze the volatile profile of the two sweet lime juices and peels and obtain a quality characterization of fruits from different varieties.

2.4. Genetic Comparison by DNA-Based Molecular Markers

The genetic similarity of the two Mediterranean sweet lime populations was finally analyzed using RAPD molecular markers [38]. This technique was preferred to DNA barcoding and phylogenic analysis because DNA barcoding works best if the sequences have sufficiently diverged. We feared that the assay could prove inconclusive in our case given the indications from chemical data that the evolutionary distance between Limmo and lemoncetta is close [39]. RAPD has instead proven useful in the identification of Citrus cultivars and the assessment of genetic relatedness for neglected or little-known citrus accessions [40,41]. Lemon (*C. limon*) cultivars of Campania (Italy) were for instance distinguished by their RAPD profiles using five arbitrary primers, confirming that RAPD markers can successfully identify lemon genotypes [42]. Iannelli et al. [43] characterized lemons by combining genome size and RAPD markers. In their work, the primer U19 utilized for distinguishing lemon genotypes.

The RAPD-PCR method allowed the genetic analysis of Neapolitan limmo and lemoncetta Locrese by simply comparing the presence/absence of bands in DNA amplification patterns, visible after electrophoresis on agarose gel, as the bands represent the numerous loci detected randomly and dispersed in their respective genomes. We used primers already successfully considered for the variety discrimination of other species [44,45] but we obtained more markers as compared to previous work. The primers used showed high reproducibility of amplification products. They also showed the absence of polymorphisms in DNA in the comparison between limmo and lemoncetta and therefore the impossibility to discriminate between the two populations (Figure 7a,b).

The RAPD profiles obtained with each primer were extremely different from each other, allowing us to explore different regions of the genome. Alleles corresponding to reproducible amplicons, given the dominant genetic nature of RAPD markers, identified a total of 80 markers or loci (Table 6). The number detected was dependent on the primer used but not on the variety, with an average of 6.7 loci per primer, going from the minimum of four bands for primers G07 and U4 and at most 12 bands for the AX08 primer. Also, the genetic analysis suggested that the Neapolitan limmo and the lemoncetta Locrese are likely synonyms of the same variety since the loci were not polymorphic and did not allow to discriminate between the two local varieties.

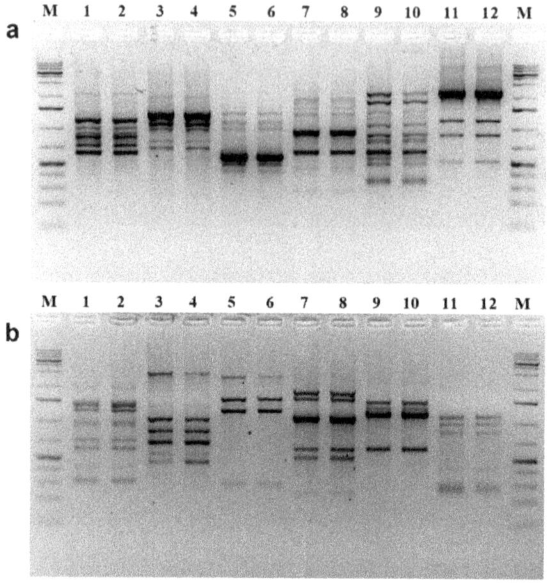

Figure 7. Genetic diversity of Neapolitan limmo and lemoncetta Locrese by RAPD molecular analysis. Comparison of the RAPD results on agarose gels of lemoncetta Locrese (odd lanes) and Neapolitan limmo genomes (even lanes), respectively, with the indicated arbitrary primers. (**a**) Lanes 1, 2: primer A05; lanes 3, 4: primer AK10; lanes 5, 6: primer AN10; lanes 7, 8: primer AX01; lanes 9, 10: primer AX08; lanes 11, 12: primer G07. (**b**) Lanes 1, 2: primer G12; lanes 3, 4: primer G19; lanes 5, 6: primer E10; lanes 7, 8: primer E11; lanes 9, 10: primer U4; lanes 11, 12: primer U19. Lanes M: GeneRuler 1 kb Plus DNA Ladder (Thermo Fisher Scientific) as molecular weight marker, containing three darkest bands consisting of 5000, 1500, and 500 bp.

3. Materials and Methods

3.1. Plant Materials

Fruits and Leaves samples of Neapolitan limmo and lemoncetta Locrese were harvested in January 2018 and 2019 from populations located in areas of Afragola (40°55′37′′ N; 14°18′42′′ E), Pozzuoli (40°49′39′′ N; 14°9′11′′ E) and Pianura (41°02′24′′ N; 14°11′09′′ E) (Campania, Italy) or Locri (38°14′19′′ N; 16°15′34′′ E) and Siderno (38°16′ N; 16°18′ E), Reggio Calabria (Calabria, Italy), respectively, and placed in a 4 °C refrigerated box to be processed.

Chemicals: All chemicals were purchased from Sigma-Aldrich (St. Louis, MO, USA) or Extrasynthes (Genay, France). An internal standard solution of camphor used for the analysis of volatile compounds in the juices was obtained from Sigma-Aldrich. The purity of all of the standards was beyond 95%. All other solvents and reagents were of analytical grade.

3.2. Preparation of the Samples for Chemical Analyses

Chemical analyses were performed on juices obtained in the laboratory from fresh fruits of Neapolitan limmo and lemoncetta Locrese. A total of 6 different Neapolitan limmo juice samples were used (2 in 2018 and 4 in 2019) and 8 lemoncetta Locrese samples (3 in 2018 and 5 in 2019). The juices were prepared using a manual squeezer, filtered through a stainless-steel filter with 1.18 mm mesh diameter, centrifuged at 18,000× *g* for 60 min at 4 °C, placed in plastic bags in 100 mL aliquots, and stored at −20° C until usage. The essential oils were extracted from fruits of limmo or lemoncetta (2 kg) through manual abrasion of the frozen flavedo at −20 °C by a small stainless steel grater with subsequent recovery and mixing 1:5 (*w*/*w*) with a saline solution (1 M NaCl) at 0 °C and subsequent centrifugation at 18,000× *g* for 60 min at 4° C. The oil recovered after centrifugation (supernatant, about 200 μL) was dried over anhydrous sodium sulfate and re-centrifuged at 12,000× *g* for 15 min at 4 °C and stored in the dark in a 1 mL vial under nitrogen at 5° C. A total of 8 samples of essential oils were prepared (all in 2019), 4 of Neapolitan limmo and 4 of lemoncetta Locrese.

3.3. Proximate Constituents

The soluble solids, expressed in Brix degrees, were determined by measurement of refractive index at 20 °C. The pH was determined by a Crison Model microTT 2050 pHmeter. Titrable acidity (total acids), expressed as citric acid monohydrate, was determined by titrating a 10 g sample with 0.1 N NaOH up to pH 8.1 according to the method reported by the International Federation of Fruit Juice Producers [46].

3.4. Organic Acid Analysis

Limmo or lemoncetta juice (20 g) were clarified by centrifugation at 12,000× *g* for 15 min. The clarified extract was filtered through a 0.45 μm Millipore filter (Darmstadt, Germany); 10 mL was chromatographed through a cation-exchange column [AG-1-X8 ($HCOO^-$) poly-prep Bio-Rad (Hercules, CA, USA)] and washed with water to a total volume of 100 mL. The organic acids were eluted with 6 M formic acid (ca. 130 mL), collected, and evaporated. The dry samples were recovered with water (10 mL) and filtered through a 0.45 μm Millipore filter before HPLC analysis. A volume of 10 μL was employed for the HPLC ThermoFinnigan Surveyor (Thermo Finnigan, Waltham, MA, USA) analysis on a Restek Allure organic acids cartridge 5 μm, 300 mm × 4.6 mm ID thermostated at 25 °C. The isocratic elution was carried out with a eluent consisting of 100 mM phosphate buffer at pH 2.5 with a flow rate of 0.5 mL/min with detection at 226 nm by a diode array detector [47] interfaced to a Dell computer Optlex gx260 with Xcalibur software for the signal acquisition and elaboration. Identification of quinic, malic, citric, fumaric, and tartaric acids was based on the retention time by co-injection of reference standards.

3.5. Flavonoid Analysis

The determination of the flavonoids (Flavanone *O*-glycosides and Flavone *O*-glycosides) in the juices was carried out by liquid chromatography according to the method of Grandi et al. [27]. Ten phenolic compounds were quantified. They included four flavanone *O*-glycosides with a rutinose (rhamnosyl-α-1,6-glucose) moiety: hesperidin, narirutin, eriocitrin and didymin (or neoponcirin). Four more were flavanone O-glycosides with neohesperidose moiety (rhamnosyl-α-1,2-glucose): naringin, neohesperidin, neoeriocitrin and poncirin.

Standard solutions of the flavonoids were prepared by weighing exactly 0.1 g of each compound and dissolving it in 100 mL of *N*,*N*-dimethylformamide. Those solutions were used to build up the calibration lines by diluting them to cover the concentration range of 1–100 mg/L.

The juices (10 mL) were shaken with 20 mL of a 1:1 (*v*/*v*) mixture of 0.25 M *N*,*N*-dimethylformamide/ ammonium oxalate and 20 mL of analytical-grade water and then filtered on 0.45 μm PTFE Pall filters. A volume of 5 μL was employed for the HPLC analysis on a Phenomenex Luna column C18 (150 × 3 mm) 5 μm thermostated at 25 °C. The elution was conducted as indicated by Grandi et al. [27]. The eluent A was made by an aqueous solution of 5 mM KH_2PO_4 adjusted at pH 3.05 with phosphoric acid. The eluent B was obtained by mixing acetonitrile/water/0.25 M KH_2PO_4, in the ratio 70:26:4 (*v*/*v*/*v*), and adding 100 μL of H_3PO_4 (87%) per liter of solution.

Finally, we quantified two flavone *O*-glycosides with a rutinose sugar moiety: rutin and diosmin. Specific wavelengths were used to identify the individual classes according to Gattuso et al. [48]: flavanones have an absorption maximum at 280–290 nm (set at 285), the flavones rutin and diosmin absorb at 304–350 nm (set at 325).

3.6. Chirospecific Analysis of Sabinene, β-Pinene, Limonene, Linalool and Linalyl Acetate in Essential Oils

The analyses were carried out by gas chromatography (GC-FID). The content in sabinene, β-pinene, limonene, linalool, and linalyl acetate of the essential oils of Neapolitan limmo and lemoncetta Locrese was determined by injecting in split modality 1:100 volume of 0.2 µL essential oil diluted 1:10 (*v/v*) in acetone. Analyses were conducted with a RTX®-5 (Resteck, Bellefonte, PA, USA) column (30 m × 0.25 mm, film 0.25 µm) at 70 °C for 1 min, 3 °C/min at 200 °C, holding for 0.3 min, 15 °C/min at 250 °C, and holding for 5 min. The metabolites content was expressed as the percentage of GC peak areas. Chirospecific analysis of the metabolites was performed with enantioselective multidimensional GC (enantio-MDGC). This technique consists in transferring part of the sample from a primary to a secondary column of different polarity or different chiral type. Our laboratory assembled a MDGC system by joining two Thermo Finnigan Trace 2000 G GC devices through a transfer line thermostated at 160 °C [36]. The first GC was equipped with a non-chiral column, the second had a column with a chiral stationary phase. A six-way rotating valve, positioned in the first GC and switchable via software, diverted the flow coming out the first column either to the first detector (FID 1) or to the second column where the substance enantiomers were separated and monitored with the second detector (FID 2) generating the chiral chromatogram. The non-chiral column employed in the first GC was a RTX®-5 column (5% diphenyl/95% dimethyl polysiloxane (30m × 0.25 mm i.d., film thickness 0.25 µm) (Resteck, Bellefonte, PA, USA). The carrier gas was helium at constant flow 1.5 mL/min. Make-up gas was nitrogen at 30 mL/min flow rate. The initial oven temperature was set at 70 °C for 10 min, then it was programmed from 70 to 85° C at 3 °C/min, from 85 to 175 °C at 5 °C/min, from 175 to 285 °C at 6 °C/min and finally at 285 °C for 5 min. The injector temperature was set at 250 °C and the detector temperature (FID) at 280 °C.

The chiral column used in the second chromatograph was a Diethyl tertbutyl silyl-BETA-Cyclodextrin column (Mega Legnano, Milan, Italy) (25m × 0.20mm i.d., 0.18 µm film thickness). The initial oven temperature was set at 35 °C for 25 min, then it was programmed from 35 to 160 °C at 4 °C/min and held at 140 °C for 2 min. The injector temperature 150 °C and detector (FID) was set at 220 °C. Carrier gas was helium at a programmed pressure. The initial pressure was 290 kpa for 30 min. then varied from 290 to 1500 kpa at 500 kpa/min. For each isomer, the enantiomeric excess (ee%) was calculated as $ee\% = ((A_{max} - A_{min})/(A_{max} + A_{min})) \times 100$ where A_{max} and A_{min} are the areas of the more and less abundant isomers respectively.

3.7. Volatile Organic Compounds (VOCs)

Determination of VOCs in the juices and peels was carried out by solid-phase micro extraction (SPME) and analyzed with a gas chromatography-mass spectrometry (GC-MS). An automatic injection autosampler CombiPal (CTC-CombiPal Analytics, Zwingen, Switzerland) was used for SPME sampling. The experiments were performed using a 50/30 µm divinylbenzene/carboxen/polydimethylsiloxane fiber (Supelco, Bellefonte, PA, USA). The fiber was conditioned according to the manufacturer's recommendation to remove contaminants. Before analysis, a fiber blank was run to confirm no contamination peak.

VOCs of juices and peels: An aliquot (8 g) of each juice diluted 20 times was weighed into a 20 mL vial and spiked with 80 µL of internal standard (camphor 3000 mg/L). Each sample was equilibrated at 40 °C for 10 min under stirring (500 rpm). After equilibration, the juice or peel were extracted by exposing the SPME fiber at 40 °C for 10 min (juice) and 2 min (peel). The analytes were desorbed at 250 °C for 15 min in the GC injection port. Measurements were always repeated at least in triplicates.

3.8. Gas Chromatography-Mass Spectrometry Analysis

The analyses were performed using a Varian 450 GC (Walnut Creek, CA, USA) coupled with Varian 300-MS mass spectrometer (Walnut Creek, CA, USA). The volatile compounds were separated using a Zebron ZB-semivolatiles capillary column (30 × 0.25 mm i.d., 0.25 µm film thickness). The oven temperature program was set as follows: initial temperature was held at 40 °C for 2 min, increased to 175 °C at 7.5 °C/min, and to 275 °C at 20 °C/min (held for 6 min). The temperatures of the transfer line and the ion source were set at 300 °C and 230 °C, respectively. Helium (99.999% purity) was used as a carrier gas at 0.9 mL/min. A split injection with a ratio of 1:100 was used. The analyses were carried out under full-scan acquisition mode. The mass range used was 30 to 450 *m/z*. The identification of volatile compounds was based on the comparison between the mass spectrum for each compound with those of two spectral libraries: NIST (National Institute of Standards and Technology, Gaithersburg, MD, USA) and WILEY (Wiley, New York, NY, USA). The data obtained were collected using the Bruker software Chemical Analysis MS Workstation version 7.0 (Karlsruhe, Germany).

3.9. Random Amplified Polymorphic DNA (RAPD-PCR) Analysis

Total genomic DNA of Neapolitan limmo and lemoncetta Locrese was isolated from 10 mg of lyophilized leaves. Plant tissue disruption was achieved at dry ice temperature through stainless beads in 2 mL tubes using a TissueLyser apparatus (Qiagen S.r.l., Milano, Italy) programmed at 30 Hz for 1 min. DNA was extracted according to the GeneJET Plant Genomic DNA Purification Mini Kit (Thermo Fisher Scientific, Carlsbad, CA, USA). The samples were dissolved in pre-heated (60 °C) lysis buffer. The suspensions were incubated at 60 °C for 10 min and subjected to the extraction procedure. RNase A treatment (5 μg/mL) was necessary to eliminate the co-extracted RNA. Finally, the DNA was eluted and diluted a final concentration of 20 ng/μL determined from the UV absorbance at 260 nm. As the purity and quality of the DNA template are crucial factors for a successful PCR, genomic DNA was checked by the 260/280 nm absorbance ratio and agarose gel electrophoresis.

The RAPD-PCR procedure was well established in our group in a previous project on *Citrus*, where sensitivity and reproducibility of the method were examined on genomic DNA from 1 to 100 ng [38,45,49]. In the present study we used 10 ng DNA template samples. The arbitrary primers tested in the PCR reaction had 60% G + C content and were 10 nucleotides long. After an initial screening of 20 arbitrary oligodeoxyribonucleotide decamer primers, 12 primers were selected for reproducibility of the band patterns: A05, AK10, AN10, AX01, AX08, G07, G12, G19, E10, E11, U4, and U19 (Table 6).

Each PCR was carried out in a 50 μL volume, containing 1X DreamTaq buffer with 2 mM $MgCl_2$, brought to 3 mM $MgCl_2$, 100 μM of each dNTP, 20 pmols of the arbitrary and unique primer, 2.0 Units of DreamTaq DNA polymerase (Thermo Fisher Scientific) and 10 ng of citrus genomic DNA. The PCR mixture was assembled on ice and transferred to a pre-cooled (6 °C) Veriti thermal cycler with a heated lid (Thermo Fisher Scientific).

Table 6. Arbitrary 10-mer primers used for RAPD analysis of Neapolitan limmo and lemoncetta Locrese.

N.	Primer Name	5′-Sequence-3′	GC (%)	Total Bands (nr. 80)	Total Bands (%)
1	A05	AGGGGTCTTG	60	8	10
2	AK10	CAAGCGTCAC	60	6	7.5
3	AN10	CTGTGTGCTC	60	6	7.5
4	AX01	GTGTGCCGTT	60	8	10
5	AX08	AGTATGGCGG	60	12	15
6	G07	GAACCTGCGG	70	4	5
7	G12	CAGCTCACGA	60	9	11.25
8	G19	GTCAGGGCAA	60	8	10
9	E10	CACCAGGTGA	60	5	6.25
10	E11	GAGTCTCAGG	60	5	6.25
11	U4	GACAGACAGG	60	4	5
12	U19	TGGGAACGGT	60	5	6.25

The DNA template was amplified by the following cycling profile: initial DNA template melting for 3 min at 95 °C, 45 cycles of denaturation for 1 min at 95 °C, primer annealing for 1 min at 40 °C, and synthesis for 1 min at 72 °C. The program ended with a final step conducted for 10 min at 72 °C. The reaction products were stored at −20 °C. Each reaction was repeated three times along with negative controls without genomic DNA. The RAPD-PCR products (25 μL) were separated by electrophoresis on 2% (*w*/*v*) agarose gel containing 0.5 μg/mL SyBr Safe and 1X TAE buffer (89 mM Tris-acetate at pH 8.4, 2 mM EDTA) at 5 V cm-1 for 1.5 h. The GeneRuler 1 kb Plus DNA ladder was used as standard marker of known molecular weights (Thermo Fisher Scientific). Amplicons were visualized under UV transilluminator and digitalized by the Electrophoresis Documentation and Analysis 120 System (Kodak ds-digital science, Rochester, NY, USA).

3.10. Statistical Analysis

All samples were analyzed in triplicates and the results were expressed as mean ± standard deviation (SD) after a normality distribution Kolmogorov-Smirnov test. Statistical analyses were performed using SPSS software ver. 21.0 (IBM, Armonk, NY, USA).

Statistical comparisons were carried out by analysis of variance (ANOVA) and post hoc Tukey-Kramer tests. A p value less than 0.05 was considered statistically significant. All tests were two tailed.

Significant differences in relative intensities of each volatile compound detected by GC-MS were analyzed by Mann-Whitney U-test between Neapoletan limmo and lemoncetta Locrese (α = 5%).

4. Conclusions

In conclusion, chemical and genetic data obtained by DNA fingerprint using RAPD markers have collectively allowed us to confirm beyond doubts that the ancient and rare sweet Mediterranean lime, known in Campania as Neapolitan limmo and in Calabria as lemoncetta Locrese are synonyms of the same citrus species.

Collectively, the obtained compositional data also indicate that, from the chemo-taxonomic point of view, both fruits belong to *Citrus lumia* Risso species despite the di stinct phenotypes. Our results might allow in the future the repopulation of limmo cultures and the reintroduction of this fruit in the essential oil and gastronomic market.

Author Contributions: D.C., A.S., S.S., and L.D.M. collected and analyzed the experiments, A.P., M.L.B. and G.F. critically evaluated the results. A.P. helped to write and finalize the paper. L.S., D.C., and D.C. formulated the original idea, supervised the research and wrote the first draft of the paper. All authors have read and agreed to the published version of the manuscript.

Funding: This work was supported by Dementia Research UK (RE1 3556).

Acknowledgments: We are grateful to Ciro Balestrieri (Università degli Studi della Campania "Luigi Vanvitelli") and to Roberta Ferrari (ProdALscarl) for inspiring this study.

Conflicts of Interest: The authors declare no conflict of interest.

References

1. Menagio, E. Le Origini della Lingua Italiana. 1685. Available online: https://archive.org/details/bub_gb_1qxS0CTglusC/page/n2 (accessed on 1 October 2019).
2. International Union for the Protection of New Varieties of Plants (UPOV). Guidelines for the Conduct of Tests for Distinctness, Uniformity and Stability CITRUS L.—Group 3 LEMONS and LIMES (TG/203/1 Rev). Available online: https://www.google.com.hk/url?sa=t&rct=j&q=&esrc=s&source=web&cd=1&ved=2ahUKEwjTue6Z7c3mAhUn7GEKHXJVAMoQFjAAegQIBRAC&url=https%3A%2F%2Fwww.upov.int%2Fedocs%2Ftgdocs%2Fen%2Ftg203.doc&usg=AOvVaw3xcpDB0URn31Vdv0HftQCj (accessed on 24 December 2019).
3. Amenta, N. Avvocato Napoletano, Parte I. 1724. Della Lingua Nobile d'Italia e del Modo di Leggiadramente Scrivere in Essa, non che di Perfettamente Parlare. Available online: https://archive.org/details/bub_gb_XMzAbS68mxIC (accessed on 24 December 2019).

4. Raimondo, F.M.; Cornara, L.; Mazzola, P.; Smeriglio, A. Le lumie di Sicilia: Note storiche e botaniche. *Quad. Bot. Amb. Appl.* **2015**, *26*, 43–50.
5. Curk, F.; Ollitrault, F.; Garcia-Lor, A.; Luro, F.; Navarro, L.; Ollitrault, P. Phylogenetic origin of limes and lemons revealed by cytoplasmic and nuclear markers. *Ann. Bot.* **2016**, *117*, 565–583. [CrossRef] [PubMed]
6. Butelli, E.; Licciardello, C.; Ramadugu, C.; Durand-Hulak, M.; Celant, A.; Reforgiato Recupero, G.; Froelicher, Y.; Martin, C. Noemi Controls Production of Flavonoid Pigments and Fruit Acidity and Illustrates the Domestication Routes of Modern Citrus Varieties. *Curr. Biol.* **2019**, *26*, 158–164. [CrossRef] [PubMed]
7. Strazzer, P.; Spelt, C.E.; Li, S.; Bliek, M.; Federici, C.T.; Roose, M.L.; Koes, R.; Quattrocchio, F.M. Hyperacidification of citrus fruits by a vacuolar proton-pumping P-ATPase Complex. *Nat. Commun.* **2019**, *10*, 744. [CrossRef]
8. Tanaka, T. Citrologia: Semicentennial Commemoration Papers on Citrus Studies. Available online: https://www.semanticscholar.org/paper/Citrologia-%3A-semi-centennial-commemoration-papers-%E9%95%B7%E4%B8%89%E9%83%8E/17726500cd7a095326ad7d8f52ce024e6ff0ee9c (accessed on 24 December 2019).
9. Chapot, H. Trois citrus Marocains. *Al Awamia* **1962**, *2*, 47–81.
10. Ferrarius, J.B. *Hesperides, Sive de Malorum Aureorum Cultura et Usu*; Sumptibus Hermani Scheus: Rome, Italy, 1646.
11. Bimbi, B. 1715—Citrus Fruit under the Medici family in Florence. Melangoli, Cedri e Limoni. An Extensive Collection of Citrus Fruit Is Depicted with Great Accuracy. The Fruit at the Centre of the Canvas (Number 14 and Circled in Red) Is Named as "Melangola", a Vernacular Term used to Define a "Limetta dolce" in the XVIII Century. Available online: https://commons.wikimedia.org/wiki/File:Bartolomeo_bimbi,_melagoli,_cedri_e_limoni,_1715,_01.JPG (accessed on 1 October 2019).
12. Benk, E.; Bergmann, R. Zur Kenntnis der Sűβlimette. *Fruchtsaft. Ind.* **1961**, *6*, 87–90.
13. Di Giacomo, A.; Costa, D.; Tita, S. Contributo alla conoscenza della limetta dolce. Nota II-Il succo. *Ess. Deriv. Agr.* **1970**, *40*, 124–128.
14. Asencio, A.D.; Serrano, M.; García-Martínez, S.; Pretel, M.T. Organic acids, sugars, antioxidant activity, sensorial and other fruit characteristics of nine traditional Spanish citrus fruits. *Eur. Food Res. Technol.* **2018**, *244*, 1497–1508. [CrossRef]
15. Brune, A.; Müller, M.; Taiz, L.; Gonzalez, P.; Etxeberria, E. Vacuolar acidification in citrus fruit: Comparison between acid lime (*Citrus Aurantifolia*) and sweet lime (*Citrus Limmetioides*) juice cells. *J. Am. Soc. Hortic. Sci.* **2002**, *127*, 171–177. [CrossRef]
16. Ministero delle Politiche Agricole e Forestali (MIPAF). "Limone Costa d'Amalfi" e "Limone di Sorrento" DM 20/07/1999. Italian Official Gazette GURI n. 177 on 30/07/1999. 2001. Available online: https://www.gazzettaufficiale.it/eli/gu/1999/07/30/177/sg/pdf (accessed on 9 October 2019).
17. Locatelli, M.; Carlucci, G.; Genovese, S.; Epifano, F. Analytical methods and phytochemistry of the typical italian liquor "Limoncello": A review. In *Phytochemicals—A Global Perspective of Their Role in Nutrition and Health*; Venketeshwer, R., Ed.; Intech: Rijeka, Croatia, 2012; pp. 93–106.
18. Ferrarini, A. *Farmacopea di Antonio Ferrarini Farmacista, Membro della Comissione di Sanità della Città e Provincia di Bologna, e già Ripetitore di Farmacia nell'Università, Dedicata a Sua Eminenza Reverendissima il Signor Cardinale Carlo Oppizzoni Arcivescovo della stessa città*; Per le stampe del Sassi: Bologna, Italy, 1825.
19. Pitrè, G. *Medicina Popolare Siciliana*; Clausen, C.: Torino-Palermo, Italy, 1896.
20. Nicolosi, E.; Deng, Z.N.; Gentile, A.; La Malfa, S.; Continella, G.; Tribulato, E. Citrus phylogeny and genetic origin of important species as investigated by molecular markers. *Theor. Appl. Genet.* **2000**, *100*, 1155–1156. [CrossRef]
21. Servillo, L.; Giovane, A.; Balestrieri, M.L.; Cautela, D.; Castaldo, D. N-methylated tryptamine derivatives in citrus genus plants: Identification of N, N, N -trimethyltryptamine in bergamot. *J. Agric. Food Chem.* **2012**, *60*, 9512–9518. [CrossRef] [PubMed]
22. Cautela, D.; Vella, F.M.; Laratta, B. The effect of processing methods on phytochemical composition in bergamot juice. *Foods* **2019**, *8*, 474. [CrossRef] [PubMed]
23. Nogata, Y.; Sakamoto, K.; Shiratsuchi, H.; Ishii, T.; Yano, M.; Ohta, H. Flavonoid composition of fruit tissues of *Citrus* species. *Biosci. Biotechnol. Biochem.* **2006**, *70*, 178–192. [CrossRef] [PubMed]

24. Mouly, P.P.; Arzouyan, C.R.; Gaydou, E.M.; Estienne, J.M. Differentiation of citrus juices by factorial discriminant analysis using liquid chromatography of flavanone glycosides. *J. Agric. Food Chem.* **1994**, *42*, 70–79. [CrossRef]
25. Dhuique-Mayer, C.; Caris-Veyrat, C.; Ollitrault, P.; Curk, F.; Amiot, M.J. Varietal and interspecific influence on micronutrient contents in *Citrus* from the Mediterranean area. *J. Agric. Food Chem.* **2005**, *53*, 2140–2145. [CrossRef] [PubMed]
26. Abad-García, B.; Garmón-Lobato, S.; Sánchez-Ilárduya, M.B.; Berrueta, L.A.; Gallo, B.; Vicente, F.; Alonso-Salces, R.M. Polyphenolic contents in *Citrus* fruit juices: Authenticity Assessment. *Eur. Food Res. Technol.* **2014**, *238*, 803–818. [CrossRef]
27. Grandi, R.; Trifirò, A.; Gherardi, S.; Calza, M.; Saccani, G. Characterization of lemon juice on the basis of flavonoid content. *Fruit Process.* **1994**, *11*, 335–359.
28. Ooghe, W.C.; Detavernier, C.M. Detection of the addition of *Citrus Reticulata* and Hybrids to *Citrus Sinensis* by flavonoids. *J. Agric. Food Chem.* **1997**, *45*, 1633–1637. [CrossRef]
29. Peterson, J.J.; Beecher, G.R.; Bhagwat, S.A.; Dwyer, J.T.; Gebhardt, S.E.; Haytowitz, D.B.; Holden, J.M. Flavanones in grapefruit, lemons, and limes: A compilation and review of the data from the analytical literature. *J. Food Compos. Anal.* **2006**, *19*, 74–80. [CrossRef]
30. Cautela, D.; Laratta, B.; Santelli, F.; Trifirò, A.; Servillo, L.; Castaldo, D. Estimating bergamot juice adulteration of lemon juice by high-performance liquid chromatography (HPLC) analysis of flavanone glycosides. *J. Agric. Food Chem.* **2008**, *56*, 5407–5414. [CrossRef]
31. Aubert, B. Contexte Historique, Scientifique et Artistique de l'Histoire Naturelle des Orangers. In *Histoire naturelle des orangers de Risso et Poiteau, Commentaires et Développements*; Aubert, B., Bové, J.M., Eds.; Connaissance & Mémoires: Paris, France, 2000; p. 40.
32. Smeriglio, A.; Cornara, L.; Denaro, M.; Barreca, D.; Burlando, B.; Xiao, J.; Trombetta, D. Antioxidant and cytoprotective activities of an ancient Mediterranean *Citrus* (*Citrus Lumia* Risso) albedo extract: Microscopic observations and polyphenol characterization. *Food Chem.* **2019**, *279*, 347–355. [CrossRef] [PubMed]
33. Da Conceicao Neta, E.R.; Johanningsmeier, S.D.; Drake, M.A.; McFeeters, R.F. A Chemical basis for sour taste perception of acid solutions and fresh-pack dill pickles. *J. Food Sci.* **2007**, *72*, 352–359. [CrossRef] [PubMed]
34. Di Giacomo, A.; Rispoli, G. Contributo alla conoscenza della limetta dolce. Nota I L'olio essenziale. *Ess. Deriv. Agr.* **1969**, *39*, 115–123.
35. El-Seedi, H.R.; El-Said, A.M.A.; Khalifa, S.A.M.; Göransson, U.; Bohlin, L.; Borg-Karlson, A.K.; Verpoorte, R. Biosynthesis, natural sources, dietary intake, pharmacokinetic properties, and biological activities of hydroxycinnamic acids. *J. Agric. Food Chem.* **2012**, *60*, 10877–10895. [CrossRef] [PubMed]
36. Dugo, G.; Bartle, K.D.; Bonaccorsi, I.; Catalfamo, M.; Cotroneo, A. Advanced analytical techniques for the analysis of *Citrus* essential oils. Part 2. Volatile fraction: LC-HRGC and MDGC. *Ess. Deriv. Agr.* **1999**, *69*, 159–217.
37. Qiu, S.; Wang, J. Application of Sensory Evaluation, HS-SPME GC-MS, E-Nose, and E-Tongue for quality detection in *Citrus* fruits. *J. Food Sci.* **2015**, *80*, 2296–2304. [CrossRef]
38. Williams, J.G.K.; Kubelik, A.R.; Livak, K.J.; Rafalski, J.A.; Tingey, S.V. DNA polymorphisms amplified by arbitrary primers are useful as genetic markers. *Nucleic Acids Res.* **1990**, *18*, 6531–6535. [CrossRef]
39. Bruni, I.; De Mattia, F.; Martellos, S.; Galimberti, A.; Savadori, P.; Casiraghi, M.; Nimis, P.L.; Labra, M. DNA barcoding as an effective tool in improving a digital plant identification system: A case study for the area of Mt. Valerio, Trieste (NE Italy). *PLoS ONE* **2012**, *7*, e43256. [CrossRef]
40. Deng, Z.N.; Gentile, A.; Nicolosi, E.; Domina, F.; Vardi, A.; Tribulato, E. Identification of in vivo and in vitro lemon mutants by RAPD Markers. *J. Hortic. Sci.* **1995**, *70*, 117–125. [CrossRef]
41. Gouri Sankar, T.; Gopi, V.; Deepa, B.; Gopal, K. Genetic diversity analysis of sweet orange (*Citrus sinensis* Osbeck) varieties/clones through RAPD markers. *Int. J. Curr. Microbiol. Appl. Sci.* **2014**, *3*, 75–84.
42. Mariniello, L.; Sommella, M.G.; Cozzolino, A.; Di Pierro, P.; Ercolini, D.; Porta, R. Identification of Campania *Citrus limon* L. by Random Amplified Polymorphic DNA markers. *Food Biotechnol.* **2004**, *18*, 289–297. [CrossRef]
43. Iannelli, D.; Cottone, C.; Viscardi, M.; D'Apice, L.; Capparelli, R.; Boselli, M. Identification of genotypes of lemon by flow cytometry and RAPD markers. *Int. J. Plant Sci.* **1998**, *159*, 864–869. [CrossRef]

44. Galderisi, U.; Cipollaro, M.; Di Bernardo, G.; De Masi, L.; Galano, G.; Cascino, A. molecular typing of italian sweet chestnut cultivars by Random Amplified Polymorphic DNA analysis. *J. Hortic. Sci. Biotechnol.* **1998**, *73*, 259–263. [CrossRef]
45. De Masi, L.; Castaldo, D.; Galano, G.; Minasi, P.; Laratta, B. Genotyping of fig (*Ficus carica* L) via RAPD markers. *J. Sci. Food Agric.* **2005**, *82*, 2235–2242. [CrossRef]
46. International Federation of Fruit Juice Producers (IFFJP). International Fruit and Vegetable Juice Association. Available online: https://ifu-fruitjuice.com/ (accessed on 24 December 2019).
47. AOAC International. *Official Methods of Analysis*, 17th ed.; AOAC International: Gaithersburg, MD, USA, 2000.
48. Gattuso, G.; Barreca, D.; Gargiulli, C.; Leuzzi, U.; Caristi, C. Flavonoid Composition of *Citrus* juices. *Molecules* **2007**, *12*, 1641–1647. [CrossRef]
49. Galtieri, V.; Laratta, B.; Minasi, P.; De Masi, L. Indagine RAPD-PCR per la tipizzazione delle *Cultivar* di bergamotto (*Citrus bergamia* Risso et Poiteau). *Ess. Deriv. Agr.* **2004**, *74*, 19–23.

molecules

MDPI

Article

The *Trichoderma atroviride* Strains P1 and IMI 206040 Differ in Their Light-Response and VOC Production

Verena Speckbacher [1], Veronika Ruzsanyi [2], Modestus Wigger [3] and Susanne Zeilinger [1,*]

1 Department of Microbiology, University of Innsbruck, 6020 Innsbruck, Austria; verena.speckbacher@uibk.ac.at
2 Breath Research Institute, University of Innsbruck, 6850 Dornbirn, Austria; veronika.ruzsanyi@uibk.ac.at
3 Umweltmonitoring und Forensische Chemie, Hochschule Hamm-Lippstadt, 59063 Hamm, Germany; modestus.wigger@stud.hshl.de
* Correspondence: susanne.zeilinger@uibk.ac.at

Academic Editor: Derek J. McPhee
Received: 9 December 2019; Accepted: 27 December 2019; Published: 3 January 2020

Abstract: *Trichoderma atroviride* is a strong necrotrophic mycoparasite antagonizing and feeding on a broad range of fungal phytopathogens. It further beneficially acts on plants by enhancing growth in root and shoot and inducing systemic resistance. Volatile organic compounds (VOCs) are playing a major role in all those processes. Light is an important modulator of secondary metabolite biosynthesis, but its influence has often been neglected in research on fungal volatiles. To date, *T. atroviride* IMI 206040 and *T. atroviride* P1 are among the most frequently studied *T. atroviride* strains and hence are used as model organisms to study mycoparasitism and photoconidiation. However, there are no studies available, which systematically and comparatively analyzed putative differences between these strains regarding their light-dependent behavior and VOC biosynthesis. We therefore explored the influence of light on conidiation and the mycoparasitic interaction as well as the light-dependent production of VOCs in both strains. Our data show that in contrast to *T. atroviride* IMI 206040 conidiation in strain P1 is independent of light. Furthermore, significant strain- and light-dependent differences in the production of several VOCs between the two strains became evident, indicating that *T. atroviride* P1 could be a better candidate for plant protection than IMI 206040.

Keywords: *Trichoderma atroviride*; mycoparasitism; secondary metabolites; volatile organic compounds (VOCs); photoconidiation; fungi; 2-octanone; injury response; light response; *Fusarium oxysporum*

1. Introduction

Filamentous fungi are prolific producers of volatile organic compounds (VOCs), small carbon-based substances that readily enter the gas phase and that are derived from both, primary and secondary metabolism. Fungal VOCs are produced as mixtures of chemically diverse compounds, with the produced VOC profile being species- and even strain-specific and greatly depending on the developmental stage of the fungus as well as environmental factors [1–3]. They are involved in various biological processes including communication with other organisms such as plants and microbes as well as self-signaling. VOC emission by fungi serves in interaction with animals as reported for 1-octen-3-ol produced by the mushroom *Clitopilus prunulus* for deterring banana slugs and by the wood-rotting fungus *Trametes gibbosa* for attracting fungus-eating beetles [4,5]. Fungal VOCs further can impact plants by activating defense responses and affecting plant growth, as well as directly inhibiting the proliferation of phytopathogens [6].

Members of the fungal genus *Trichoderma* are efficient mycoparasites that antagonize a wide range of phytopathogenic fungi by direct parasitism employing secreted antifungal hydrolytic enzymes and metabolites [7]. At least 480 different VOCs have been identified from *Trichoderma* species yet, with 6-pentyl-2*H*-pyran-2-one (6PP) being one of the first volatiles isolated from this fungal genus [8,9]. *Trichoderma*-emitted VOCs have been connected to the antagonistic activity of these fungi, as they were shown to reduce the mycelial growth of fungal plant pathogens [8,10] as well as wood decay fungi [11]. However, only recently VOC production in co-cultivation was addressed, showing that interaction of *Trichoderma* spp. such as *T. harzianum, T. hamatum* and *T. velutinum* with the ectomycorrhizal fungus *Laccaria bicolor* dramatically altered the VOC emission patterns [12]. In addition to VOCs with bioactivity against fungi, *Trichoderma* spp. release volatiles that affect plant development and immunity. *Arabidopsis* plants exposed to the pool of VOCs emitted by *T. viride* showed increased lateral root formation and growth, and comparable results were obtained with *T. virens* and *T. pseudokoningii* volatiles [13–15]. Similarly, *T. atroviride*-derived 6PP stimulated tomato plant growth and systemic defense; however, the effects were concentration-dependent [8]. In addition to 6-PP, 24 members of the compound classes of alkanes, alcohols, ketones, lactones, furanes, monoterpenes, and sesquiterpenes—including typical fungal C8-compounds—as well as several volatiles not known to be produced by *Trichoderma* before, were identified in the headspace of cultures of *T. atroviride* strain P1 (ATCC 74058) [16]. C8-compounds such as 1-octen-3-ol, 3-octanol and 3-octanone are end-products of fatty acid metabolism [17] and act as signaling molecules regulating fungal development and inter-colony communication. In *T. atroviride*, these volatiles are increasingly produced by conidiating cultures and were described to act as elicitors of conidiation [18]. In addition, 1-octen-3-ol production by *T. atroviride* was up-regulated upon treatment of the fungus with the *Fusarium*-derived mycotoxin fusaric acid, while levels of other VOCs such as 6PP, alpha-phellandrene, and beta-phellandrene decreased [16].

Light, especially in the UV/blue spectrum, is an important environmental cue that triggers asexual reproduction in many fungal species [19]. In *Trichoderma*, most studies on photoconidiation have been performed with *T. atroviride* strain IMI 206040 as a model [20–26]. In complete darkness, *T. atroviride* IMI 206040 has been reported to grow infinitely as mycelium, while exposure to light induces the formation of green conidia [27,28].

In the present study, two different strains (*T. atroviride* P1, ATCC 74058 and *T. atroviride* IMI 206040) of the strong mycoparasite *T. atroviride* were analyzed for their differences in VOC biosynthesis by an in-house made high-resolution ion mobility spectrometer (IMS) with gas chromatographic (GC) pre-separation. Despite the fact that *T. atroviride* is a model to study photoconidiation, no studies have systematically and comparatively analyzed putative strain-, or light-dependent differences in the composition of VOC mixtures released by these fungi. We hence explored and compared their VOC profiles along a cultivation period of 120 h in complete darkness and upon exposure to light, as well as during the mycoparasitic interaction with the host fungi *Rhizoctonia solani* and *Fusarium oxysporum*. Remarkable differences not only in VOC production, but also in light-dependent conidiation and in the mycoparasitism on *F. oxysporum* became evident.

2. Results

2.1. The Vegetative Growth Rate of *T. atroviride* Is Strain- and Light- Dependent

Upon cultivation on PDA plates, the radial growth rate differed between *T. atroviride* strains P1 and IMI 206040. *T. atroviride* IMI 206040 exhibited a higher radial growth rate than *T. atroviride* P1, irrespective of the applied light regime. However, both strains showed enhanced radial growth upon cultivation in complete darkness compared to light-dark conditions (Figure 1).

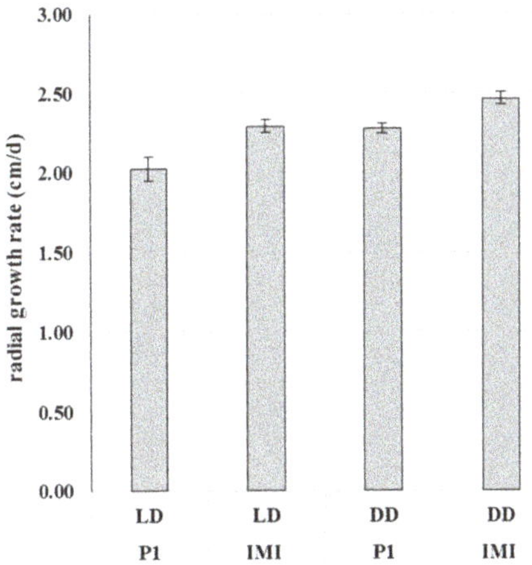

Figure 1. Strain-specific differences in radial growth of *Trichoderma atroviride*. The radial growth rate (cm/d) of *T. atroviride* P1 (P1) and IMI 206040 (IMI) after three days of cultivation on PDA at 25 °C under light-dark (LD) conditions or complete darkness (DD). Results shown are means ± SD (n = 4).

2.2. Asexual Sporulation in *T. atroviride* Is Strain- and Light-Dependent

Comparative analysis of *T. atroviride* P1 and IMI 206040 under conidiation-inducing conditions revealed significant differences between the two strains. In *T. atroviride* IMI 206040, asexual sporulation only occurred under light-dark conditions, while conidia were not formed upon cultivation in complete darkness. According to previous reports [29], conidiation could further be triggered in dark-grown *T. atroviride* IMI 206040 by mechanical injury or a pulse of blue light, respectively. In strain IMI 206040 injury resulted in low conidiation along the cutting sites only, whereas blue light treatment led to the production of massive amounts of heavily pigmented conidia. In contrast, *T. atroviride* P1 fully conidiated even upon growth in complete darkness. Mechanical injury led to strong conidiation and the generation of scarring tissue along the cutting sites in this strain (Figure 2).

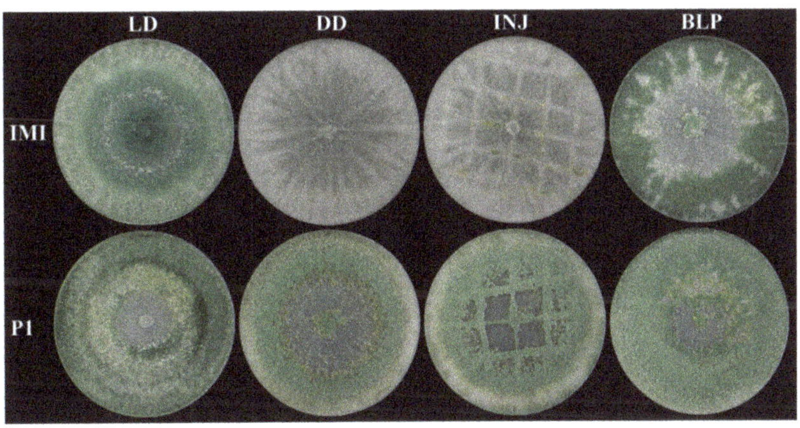

Figure 2. Strain-specific differences in conidiation upon growth under different light regimes and upon mechanical injury. *T. atroviride* P1 (P1) and *T. atroviride* IMI 206040 (IMI) were grown on PDA at 25 °C for five days under light-dark (LD) conditions or in complete darkness (DD). For induction of conidiogenesis, the fungi were grown in complete darkness for two days, treated by either mechanical injury (INJ) or a 10 min blue-light pulse (BLP) followed by incubation for further three days in complete darkness. A representative image of four biological replicates (n = 4) is shown.

2.3. The Mycoparasitic Activity of T. atroviride Is Strain- and Light-Dependent

In direct confrontation assays, the antagonistic activities against the tested host fungi differed between the two *T. atroviride* strains and turned out to be influenced by light. Upon growth in light–dark conditions, neither *T. atroviride* IMI 206040 nor strain P1 were able to fully overgrow and mycoparasitize the two tested host fungi *R. solani* and *F. oxysporum*. However, maybe due to its higher growth rate, *T. atroviride* IMI 206040 made better progress in overgrowing both hosts compared to *T. atroviride* P1. Upon growth in complete darkness both *T. atroviride* strains showed significantly increased antagonistic activities compared to light-dark conditions; they were able to fully overgrow and lyse *R. solani* within five days. Interestingly, *F. oxysporum* could only be fully overgrown and mycoparasitized by *T. atroviride* P1, which was even conidiating under these conditions, while *T. atroviride* IMI 206040 was less active against this host fungus and did not conidiate in darkness. These data suggest that upon growth in complete darkness, *T. atroviride* P1 is a stronger mycoparasite of *F. oxysporum* than *T. atroviride* IMI 206040 (Figure 3).

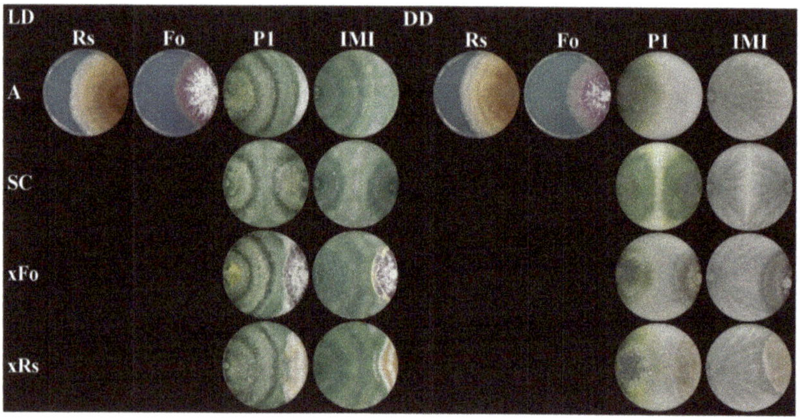

Figure 3. Strain-specific differences in the mycoparasitic behavior against *Fusarium oxysporum* and *Rhizoctonia solani* upon incubation under different light regimes. Plate confrontation assays of the mycoparasites *T. atroviride* P1 (P1) and *T. atroviride* IMI 206040 (IMI) against the host fungi *F. oxysporum* (Fo) and *R. solani* (Rs). As controls, all strains were grown alone (A). The mycoparasites were as well grown in self confrontation (SC) and in direct confrontation with *F. oxysporum* (xFo) or *R. solani* (xRs) on PDA at 25 °C for five days under either light-dark (LD) conditions or complete darkness (DD). A representative image of four biological replicates (n = 4) is shown.

2.4. VOC Biosynthesis Is Mostly Strain-Specific but Also Light- and Host Dependent

VOC analyses of *T. atroviride* P1 and IMI 206040 along a period of 120 h upon axenic cultivation under light–dark conditions or in complete darkness as well as upon interaction with host fungi revealed more than 50 peaks in the GC–IMS chromatograms, which corresponded to approx. 20 to 25 VOCs. Since for IMS no commercial database, such as the NIST database for GC–MS, is available, we had to measure VOC standards and confirm the detected substances based on the characteristic IMS and GC data (drift and retention times, respectively). This resulted in the identification of a total of ten volatiles (Table 1). Except 2-octanone, which was solely detected in direct confrontation between *T. atroviride* and *F. oxysporum*, and 3-octanone, which was released by strain IMI 206040 solely upon light–dark conditions, all of the identified VOCs were produced by both strains irrespective of the cultivation condition. However, their amounts and course of release strongly differed between the different experimental setups. The PDA media itself did not emit any of the mentioned VOCs in significant concentrations.

Table 1. Volatile organic compounds (VOCs) identified via gas chromatography–ion mobility spectrometry (GC–IMS) in the headspace of *T. atroviride* P1 and IMI 206040.

Formula	Name of Compound	CAS-Number
C_2H_6O	ethanol	64-17-5
C_3H_8O	1-propanol	71-23-8
$C_4H_{10}O$	2-methyl-propanol	78-83-1
$C_5H_{10}O$	3-methylbutanal	590-86-3
$C_5H_{12}O$	2-methyl-butanol	137-32-6
$C_5H_{12}O$	3-methyl-1-butanol	123-51-3
$C_7H_{14}O$	2-heptanone	110-43-0
$C_8H_{16}O$	1-octen-3-ol	3391-86-4
$C_8H_{16}O$	2-octanone	111-13-7
$C_8H_{16}O$	3-octanone	106-68-3

2.4.1. VOCs Emitted in Strain-Specific Amounts

Whereas 2-heptanone and ethanol were released in similar amounts to the headspace of the two tested *T. atroviride* strains, the amounts of 1-propanol, 2-methyl-propanol, 3-methylbutanal, 2-methyl-butanol, 3-methyl-1-butanol, 3-octanone and 1-octen-3-ol significantly varied between the two *T. atroviride* strains with strain P1 emitting considerably higher amounts than IMI 206040.

In *T. atroviride* P1, the release of 1-octen 3-ol to the headspace strongly increased from 68.5 h on and reached a maximal concentration of approx. 375 parts per billion (ppb) after 96.5 h of cultivation. In *T. atroviride* IMI 206040, 1-octen-3-ol emission peaked at the same time point with a maximal concentration of approx. 60 ppb meaning an 84% reduction compared to strain P1 (Figure 4A). The emission of 1-propanol increased during the early cultivation period, peaked at 43.5 h with approx. 50 ppb in *T. atroviride* P1 and afterwards decreased again. In *T. atroviride* IMI 206040, 1-propanol levels peaked at the same time point, but with a maximal concentration of approx. 10 ppb, which is equivalent to an 80% reduction compared to strain P1 (Figure 4B). 2-heptanone was increasingly released to the headspace over time, reaching and staying at a maximum of ≥1000 ppb at the end of the cultivation period in both *T. atroviride* strains. However, this volatile was already produced in considerable amounts of approx. 275 ppb after 68.5 h of cultivation in *T. atroviride* IMI 206040, while 2-heptanone emission by strain P1 was delayed and increased not before 96.5 h (Figure 4C). 2-methyl-propanol emission followed a similar time course in both *T. atroviride* strains. However, after a steep increase in the early phase of the cultivation, *T. atroviride* P1 produced maximal levels of approx. 650 ppb at the 68.5 h time-point, whereas *T. atroviride* IMI 206040 cultures only emitted up to approx. 175 ppb, which is equivalent to a 73% reduction compared to strain P1 (Figure 4D). 3-methyl-butanal emission by *T. atroviride* P1 reached 45 ppb after 43.5 h and afterwards decreased to 20 ppb until the end of cultivation, whereas in *T. atroviride* IMI 206040 approx. 20 ppb of this volatile were constantly emitted along the whole cultivation period (Figure 4E). The release of ethanol to the headspace increased at the early phases of the cultivation period, peaked at 24.5 h with approx. 65 ppb in *T. atroviride* P1 and afterwards decreased again to approx. 25 ppb. Young *T. atroviride* IMI 206040 cultures produced similar amounts of this volatile with a maximal level of approx. 60 ppb after 43.5 h of cultivation, but with a successive steeper decrease down to approx. 5 ppb at the end of the cultivation period (Figure 4F).

2.4.2. VOCs Emitted in Strain-Specific Amounts in a Light-Dependent Manner

3-octanone, 2-methyl-butanol, and 3-methyl-1-butanol were not only secreted in a strain- but also in a light-dependent manner.

Upon growth in complete darkness, *T. atroviride* IMI 206040 did not emit any 3-octanone. However, up to 25 ppb were measured after 96.5 h of cultivation under light-dark conditions. As the strain already conidiated at this time point in the presence of light, 3-octanone production could not only be light-induced but also conidiation-associated in *T. atroviride* IMI 206040. In contrast, 3-octanone emission by *T. atroviride* P1 was, similar to conidiation, light-independent and increased with cultivation time to a maximum of 80–95 ppb after 115.5 h of growth (Figure 5; row 1).

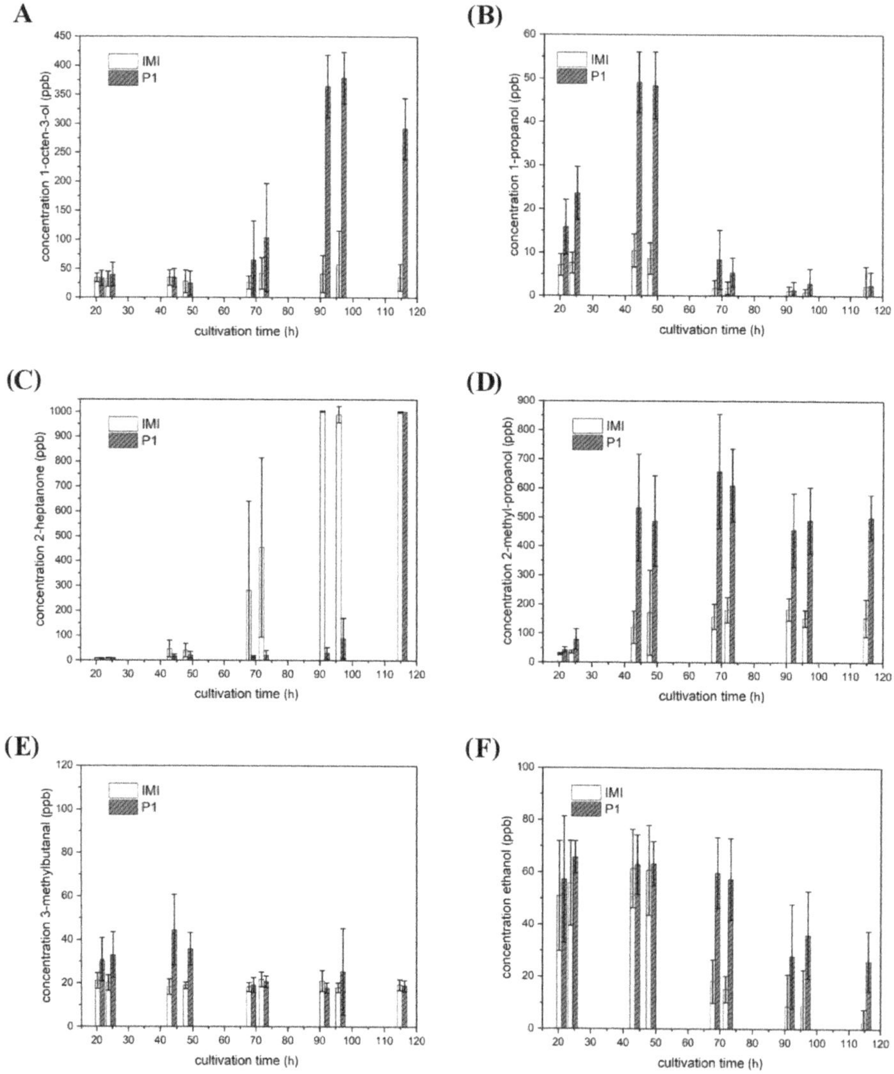

Figure 4. VOCs produced by *T. atroviride* P1 and IMI 206,040 in a strain-specific but light-independent manner. Levels (ppb) of (**A**) 1-octen 3-ol, (**B**) 1-propanol, (**C**) 2-heptanone, (**D**) 2-methyl-propanol, (**E**) 3-methylbutanal, and (**F**) ethanol detected in the headspace of *T. atroviride* P1 (P1) and *T. atroviride* IMI 206040 (IMI) cultures grown on PDA in Schott bottles. GC–IMS measurements were conducted along a cultivation period of 120 h. As the given volatiles were produced in similar amounts under light-dark conditions and in complete darkness, results shown are means ± SD (n = 8).

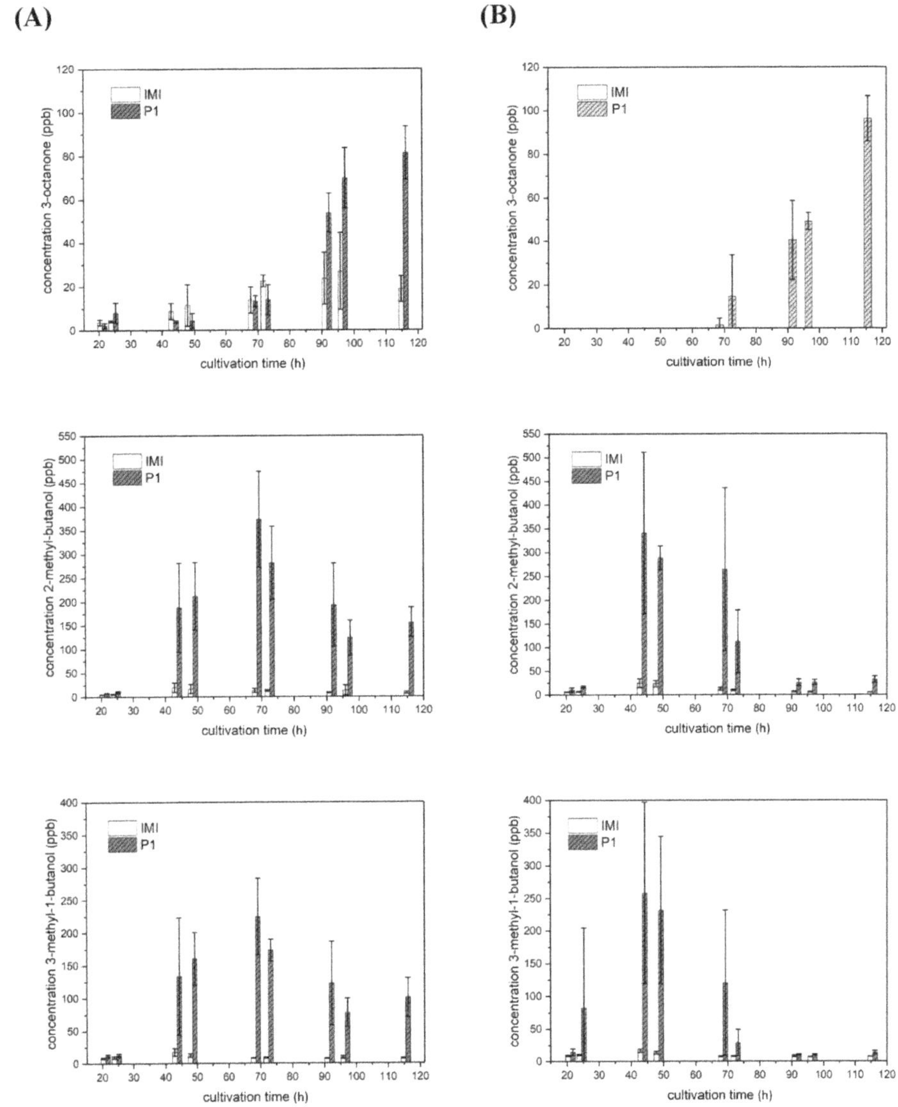

Figure 5. VOCs produced by *T. atroviride* P1 and IMI 206,040 in a strain-specific and light-dependent manner. Levels (ppb) of 3-octanone (row 1), 2-methyl-butanol (row 2), and 3-methyl-1-butanol (row 3) detected in the headspace of *T. atroviride* P1 (P1) and *T. atroviride* IMI 206040 (IMI) cultures grown on PDA in Schott bottles at 25 °C under (**A**) light-dark conditions or (**B**) in complete darkness. GC–IMS measurements were conducted along a cultivation period of 120 h. Results shown are means ± SD (n = 4).

2-methyl-butanol and 3-methyl-1-butanol were produced light-independently by *T. atroviride* IMI 206040. Low amounts of approx. 25 ppb of both volatiles were detected under both light-dark conditions and complete darkness along the whole cultivation period of this strain (Figure 5 rows 2 and 3). In contrast, *T. atroviride* P1 emitted 2-methyl-butanol and 3-methyl-1-butanol in a slightly light-dependent manner. Upon cultivation under light-dark conditions, increasing amounts of both substances were released to the headspace with a peak of approx. 375 ppb of 2-methyl-butanol and approx. 225 ppb of 3-methyl-1-butanol after 68.5 h of cultivation. Similar maximal amounts were emitted by strain P1 upon growth in complete darkness; however, production and decline started earlier compared to light–dark conditions (Figure 5; rows 2 and 3).

2.4.3. VOC Emission in Co-Culture

To assess the influence of a host fungus on the emission of VOCs by *T. atroviride*, both strains—IMI 206040 and P1—were co-cultured with *R. solani* or *F. oxysporum*.

Compared to axenic cultures of both *T. atroviride* strains and the fungal host, the amounts of 2-methyl-propanol and 3-methylbutanal strongly differed in co-cultures with *R. solani*. The host released a maximal amount of approx. 260 ppb 2-methyl-propanol after 91.5 h in a light-independent way, whereas it did not produce 3-methylbutanal upon axenic cultivation. The course of 2-methyl-propanol and 3-methylbutanal emission upon confrontation of *T. atroviride* with *R. solani* was similar to the axenic *T. atroviride* cultures. However, 2-methyl-propanol levels were enhanced in co-cultures involving *T. atroviride* P1 compared to the respective axenic cultures (Figure 6; row 1). In contrast, decreased amounts of 3-methylbutanal were released to the headspace by both *T. atroviride* strains upon confrontation with *R. solani* compared to axenic cultures (Figure 6; row 2).

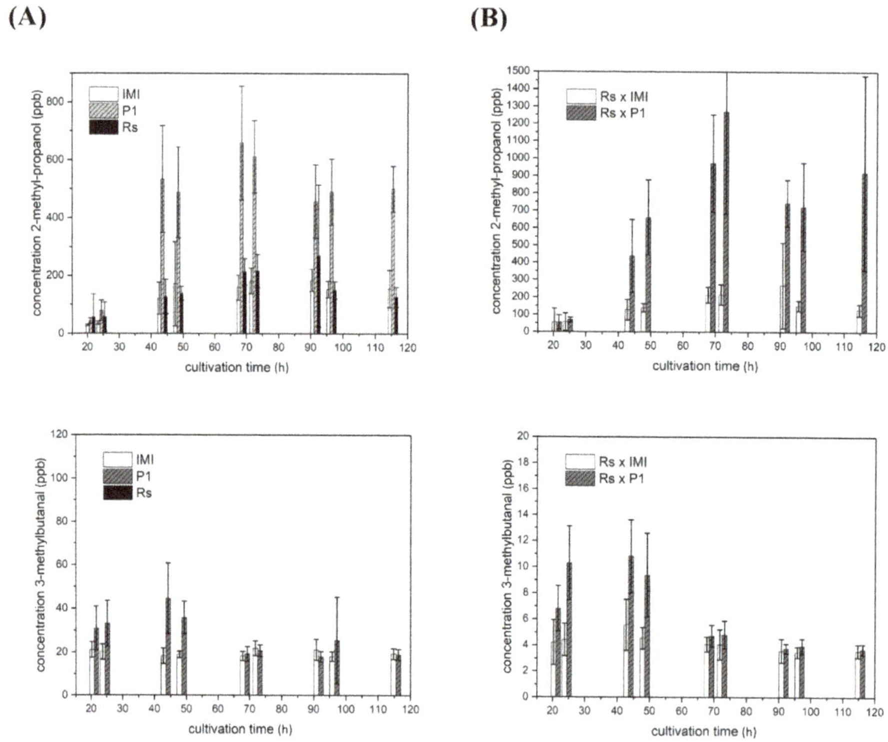

Figure 6. Emission levels of light-independently produced VOCs upon co-cultivation of *T. atroviride* P1 or IMI 206040 with *R. solani*. Levels (ppb) of 2-methyl-propanol (row 1) and 3-methylbutanal (row **2**) detected in the headspace of (**A**) axenic cultures of *T. atroviride* P1 (P1), *T. atroviride* IMI 206040 (IMI) and *R. solani* (Rs) or of (**B**) co-cultures of *T. atroviride* P1 or IMI 206040 confronted with *R. solani* (Rs × P1; Rs × IMI) upon cultivation on PDA in Schott bottles at 25 °C. GC–IMS measurements were conducted along a cultivation period of 120 h. As the given volatiles were produced in similar amounts under light-dark conditions and in complete darkness, results shown are means ± SD (n = 4).

Upon co-cultivation of *T. atroviride* with *F. oxysporum*, 1-propanol, 2-heptanone, and 3-octanone were produced in different amounts compared to the respective axenic cultures. In addition, 2-octanone was exclusively detected upon co-cultivation.

When grown in axenic culture, *F. oxysporum* constantly released approx. 55 ppb 1-propanol in a light-independent manner, whereas 2-heptanone, 3-octanone, and 2-octanone could not be detected under these conditions. However, confrontations of both *T. atroviride* strains with *F. oxysporum* led to the release of 2-octanone, which was not produced under any other condition tested in this study. Interestingly, 2-octanone emission during co-cultivation of strain P1 with *F. oxysporum* was significantly higher compared to the IMI 206040—*F. oxysporum* pairing. 2-octanone release started after 68.5 h of growth, peaked after 91.5 h with approx. 25 ppb in the co-culture involving strain P1 and with 4 ppb in the co-culture with strain IMI 206040 and afterwards decreased again (Figure 7). 1-propanol levels were increased, especially between 68.5 and 96.5 h of growth upon co-cultivation of *T. atroviride* P1 or IMI 206040 and *F. oxysporum* compared to the respective axenic cultures. In contrast to axenic cultures, which peaked between 40 and 50 h of cultivation, all co-cultivations with *F. oxysporum* resulted in high initial levels, which then constantly decreased over the cultivation period. (Figure 8; row 1).

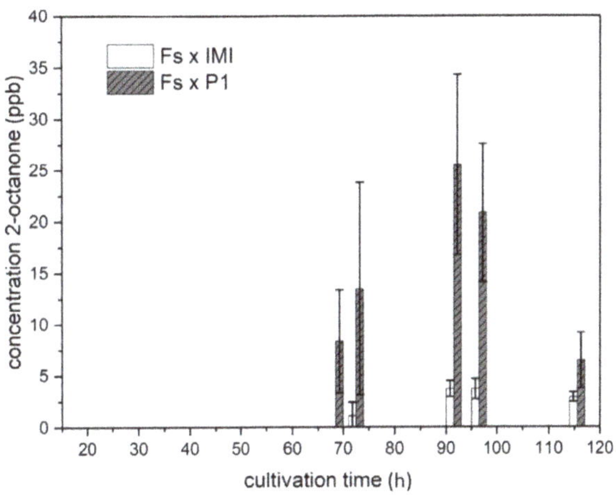

Figure 7. Exclusive, light-independent production of 2-octanone in direct confrontation of *T. atroviride* P1 and IMI 206040 with *F. oxysporum*. Levels (ppb) of 2-octanone detected in the headspace of co-cultures of *T. atroviride* P1 or IMI 206040 with *F. oxysporum* (Fo × P1; Fo × IMI) grown on PDA in Schott bottles at 25 °C. GC–IMS measurements were conducted along a cultivation period of 120 h. As 2-octanone emission in co-culture reached similar levels under light-dark conditions and in complete darkness, results shown are means ± SD (n = 4).

Interestingly, 2-heptanone emission by *T. atroviride* P1 in co-culture with *F. oxysporum* was light-dependent. In dark-grown co-cultures, 2-heptanone production by strain P1 followed a completely different course compared to co-cultivation under light–dark conditions, IMI 206040—*F. oxysporum* pairings, and axenic cultivations. 2-heptanone levels peaked with approx. 550 ppb after 68.5 h of co-cultivation in darkness and afterwards rapidly declined again. In contrast, 2-heptanone concentrations in the headspace of co-cultures grown under light–dark conditions increased steadier and earlier than upon axenic growth and finally reached and stayed at ≥1000 ppb (Figure 8; row 2). A similar trend was evident in the co-culture of strain IMI 206040 and *F. oxysporum*.

Similar to axenic cultivation, 3-octanone was solely produced in the presence of light by strain IMI 206040, irrespective of the presence of *F. oxysporum*. In axenic cultures, 3-octanone release increased over the whole cultivation period. In contrast, co-cultivation with *F. oxysporum* resulted in maximal 3-octanone production after 91.5 h, although 3-octanone levels were about 50% reduced compared to axenic cultivation. A massive reduction in 3-octanone emission upon interaction with *F. oxysporum* was also evident in strain P1, especially upon cultivation in darkness. However, 3-octanone production in co-culture was triggered by light as strain P1 produced nearly the double amount of this volatile in co-culture with *F. oxysporum* under light-dark conditions compared to constant darkness (Figure 8; row 3).

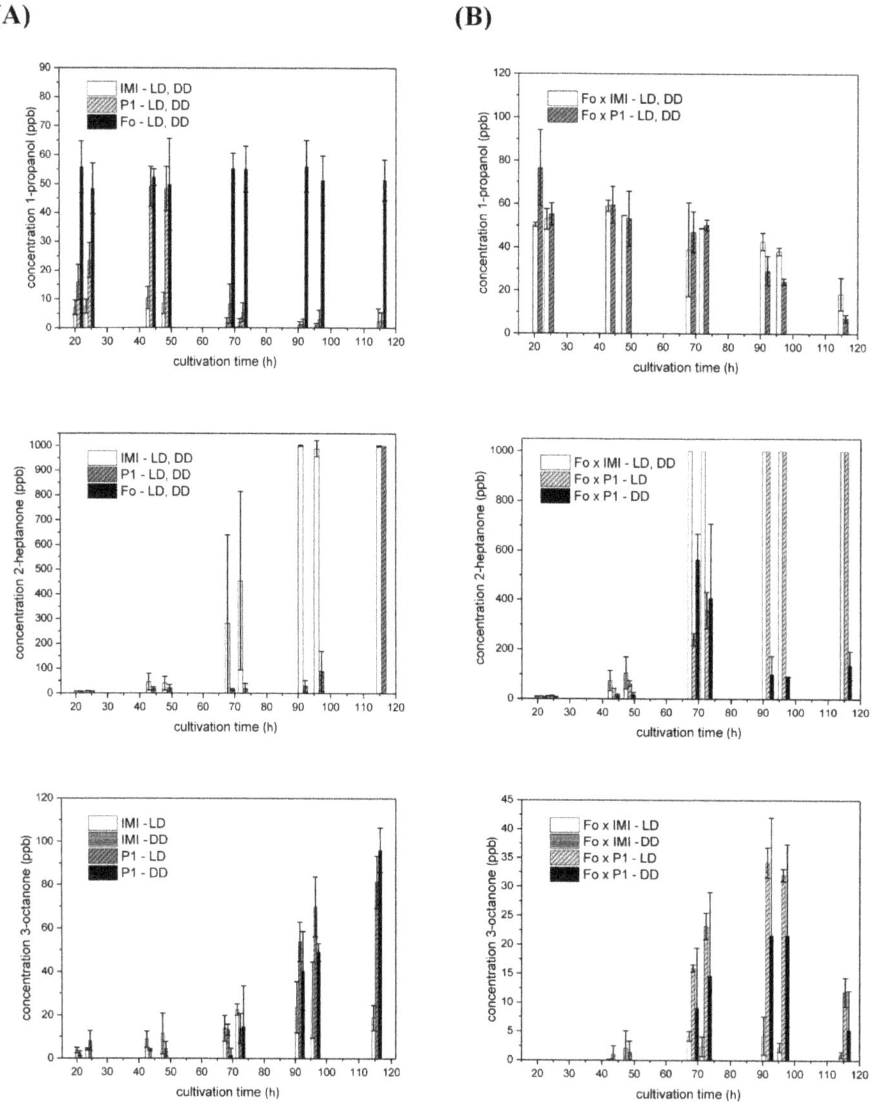

Figure 8. VOC production by *T. atroviride* P1 and IMI 206,040 upon interaction with *F. oxysporum*. Levels (ppb) of 1-propanol (row 1), 2-heptanone (row 2), and 3-octanone (row 3) detected in the headspace of (**A**) axenic cultures of *T. atroviride* P1 (P1), *T. atroviride* IMI 206040 (IMI), or *F. oxysporum* (Fo) or of (**B**) co-cultures of *T. atroviride* P1 or IMI 206040 with *F. oxysporum* (Fo × P1; Fo × IMI) on PDA in Schott bottles at 25 °C. Cultures were incubated either under light-dark conditions (LD) or in complete darkness (DD) as indicated. GC–IMS measurements were conducted along a cultivation period of 120 h. Results shown are means ± SD ($n \geq 3$).

3. Discussion

Trichoderma atroviride's natural habitat is soil and soil itself is penetrated by light in its upmost layer (with the exception of far-red light) solely. From the ecological point of view, the observed enhanced radial growth of *T. atroviride* colonies upon incubation in complete darkness indicates that dark conditions were closer to the conditions prevailing in the natural habitat than direct exposure to light. In accordance with this assumption, both *T. atroviride* strains exhibited a higher ability of antagonizing *R. solani* and particularly *F. oxysporum* in darkness compared to light–dark conditions, under which the latter fungal host could not be overgrown. The faster growth of *T. atroviride* IMI 206040 compared to strain P1 could account for its higher mycoparasitic activity in light. On the other hand, the observed higher antagonistic potential of *T. atroviride* P1 against *F. oxysporum* under complete darkness indicates that this strain is a stronger mycoparasite of *Fusarium* than *T. atroviride* IMI 206040, especially since its radial growth rate was lower. Our results point out that the frequently applied light–dark cycle, exposing the fungi for 12 h to illumination with white light, is an artificial and stressful condition that results in reduced growth and reduced mycoparasitic activity. Since under natural conditions the contact with light usually is accompanied by higher temperatures, drought and exposure to mutagenic ultraviolet light, paralleled by an extensive onset of reactive oxygen species production [29,30], the observed light-triggered growth reduction seems plausible. A strategy to react to and protect from light is conidiation, which was reported to be induced by (a pulse of blue-) light as well as mechanical injury in *T. atroviride* IMI 206040 [31]. Surprisingly, we found that *T. atroviride* P1 similarly produces asexual spores in the presence as well as absence of light, which is in massive contrast to the lack of sporulation in darkness in strain IMI 206040. Differences between the two tested *T. atroviride* strains were also observed regarding their response to mycelial injury. While injury led to the formation of strongly conidiating hypertrophic scarring tissue along the cutting sites in strain P1, the injury response of strain IMI 206040 resulted in only slight conidiation along the cutting lines without an obvious formation of scarring tissue. In combination with its high resistance against several fungicides [32], the light-independent conidiation behavior of *T. atroviride* P1 could be a major advantage for its application as a biocontrol agent in the field, since conidia, which are the main form of application, can be produced in a straightforward manner. Furthermore, belowground conidiation could also be a major advantage for long-term persistence of the mycoparasite in soil.

VOCs are playing important and versatile ecological roles in various intra- and inter-species or even inter-kingdom interactions [2,6]. A well-known example of such "multi-purpose" fungal VOCs are the eight-carbon volatiles 1-octene-3-ol and 3-octanone, which are contributing to the characteristic mushroom aroma and therefore are applied as food odorants in industry [33]. Furthermore, 1-octen-3-ol is known to act as mosquito attractant [34] and both eight-carbon volatiles were found to endogenously regulate conidial germination as well as the induction and inhibition of conidiation in a concentration-dependent manner [18,35–37]. Since *T. atroviride* IMI 206040 does not conidiate under dark conditions, the complete absence of 3-octanone in all samples obtained from dark-grown cultures of this strain indicates that 3-octanone is exclusively produced by conidia or under conidiation-inducing conditions. The substantially higher release of the majority of VOCs detected in this study by *T. atroviride* P1 suggests that this strain has a higher potential for biotechnological VOC production than strain IMI 206040. As described above, the versatile ecological roles of several fungal VOCs are well known; however, the regulation of their production along cultivation, like in dependence of culture age, and in response to stress is sparsely understood. In plants, VOC biosynthesis is regulated by abiotic stress, in particular by light and temperature stress, as well as biotic stress [38,39]. Similar to plants, fungi are relatively immobile and a regulation of VOC production as a stress response or coping strategy therefore is very likely. In this study, we found the three VOCs 3-octanone, 2-methyl-butanol, and 3-methyl-1-butanol to be released by *T. atroviride* axenic cultures in a light-dependent manner. Since the natural habitat of *T. atroviride* is soil, light and in particular UV irradiation is a stressful condition leading to the release of reactive oxygen species and to asexual reproduction [28,29]. In that respect, the light-dependent emission of 3-octanone, 2-methyl-butanol, and 3-methyl-1-butanol could play a role in the abiotic stress response to light in *T. atroviride*. Interestingly, 2-heptanone production by *T. atroviride* P1 during co-culture with *F. oxysporum* was also regulated in a light-dependent manner. Compared to *R. solani*, *F. oxysporum* is a more challenging host fungus for *T. atroviride*, which in our study could only be fully overgrown by strain P1 under dark conditions. In contrast, strain IMI 206040 was unable to fully parasitize *F. oxysporum*, irrespective of the applied light regime. In dark-grown co-cultures of *T. atroviride* P1 and *F. oxysporum*, only low 2-heptanone levels were emitted, which peaked and then rapidly decreased again. In contrast, confrontation of strain P1 with *F. oxysporum* in the presence of light as well as IMI 206040—*F. oxysporum* co-cultures and axenic cultivation of both *T. atroviride* strains in darkness and light resulted in a constant increase of 2-heptanone emission reaching and staying at levels of ≥1000 ppb. Low 2-heptanone levels hence paralleled the successful parasitism of *F. oxysporum* by *T. atroviride* P1, while the light-triggered emission of high levels of this volatile was associated with difficulties of *T. atroviride* to overgrow and lyse this host. *F. oxysporum* spp. are known to produce a plethora of toxic secondary metabolites [40,41], of which e.g., fumonisin is increasingly produced in the presence of light [42]. 2-hepanone emission by *T. atroviride* could be part of a coping strategy against abiotic (light) and biotic (*F. oxysporum* mycotoxins) stress.

F. oxysporum—*T. atroviride* co-cultivation further resulted in the production of 2-octanone, a volatile that was not emitted by any of the interaction partners upon axenic growth. Interestingly, a recent study revealed that deletion of the histone deacetylase-encoding gene *hda-2* in *T. atroviride* IMI 206040 leads to the biosynthesis of 2-octanone as a new volatile in this strain [43]. This report together with our results points out that 2-octanone biosynthesis genes could be shut-down by a repressive chromatin structure and could be activated epigenetically as a consequence of environmental stimuli, in this case the interaction with *F. oxysporum*. Accordingly, the bacterium *Streptomyces rapamycinicus* triggered the expression of the silent gene cluster responsible for orsellinic acid production in *Aspergillus nidulans* by activation of the histone acetyltransferase GcnE [44].

In summary, our study revealed significant differences between the two tested *T. atroviride* strains regarding injury-response, light-dependent regulation of conidiation and mycoparasitic activity as well as VOC production. Besides volatiles with strain- and light-dependent biosynthesis patterns, we found an interaction-dependent regulation of the biosynthesis of individual substances and obtained evidence, that the biosynthesis of certain VOCs is specifically linked to abiotic and/or biotic stress responses in *T. atroviride*. Additional effort should be undertaken to further decipher the multifactorial regulation of secondary metabolite biosynthesis in *Trichoderma* mycoparasites.

4. Materials and Methods

4.1. Fungal Strains and Culture Conditions

Trichoderma atroviride P1 (ATCC 74058), *T. atroviride* IMI 206040, *Fusarium oxysporum* f. sp. *lycopersici* strain 4287 and *Rhizoctonia solani* strain AG-5 were used in this study.

Pre-cultivation was done by passaging a 6 mm diameter agar plug of the actively growing colony margin at least two times after 1.5 days each upside down to the centre of a fresh petri dish (94 mm × 16 mm, Greiner Bio-One GmbH, Kremsmünster, Austria) containing 25 mL potato dextrose agar (PDA; Becton, Dickinson and Company, Le Pont De Claix, France). Plates were incubated at 25 °C under light-dark conditions (12:12 h, 300 Lux; Snijders Micro Clima-Series TM Labs Economic Lux Chamber; Snijders Labs, Tiburg, The Netherlands) or under complete darkness.

For determination of the radial growth rate of the two *T. atroviride* strains, 6 mm diameter agar plugs of the actively growing colony margins from the pre-cultures were inoculated in quadruplicates at the outmost margin of fresh agar plates. Plates were incubated at 25 °C under light-dark conditions (12:12 h; 300 Lux) or under complete darkness for three days. The radial growth rate was measured and calculated (cm/d) after three days of growth.

4.2. Mechanical Injury- and Blue-Light-Induced Conidiation Assays

Mechanical injury- and blue-light-induced conidiation assays were performed in quadruplicates according to [31] with slight modifications. Six millimeter diameter agar plugs of the actively growing colony margins from the pre-cultures were inoculated upside-down at the centre of fresh agar plates. Cultures were incubated at 25 °C for five days under light-dark conditions or under complete darkness. After two days of incubation in complete darkness, conidiation was induced by mycelial injury, which was set by cutting five horizontal and five vertical lines with a scalpel under red safety light. Alternatively, a 10 min pulse of blue light was set with a blue light source (Blacklight Blue UV-A Lamp: Supratec L18 W/73, 300–400 nm; 25 Lux; Osram GmbH, Garching, Germany; distance 19 cm). Plates were further incubated under complete darkness for a total of five days. Photos were taken after five days of growth.

4.3. Dual Plate Confrontation Assays

Dual plate confrontation assays were performed in quadruplicates according to [30] with slight modifications. Six millimeter diameter agar plugs of the actively growing colony margins from the pre-cultures of *T. atroviride* and a respective host fungus were inoculated on opposite sides at the outmost margin of fresh agar plates. As controls, all strains were additionally grown alone; the mycoparasites were also grown in self-confrontations. Due to its slower radial growth rate, *F. oxysporum* was inoculated 24 h earlier than the other strains. Fungi were grown at 25 °C under light-dark conditions or under complete darkness for five and six days, respectively. Pictures were taken at the end of the cultivation period to document the progress of the mycoparasitic attack.

4.4. Gas Chromatography–Ion Mobility Spectrometry (GC–IMS) Analysis of Headspace VOCs

For gas chromatography–ion mobility spectrometry (GC–IMS) analysis, a 6 mm diameter agar plug of the actively growing colony margin from the pre-culture was inoculated upside-down at the outmost margin of a 150 mL glass bottle (Duran GmbH, Mainz, Germany) filled with 25 mL PDA at the bottom. Fungi were grown in axenic culture or direct confrontation in the glass bottles according to the description above. As a control, PDA was also measured alone. The bottles were closed with Teflon® screw caps (Bohlender™, Merck, Vienna, Austria) with two openings for air in- and outlet. For dark conditions, the bottles were covered with several layers of aluminum foil. Due to a slower radial growth rate, *F. oxysporum* was pre-grown for 24 h. The two *T. atroviride* strains and *R. solani* were pre-grown for 20 h at 25 °C under light-dark conditions (12:12 h; 300 Lux) or under complete darkness before flasks were connected to the MS device.

For headspace measurements, cultures in glass bottles were held in an incubator and connected in a gastight way, parallel to each other. To avoid condensation, the inner temperature of the oven (so the temperature of the headspace air) was held at 40 °C, while the water bath was kept at 23 ± 2 °C. For ventilation, purified air with 5 mL/min flow was continuously streaming through the flasks and regulated by four mass flow controllers (MCF, Bronkhorst, Ruurlo, The Netherlands). Additionally, 5 mL/min purified air was connected as dilution flow to the flasks to reduce moisture in the samples. Headspace air samples were collected in the incubator at 40 °C in 100 mL and 250 mL glass syringes (Socorex, Ecublens, Switzerland). Sampling and measurement were performed 21, 24.5, 43.5, 48.5, 68.5, 72.5, 91.5, 96.5 and 115.5 h after inoculation.

A high resolution GC–IMS developed at Leibniz Universität, Hannover was applied to monitor the emitted VOCs. Samples were injected immediately after collection through a heated inlet (40 °C) into the GC column using a stainless steel sample loop (200 µL) installed on a six-way valve (VICI AG International, Schenkon, Switzerland). Volatiles were separated using a RTX volatiles column (10 m × 0.53 mm × 2 µm, Restek GmbH, Bad Homburg, Germany) working at a constant temperature of 50 °C. The carrier-gas-flow-rate program was as follows: 3 mL/min for 10 min and then 10 mL/min for another 10 min, resulting in a total GC-runtime of 20 min. The IMS, with a drift tube length of 7.5 cm, provided a resolving power of R = 90 using a drift voltage of 5 kV. The instrument operated at 40 °C with purified air as drift gas at the flow of 150 mL/min and 10 mbar above the ambient pressure. For ionization of the volatiles, a radioactive β-emitter ^3H (300 MBq) was used. A detailed description of the system can be found in [31]. Chromatographic data was acquired using the Agilent Chemstation Software (GC-MS Data Analysis from Agilent, Waldbronn, Germany). Data were analyzed using the software OpenChrom (vers. 1.4.0, Lablicate GmbH, Hamburg, Germany) and the mass spectrum library NIST 2008 (Gatesburg, PA, USA) was applied for identification.

Author Contributions: V.S. conceived and directed this study. V.S. and S.Z. drafted the manuscript. V.S., M.W. and V.R. conducted the experiments and analyzed the data. All authors have read and agreed to the published version of the manuscript.

Funding: This research received no external funding.

Acknowledgments: We thank Christopher Mayhew and Helmut Wiesenhofer for assisting in the GC–IMS analysis and Stefan Zimmermann from the Institute of Electrical Engineering and Measurement Technology, University of Hannover for providing the GC–IMS.

Conflicts of Interest: The authors declare no conflict of interest.

References

1. Korpi, A.; Järnberg, J.; Pasanen, A.-L. Microbial volatile organic compounds. *Crit. Rev. Toxicol.* **2009**, *39*, 139–193. [CrossRef]
2. Schulz-Bohm, K.; Martín-Sánchez, L.; Garbeva, P. Microbial Volatiles: Small Molecules with an Important Role in Intra- and Inter-Kingdom Interactions. *Front. Microbiol.* **2017**, *8*, 2484. [CrossRef] [PubMed]
3. Morath, S.U.; Hung, R.; Bennett, J.W. Fungal volatile organic compounds: A review with emphasis on their biotechnological potential. *Fungal Biol. Rev.* **2012**, *26*, 73–83. [CrossRef]
4. Wood, W.F.; Archer, C.L.; Largent, D.L. 1-Octen-3-ol, a banana slug antifeedant from mushrooms. *Biochem. Syst. Ecol.* **2001**, *29*, 531–533. [CrossRef]
5. Thakeow, P. Development of a basic biosensor system for wood degradation using volatile organic compounds. Ph.D. Dissertation, Universität Göttingen, Göttingen, Germany, 2008.
6. Bitas, V.; Kim, H.-S.; Bennett, J.W.; Kang, S. Sniffing on microbes: Diverse roles of microbial volatile organic compounds in plant health. *Mol. Plant Microbe Interact.* **2013**, *26*, 835–843. [CrossRef] [PubMed]
7. Druzhinina, I.S.; Seidl-Seiboth, V.; Herrera-Estrella, A.; Horwitz, B.A.; Kenerley, C.M.; Monte, E.; Mukherjee, P.K.; Zeilinger, S.; Grigoriev, I.V.; Kubicek, C.P. *Trichoderma*: The genomics of opportunistic success. *Nat. Rev. Microbiol.* **2011**, *9*, 749–759. [CrossRef]
8. Vinale, F.; Sivasithamparam, K.; Ghisalberti, E.L.; Marra, R.; Barbetti, M.J.; Li, H.; Woo, S.L.; Lorito, M. A novel role for *Trichoderma* secondary metabolites in the interactions with plants. *Physiol. Mol. Plant Pathol.* **2008**, *72*, 80–86. [CrossRef]
9. Siddiquee, S. Recent Advancements on the Role and Analysis of Volatile Compounds (VOCs) from *Trichoderma*. In *Biotechnology and Biology of Trichoderma*; Gupta, V., Schmoll, M., Herrera-Estrella, A., Upadhyay, R., Druzhinina, I., Tuohy, M., Eds.; Elsevier: Amsterdam, The Netherlands, 2014; pp. 139–175. ISBN 9780444595768.
10. Amin, F.; Razdan, V.K.; Mohiddin, F.A.; Bhat, K.A.; Sheikl, P.A. Effect of volatile metabolites of *Trichoderma* species against seven fungal plant pathogens in-vitro. *J. Phytol.* **2010**, *2*, 34–37.
11. Bruce, A.; Wheatley, R.E.; Humphris, S.N.; Hackett, C.A.; Florence, M.E.J. Production of volatile organic compounds by *Trichoderma* in media containing different amino acids and their effect on selected wood decay fungi. *Holzforschung* **2000**, *54*, 481–486. [CrossRef]
12. Guo, Y.; Ghirardo, A.; Weber, B.; Schnitzler, J.-P.; Benz, J.P.; Rosenkranz, M. *Trichoderma* Species Differ in Their Volatile Profiles and in Antagonism Toward Ectomycorrhiza *Laccaria bicolor*. *Front. Microbiol.* **2019**, *10*, 891. [CrossRef]
13. Contreras-Cornejo, H.A.; Macías-Rodríguez, L.; Herrera-Estrella, A.; López-Bucio, J. The 4-phosphopantetheinyl transferase of *Trichoderma virens* plays a role in plant protection against *Botrytis cinerea* through volatile organic compound emission. *Plant Soil* **2014**, *379*, 261–274. [CrossRef]

14. Lee, S.; Yap, M.; Behringer, G.; Hung, R.; Bennett, J.W. Volatile organic compounds emitted by *Trichoderma* species mediate plant growth. *Fungal Biol. Biotechnol.* **2016**, *3*, 7. [CrossRef] [PubMed]
15. Hung, R.; Lee, S.; Bennett, J.W. *Arabidopsis thaliana* as a model system for testing the effect of *Trichoderma* volatile organic compounds. *Fungal Ecol.* **2013**, *6*, 19–26. [CrossRef]
16. Stoppacher, N.; Kluger, B.; Zeilinger, S.; Krska, R.; Schuhmacher, R. Identification and profiling of volatile metabolites of the biocontrol fungus *Trichoderma atroviride* by HS-SPME-GC-MS. *J Microbiol. Methods* **2010**, *81*, 187–193. [CrossRef]
17. Schnürer, J.; Olsson, J.; Börjesson, T. Fungal volatiles as indicators of food and feeds spoilage. *Fungal Genet. Biol.* **1999**, *27*, 209–217. [CrossRef]
18. Nemcovic, M.; Jakubíková, L.; Víden, I.; Farkas, V. Induction of conidiation by endogenous volatile compounds in *Trichoderma* spp. *FEMS Microbiol. Lett.* **2008**, *284*, 231–236. [CrossRef]
19. Yu, Z.; Fischer, R. Light sensing and responses in fungi. *Nat. Rev. Microbiol.* **2019**, *17*, 25–36. [CrossRef]
20. Casas-Flores, S.; Rios-Momberg, M.; Bibbins, M.; Ponce-Noyola, P.; Herrera-Estrella, A. BLR-1 and BLR-2, key regulatory elements of photoconidiation and mycelial growth in *Trichoderma atroviride*. *Microbiology* **2004**, *150*, 3561–3569. [CrossRef]
21. Casas-Flores, S.; Rios-Momberg, M.; Rosales-Saavedra, T.; Martínez-Hernández, P.; Olmedo-Monfil, V.; Herrera-Estrella, A. Cross talk between a fungal blue-light perception system and the cyclic AMP signaling pathway. *Eukaryot. Cell* **2006**, *5*, 499–506. [CrossRef]
22. Esquivel-Naranjo, E.U.; Herrera-Estrella, A. Enhanced responsiveness and sensitivity to blue light by *blr-2* overexpression in *Trichoderma atroviride*. *Microbiology* **2007**, *153*, 3909–3922. [CrossRef]
23. Friedl, M.A.; Schmoll, M.; Kubicek, C.P.; Druzhinina, I.S. Photostimulation of *Hypocrea atroviridis* growth occurs due to a cross-talk of carbon metabolism, blue light receptors and response to oxidative stress. *Microbiology* **2008**, *154*, 1229–1241. [CrossRef] [PubMed]
24. García-Esquivel, M.; Esquivel-Naranjo, E.U.; Hernández-Oñate, M.A.; Ibarra-Laclette, E.; Herrera-Estrella, A. The *Trichoderma atroviride* cryptochrome/photolyase genes regulate the expression of blr1-independent genes both in red and blue light. *Fungal Biol.* **2016**, *120*, 500–512. [CrossRef] [PubMed]
25. Cetz-Chel, J.E.; Balcázar-López, E.; Esquivel-Naranjo, E.U.; Herrera-Estrella, A. The *Trichoderma atroviride* putative transcription factor Blu7 controls light responsiveness and tolerance. *BMC Genomics* **2016**, *17*, 327. [CrossRef] [PubMed]
26. Osorio-Concepción, M.; Cristóbal-Mondragón, G.R.; Gutiérrez-Medina, B.; Casas-Flores, S. Histone Deacetylase HDA-2 Regulates *Trichoderma atroviride* Growth, Conidiation, Blue Light Perception, and Oxidative Stress Responses. *Appl. Environ. Microbiol.* **2017**, *83*, e02922-16. [CrossRef] [PubMed]
27. Lilly, V.G.; Barnett, H.L. *Physiology of the Fungi*; McGraw-Hill: New York, NY, USA, 1951.
28. Schmoll, M.; Esquivel-Naranjo, E.U.; Herrera-Estrella, A. *Trichoderma* in the light of day-physiology and development. *Fungal Genet. Biol.* **2010**, *47*, 909–916. [CrossRef]
29. Rodriguez-Romero, J.; Hedtke, M.; Kastner, C.; Müller, S.; Fischer, R. Fungi, hidden in soil or up in the air: Light makes a difference. *Annu. Rev. Microbiol.* **2010**, *64*, 585–610. [CrossRef]
30. Fuller, K.K.; Loros, J.J.; Dunlap, J.C. Fungal photobiology: Visible light as a signal for stress, space and time. *Curr. Genet.* **2015**, *61*, 275–288. [CrossRef]
31. Hernández-Oñate, M.A.; Esquivel-Naranjo, E.U.; Mendoza-Mendoza, A.; Stewart, A.; Herrera-Estrella, A.H. An injury-response mechanism conserved across kingdoms determines entry of the fungus *Trichoderma atroviride* into development. *Proc. Natl. Acad. Sci. USA* **2012**, *109*, 14918–14923. [CrossRef]
32. Tronsmo, A. Biological and integrated controls of *Botrytis cinerea* on apple with *Trichoderma harzianum*. *Biol. Control* **1991**, *1*, 59–62. [CrossRef]
33. Combet, E.; Eastwood, D.C.; Burton, K.S.; Henderson, J. Eight-carbon volatiles in mushrooms and fungi: Properties, analysis, and biosynthesis. *Mycoscience* **2006**, *47*, 317–326. [CrossRef]
34. Kline, D.L.; Allan, S.A.; Bernier, U.R.; Welch, C.H. Evaluation of the enantiomers of 1-octen-3-ol and 1-octyn-3-ol as attractants for mosquitoes associated with a freshwater swamp in Florida, U.S.A. *Med. Vet. Entomol.* **2007**, *21*, 323–331. [CrossRef] [PubMed]
35. Chitarra, G.S.; Abee, T.; Rombouts, F.M.; Posthumus, M.A.; Dijksterhuis, J. Germination of *Penicillium paneum* Conidia is regulated by 1-octen-3-ol, a volatile self-inhibitor. *Appl. Environ. Microbiol.* **2004**, *70*, 2823–2829. [CrossRef]

36. Chitarra, G.S.; Abee, T.; Rombouts, F.M.; Dijksterhuis, J. 1-Octen-3-ol inhibits conidia germination of *Penicillium paneum* despite of mild effects on membrane permeability, respiration, intracellular pH, and changes the protein composition. *FEMS Microbiol. Ecol.* **2005**, *54*, 67–75. [CrossRef] [PubMed]
37. Herrero-Garcia, E.; Garzia, A.; Cordobés, S.; Espeso, E.A.; Ugalde, U. 8-Carbon oxylipins inhibit germination and growth, and stimulate aerial conidiation in *Aspergillus nidulans*. *Fungal Biol.* **2011**, *115*, 393–400. [CrossRef] [PubMed]
38. Vickers, C.E.; Gershenzon, J.; Lerdau, M.T.; Loreto, F. A unified mechanism of action for volatile isoprenoids in plant abiotic stress. *Nat. Chem. Biol.* **2009**, *5*, 283–291. [CrossRef] [PubMed]
39. Loreto, F.; Barta, C.; Brilli, F.; Nogues, I. On the induction of volatile organic compound emissions by plants as consequence of wounding or fluctuations of light and temperature. *Plant Cell Environ.* **2006**, *29*, 1820–1828. [CrossRef] [PubMed]
40. Manici, L.M.; Caputo, F.; Saccà, M.L. Secondary metabolites released into the rhizosphere by *Fusarium oxysporum* and *Fusarium* spp. as underestimated component of nonspecific replant disease. *Plant Soil* **2017**, *415*, 85–98. [CrossRef]
41. Waskiewicz, A.; Golinski, P.; Karolewski, Z.; Irzykowska, L.; Bocianowski, J.; Kostecki, M.; Weber, Z. Formation of fumonisins and other secondary metabolites by *Fusarium oxysporum* and *F. proliferatum*: A comparative study. *Food Addit. Contam. Part A Chem. Anal. Control Expo. Risk Assess.* **2010**, *27*, 608–615. [CrossRef]
42. Matić, S.; Spadaro, D.; Prelle, A.; Gullino, M.L.; Garibaldi, A. Light affects fumonisin production in strains of *Fusarium fujikuroi*, *Fusarium proliferatum*, and *Fusarium verticillioides* isolated from rice. *Int. J. Food Microbiol.* **2013**, *166*, 515–523. [CrossRef]
43. Estrada-Rivera, M.; Rebolledo-Prudencio, O.G.; Pérez-Robles, D.A.; Rocha-Medina, M.D.C.; González-López, M.D.C.; Casas-Flores, S. *Trichoderma* Histone Deacetylase HDA-2 Modulates Multiple Responses in *Arabidopsis*. *Plant Physiol.* **2019**, *179*, 1343–1361. [CrossRef]
44. Nützmann, H.-W.; Reyes-Dominguez, Y.; Scherlach, K.; Schroeckh, V.; Horn, F.; Gacek, A.; Schümann, J.; Hertweck, C.; Strauss, J.; Brakhage, A.A. Bacteria-induced natural product formation in the fungus *Aspergillus nidulans* requires Saga/Ada-mediated histone acetylation. *Proc. Natl. Acad. Sci. USA* **2011**, *108*, 14282–14287. [CrossRef] [PubMed]

Sample Availability: Not available.

molecules

MDPI

Article

A Preliminary Study on Metabolome Profiles of Buffalo Milk and Corresponding Mozzarella Cheese: Safeguarding the Authenticity and Traceability of Protected Status Buffalo Dairy Products

Angela Salzano [1], Gelsomina Manganiello [2], Gianluca Neglia [1,*], Francesco Vinale [1,3], Donato De Nicola [1], Michael D'Occhio [4] and Giuseppe Campanile [1]

[1] Dipartimento di Medicina Veterinaria e Produzioni Animali, University of Naples "Federico II", 80137 Naples, Italy; angela.salzano@unina.it (A.S.); frvinale@unina.it (F.V.); denicolavet@gmail.com (D.D.N.); giucampa@unina.it (G.C.)

[2] Dipartimento di Agraria, University of Naples "Federico II", 80055 Portici (NA), Italy; gelsomina.manganiello@hotmail.it

[3] Istituto per la Protezione Sostenibile delle Piante, National Research Council, 80100 Naples, Italy

[4] Faculty of Science, The University of Sydney, Sydney, NSW 2006, Australia; michael.docchio@sydney.edu.au

* Correspondence: neglia@unina.it; Tel.: +39-0812536063

Academic Editor: Derek J. McPhee

Received: 12 November 2019; Accepted: 10 January 2020; Published: 12 January 2020

Abstract: The aim of this study is to combine advanced GC-MS and metabolite identification in a robust and repeatable technology platform to characterize the metabolome of buffalo milk and mozzarella cheese. The study utilized eleven dairies located in a protected designation of origin (PDO) region and nine dairies located in non-PDO region in Italy. Samples of raw milk (100 mL) and mozzarella cheese (100 g) were obtained from each dairy. A total of 185 metabolites were consistently detected in both milk and mozzarella cheese. The PLS-DA score plots clearly differentiated PDO and non-PDO milk and mozzarella samples. For milk samples, it was possible to divide metabolites into two classes according to region: those with lower concentrations in PDO samples (galactopyranoside, hydroxybuthyric acid, allose, citric acid) and those with lower concentrations in non-PDO samples (talopyranose, pantothenic acid, mannobiose, etc.,). The same was observed for mozzarella samples with the proportion of some metabolites (talopyranose, 2, 3-dihydroxypropyl icosanoate, etc.,) higher in PDO samples while others (tagatose, lactic acid dimer, ribitol, etc.,) higher in non-PDO samples. The findings establish the utility of GC-MS together with mass spectral libraries as a powerful technology platform to determine the authenticity, and create market protection, for "Mozzarella di Bufala Campana."

Keywords: metabolome; GC-MS; buffalo; milk; mozzarella; authenticity

---

1. Introduction

The authenticity, integrity, and traceability of food products is very important for market protection in global food systems [1–4]. This applies particularly to food that has a market niche and for which consumers pay a high price. The premium price of niche food makes it susceptible to substitution with falsely-labelled, non-authentic food [5]. Examples are substitution in olive oil [6,7] and dairy products [8–10]. Italian mozzarella cheese is made from the milk of Italian Mediterranean water buffalo (*Bubalus bubalis*) and is recognized globally for its exceptional eating qualities. The water buffalo undergo intense selection for efficiency of production and product quality, and today has an important commercial and cultural niche in several regions of Italy. Genuine Italian mozzarella is known as

"Mozzarella di Bufala Campana." It has European Union protected designation of origin (PDO) and protected geographical indication (PGI) status. This identifies genuine mozzarella with four regions in Italy (Apulia, Campania, Lazio, Molise) [11,12]. Attempts are regularly made to place non-authentic mozzarella in premium markets as a substitute for buffalo mozzarella cheese [13–15]. Consumers can be denied a genuine product and confidence in certified farmers is threatened. The potential for product substitution has led to interest in the development of screening technology that ensures the authenticity of buffalo mozzarella and traceability to geographical origin [14,16,17].

The characterization of the metabolome of food is a promising approach for establishing authenticity [18]. The metabolome signature of biological material can be obtained using gas chromatography/mass spectrometry (GC-MS) [19,20], NMR spectroscopy [21,22], and high-resolution magic angle spinning (HRMAS) NMR spectroscopy [23]. Some initial screening of buffalo mozzarella cheese has used NMR [16], HRMAS NMR [16], and GC-MS/LC-MS [14,15,17]. To date, individual studies have looked at the metabolome of either buffalo milk or mozzarella. Substitution can occur at different stages of the supply chain and it is important that authenticity can be established from primary product (milk) through to secondary product (mozzarella). There are also limited studies in dairy on the impacts of breeding, diet, and geographical location on product quality [24]. The present study sought to compare, for the first time, the metabolomes of both unprocessed milk and corresponding mozzarella cheese for buffalo in PDO and non-PDO regions in Italy. Differences in the milk and/or mozzarella metabolomes between buffalo in PDO and non-PDO regions have potential practical application in safeguarding the authenticity and traceability of protected status buffalo dairy products.

2. Results

No differences were observed in the quality of milk and mozzarella cheese from PDO and non-PDO regions (Tables 1 and 2). A total of 185 metabolites were detected consistently. In particular, 113 compounds were detected in milk, 102 in mozzarella cheese, and 30 compounds out of 185 were found in both matrices. The PLS-DA score plots clearly differentiated PDO and non-PDO milk and mozzarella samples (Figure 1(A1,B1)). The 15 highest scoring variable importance in projection VIP variables (VIP score > 1.5) identified by PLS-DA are shown in Figure 1(A2,B2).

Table 1. Average composition of milk samples from different farms located in protected destination of origin (PDO) and non-PDO regions in Italy. Data are expressed as means ± ES.

BUFFALO MILK					
Nutrition Facts for 100 g of Product (Reg. UE 1169/2011)	**NON-PDO REGION (n = 9)**		**PDO REGION (n = 11)**		***p* Value**
	U.M	**VALUE**	**U.M**	**VALUE**	
Energy net value	KJ/100 g	473 ± 8.4	KJ/100 g	479 ± 7.9	0.82
	Kcal/100 g	112 ± 3.5	Kcal/100 g	116 ± 3.7	0.77
Total protein	g/100 g	4.5 ± 0.7	g/100 g	4.6 ± 0.6	0.85
Total fat	g/100 g	8.4 ± 0.9	g/100 g	8.6 ± 0.8	0.69
Saturated fat	g/100 g	4.9 ± 0.5	g/100 g	4.9 ± 0.5	0.91
Total carbohydrates	g/100 g	5.2 ± 0.4	g/100 g	5.2 ± 0.5	0.82
Sugars	g/100 g	0.8 ± 0.1	g/100 g	0.9 ± 0.1	0.86
Salt	g/100 g	Nd	g/100 g	Nd	
Ashes	g/100 g	0.8 ± 0.0	g/100 g	0.8 ± 0.1	0.90

U.M. Unit measure.

Table 2. Average mozzarella cheese composition of samples from different farms located in protected designation of origin (PDO) and non-PDO regions in Italy. Data are expressed as means ± ES.

BUFFALO MOZZARELLA CHEESE					
Nutrition Facts for 100 g of Product (Reg. UE 1169/2011)	**NON-PDO REGION (n = 9)**		**PDO REGION (n = 11)**		***p* Value**
	U.M.	**VALUE**	**U.M.**	**VALUE**	
Energy net value	KJ/100 g	1129.9 ± 9.4	KJ/100 g	1131.1 ± 8.2	0.72
	Kcal/100 g	268.6 ± 4.0	Kcal/100 g	269.2 ± 3.2	0.84
Total protein	g/100 g	13.4 ± 1.0	g/100 g	13.5 ± 0.9	0.86
Total fat	g/100 g	23.3 ± 2.6	g/100 g	23.5 ± 2.9	0.73
Saturated fat	g/100 g	14.6 ± 0.4	g/100 g	14.6 ± 0.6	0.79
Total carbohydrates	g/100 g	0.8 ± 0.0	g/100 g	0.8 ± 0.0	0.91
Sugars	g/100 g	0.8 ± 0.1	g/100 g	0.8 ± 0.0	0.88
Salt	g/100 g	0.9 ± 0.1	g/100 g	0.9 ± 0.1	0.90
Ashes	g/100 g	1.4 ± 0.2	g/100 g	1.4 ± 0.2	0.93

U.M. Unit measure.

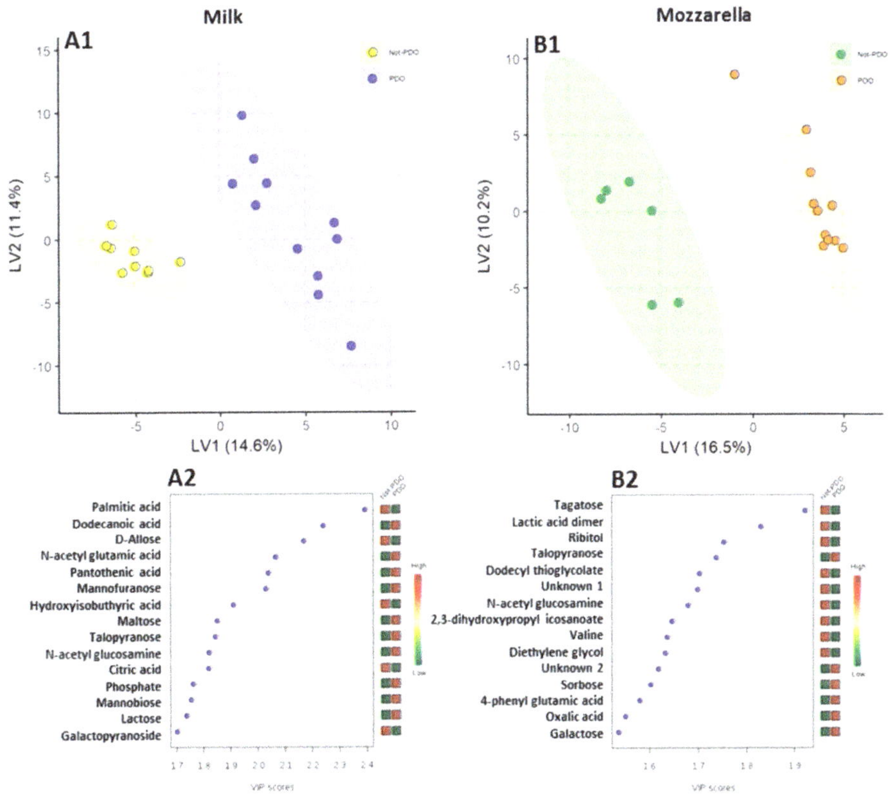

Figure 1. Partial least square discriminant analysis (PLS-DA) models to discriminate PDO and non-PDO buffalo milk (**A1**) and mozzarella (**B1**) samples. The explained variance of each component is shown in parentheses on the corresponding axis. (**A2**) and (**B2**) panels show the 15 top-scoring variable importance in projection (VIP) metabolites (VIP-score ≥ 1.5) for milk (A2) and mozzarella (B2) samples. The colored boxes on the right indicate the relative amount of the corresponding metabolite in each group under study.

Differences in metabolite concentrations between PDO and non-PDO raw milk samples allowed the metabolites to be separated into two classes: those with lower ($p < 0.05$) concentrations in PDO milk (galactopyranoside, hydroxybuthyric acid, allose, citric acid) and those with higher ($p < 0.05$) concentrations in PDO milk (talopyranose, pantothenic acid, mannobiose, maltose, phosphate, mannofuranose, dodecanoic acid, lactose, palmitic acid, n-acetyl glutamic acid, *n*-acetyl glucosamine) (Figure 2).

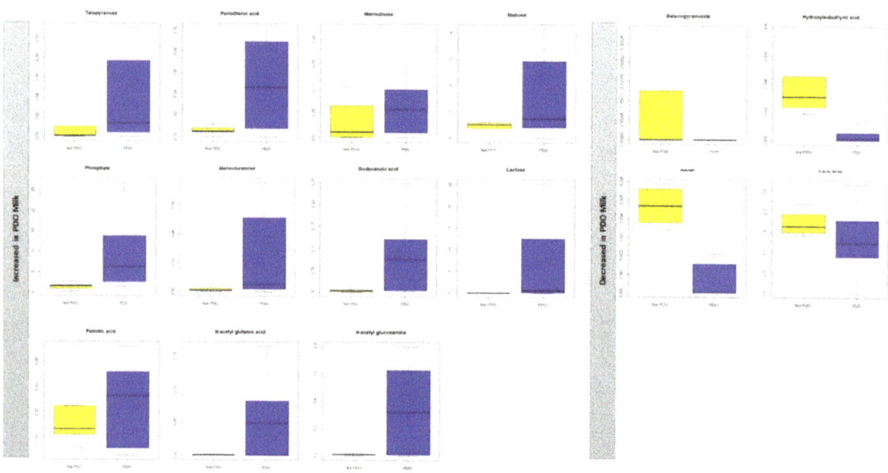

Figure 2. Box and Whisker plot of the VIP metabolites in buffalo milk samples. Boxes represent non-PDO (yellow) and PDO (blue) milk samples. The vertical axis reports the log of the gas chromatography mass spectrometry values of the normalized area of each metabolite.

Metabolites could also be separated into two classes for mozzarella cheese samples: those with higher ($p < 0.05$) concentrations in PDO mozzarella (talopyranose, 2, 3-dihydroxypropyl icosanoate, sorbose, 4-pnehyl glutamic acid, oxalic acid, galactose) and those with higher ($p < 0.05$) concentrations in non-PDO mozzarella (tagatose, lactic acid dimer, ribitol, dodecyl thioglycolate, *n*-acetyl glucosamine, valine, diethylene glycol) (Figure 3).

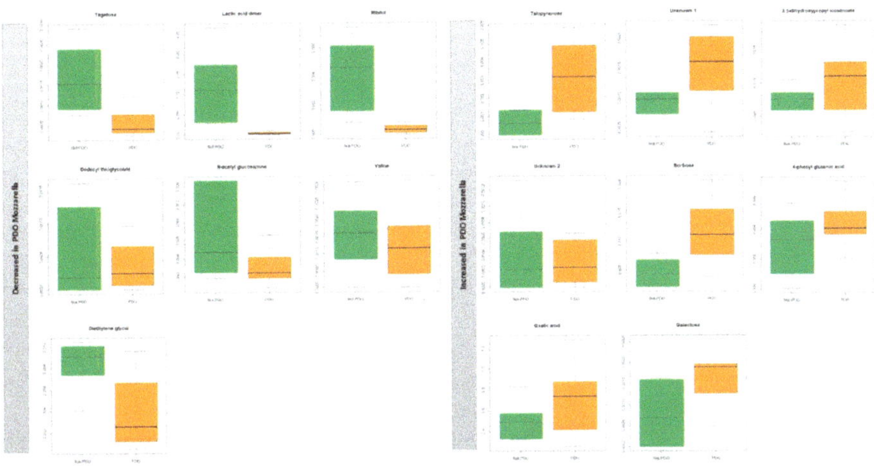

Figure 3. Box and Whisker plot of the VIP metabolites in buffalo mozzarella samples. Boxes represent non-PDO (green) and PDO (orange) mozzarella samples. The vertical axis reports the log of the gas chromatography mass spectrometry values of the normalized area of each metabolite.

3. Discussion

The present study sought to compare the metabolomes of unprocessed milk and corresponding mozzarella cheese for buffalo in PDO and non-PDO regions in Italy. It was found that the milk metabolome differed between buffalo in PDO and non-PDO regions. The mozzarella metabolome also differed between buffalo in PDO and non-PDO regions. This is the first time that some metabolites have been detected in both the metabolome of unprocessed milk and corresponding mozzarella cheese in buffalo. It is also the first time that differences have been found between buffalo in PDO and non-PDO regions in both milk and mozzarella metabolomes. The number of individual farms from PDO (n = 11) and non-PDO (n = 9) regions might be considered a relatively small sample size, although it was comparable to numbers in previous reports that looked at the metabolome [24–26]. Notwithstanding the sample size, the combination of GC-MS and mass spectral libraries (NIST library) proved to be a robust technology platform for determining the metabolomes of buffalo milk and mozzarella. This platform, together with a rigorous analysis of the data, has provided a sound foundation to inform further studies. In particular, a number of notable metabolites identified in unprocessed milk (n = 15) and mozzarella cheese (n = 13) could be used to validate the utility of the metabolome for safeguarding the protected status buffalo dairy products. Candidate metabolites that differentiated PDO from non-PDO milk included several carbohydrates (D-allose, mannofuranose, maltose, and talopyranose). Candidate metabolites that differentiated both milk and mozzarella between PDO and non-PDO origin were talopyranose and *N*-acetyl glucosamine. Talopyranose is the pyranose form of talose and an epimer of glucose. Talose was already reported to be a "differential" marker between bovine milk and goat milk [27]. *N*-acetylglucosamine is a derivative amide of the monosaccharide glucose and a secondary amide between glucosamine and acetic acid. It is significant in several biological systems and a major component of the cell walls of most fungi [28]. Talopyranose is of particular interest as amounts were substantially higher in both milk and mozzarella of PDO origin. A number of saccharides (tagatose, talopyranose, sorbose, galactose) differentiated mozzarella of DPO and non-DPO origin. Two metabolites in mozzarella cheese had markedly different concentrations between PDO and non-PDO samples. These metabolites could not be identified from public mass spectral libraries. However, they require further study as they may emerge as highly valuable in establishing the authenticity and integrity of protected status buffalo mozzarella cheese.

Irrespective of PDO or non-PDO origin, the milk aqueous fraction was rich in short-chain saturated carboxylic acids (carbonic acid, acetic acid, propanoic acid, butanoic acid, octanoic acid, decanoic acid) and long-chain saturated carboxylic acids (lauric acid, palmitic acid, stearic acid). Free amino acids in milk included serine, threonine, and valine. Compared with milk, mozzarella cheese contained lower amounts of some short-chain saturated carboxylic acids (acetic, propanoic, nonanoic acid) and only two long-chain saturated carboxylic acids were found in mozzarella (palmitic acid, stearic acid). Mozzarella was, however, richer in free amino acids (serine, leucine, isoleucine, alanine, proline, valine, norvaline). Some of the metabolites found in mozzarella samples in the present study (lactic acid dimer, valine, oxalic acid, and galactose) were also identified in a study that looked at the metabolomic and microbiological differences of Italian mozzarella cheese produced with buffalo or cow milk [29].

Processing alters milk constituents and their concentrations. For example, milk monosaccharide levels change in response to heat treatment [30] and storage [31]. The composition of the starter, which can be influenced by the environment, management and cheese-making technologies, also affects cheese characteristics by altering the milk quality. [32–34]. In PDO mozzarella, there may have been a synergistic effect between different lactic acid bacteria and yeast species, which ferment residual galactose and lactose and thereby increase lactic acid production. The processed cheese industry uses citrate or phosphate salts to sequester Ca^{2+} from residual colloidal calcium phosphate. This solubilizes caseins which can then emulsify fat globules. The acidity of whey at drainage, and the rate of acid development, are important parameters that determine the mineral content, acidity, and quality of cheese [35]. Milk of PDO origin had higher phosphate content and lower citric acid compared to non-PDO milk. Differences in milk salt composition most likely contributed to the differences in metabolites between PDO and non-PDO mozzarella cheese. Differences in packaging could also modify the metabolomic profile of mozzarella cheese. The higher diethylene glycol concentration in samples of non-PDO mozzarella compared to PDO was attributed to packaging [36].

Climatic conditions influence the rate of uptake of metabolites by plants [37,38]. Also, lower temperatures and frequent rainfall can affect the drying process of forages; essentially, reducing sugars, mineral salts, and soluble nitrogenous substances, while increasing fermentation [39]. Rainfall is usually less frequent in the PDO regions in the present study compared with the non-PDO regions, and the higher concentrations of some sugars (maltose, lactose, mannobiose, talopyranose) in milk of PDO origin may have been due to differences in climatic conditions. The climate and soils of a region directly and indirectly impact the biochemical and biophysical properties of food products. For example, the isotopic and elemental composition of milk is closely related to geographical origin and this carries over to the properties of cheese [39]. To have PDO certification, at least 70% of the dry matter of fodder, or 40% of the dry matter of the ration, must come from the PDO region. The use of local fodder helps to maintain the strict relationship between product and region for protected status of buffalo mozzarella cheese. There are close relationships between the environment, rumen microbiome, and animal metabolome. While the major families of microbes in the rumen are broadly similar across diverse landscapes [40], the relative populations of different microbes change according to local conditions of climatic, soil, feed, and management [40–42]. The amount of crude protein, neutral-detergent fiber, and acid detergent lignin are higher in cultivars from the PDO region [43]. It can be assumed that this would have influenced the rumen microbial populations and ruminal metabolome, which could have contributed, at least in part, to differences in the milk and mozzarella metabolomes between PDO and non-PDO regions.

In conclusion, a robust GC-MS and mass spectral library technology platform was used to identify for the first time the metabolome of unprocessed milk and corresponding mozzarella cheese in buffalo. Differences in both the milk and mozzarella metabolomes between buffaloes in PDO and non-PDO regions were also shown for the first time. A number of candidate metabolites in milk and mozzarella were identified that will be important in validation studies that aim to develop practical protocols to distinguish between PDO and non-PDO buffalo milk and mozzarella. Talopyranose was a particularly notable candidate metabolite as it differed substantially between PDO and non-PDO buffalo, for both milk and mozzarella. The development of quality assurance and certification protocols for milk and cheese will help to ensure the authenticity and traceability of primary (milk) and secondary (mozzarella) protected status buffalo dairy products. This is necessary to ensure that the investment in breeding, feeding, and management of buffalo in PDO regions is safeguarded.

4. Materials and Methods

4.1. Sample Collection

The study utilized 20 commercial buffalo dairies. Eleven dairies were located in a protected designation of origin (PDO, Campania, Italy) region and nine dairies were in non-PDO regions in Italy (). All dairies had a processing facility that produced mozzarella cheese exclusively from their own milk. Milk and mozzarella cheese quality was assessed the day of the sampling and the average results from PDO and non-PDO areas are reported in Tables 1 and 2. Pooled samples of raw milk (100 mL) and mozzarella cheese (100 g) were obtained from each dairy. The samples underwent similar processing and were obtained approximately 2 h after preparation. Mozzarella samples were immersed in mozzarella whey and stored at 25 °C until analysis.

4.2. Metabolomics Analysis

4.2.1. Metabolite Extraction and Derivatization

Metabolite extraction, purification, and derivatization were carried out using the MetaboPrep GC kit according to the manufacturer's instructions (Theoreo srl, Montecorvino Pugliano [SA], Italy).

From each mozzarella sample 10 ± 1 mg was transferred to an Eppendorf microcentrifuge tube containing the extraction solution. The samples were then centrifuged at 800× *g* for 30 min, before putting the samples in an ultrasonic bath at 30 °C for 30 min. The samples were then centrifuged for 5 min at 10,000× *g* at 4 °C. From the supernatant, 200 µL was removed and transferred to an Eppendorf microcentrifuge tube containing a purification mixture, and then vortexed at 800× *g* for 30 sec. The sample was again centrifuged at 10,000× *g* (at 4 °C). Finally, 175 µL supernatant was transferred into a 2-mL glass autosampler vial and freeze-dried overnight.

To facilitate derivatization, "50 µL of pyridine/methoxamine (1/1 *v:v*) were added to the sample and centrifuged at 800× *g* (25 °C) for 90 min. Then 25 µL of the second derivatization mixture containing *N,O-bis*(trimethylsilyl)trifluoroacetamide (BSTFA) and trimethylchlorosilane (TMCS) were added and vortexed at 800× *g* (25 °C) for 90 min. The solution was centrifuged for 5 min at 10,000 rpm× *g* (4 °C) before injecting into the GC-MS."

4.2.2. GC-MS Analysis

Samples (2 µL) of the derivatized solution were injected into the GC-MS system (GC-2010 Plus gas chromatograph coupled to a 2010 Plus single quadrupole mass spectrometer; Shimadzu Corp., Kyoto, Japan). Chromatographic separation was achieved with a 30 m 0.25 mm CP-Sil 8 CB fused silica capillary GC column with 1.00 µm film thickness (Agilent, J&W Scientific, Folsom, CA, USA), with helium as carrier gas. The initial oven temperature of 100 °C was maintained for 1 min and then raised by 6 °C/min to 320 °C with a further 2 min of holding time. The gas flow was set to obtain a constant linear velocity of 39 cm/s and the split flow was set at 1:5. The mass spectrometer was operated with electron impact ionization (70 eV) in full scan mode in the interval of 35–600 *m/z* with a scan velocity of 3333 AMU/sec and a solvent cut time of 4.5 min. The complete GC program duration was 40 min.

4.3. Metabolites Identification

Metabolite identification was performed according to Troisi et al. [44]. Briefly, untargeted metabolites were identified by comparing the mass spectrum of each peak with the NIST library collection (NIST, Gaithersburg, MD, USA). The linear index difference max tolerance was set to 10, while the minimum matching spectra library search was set to 85% (level 2 identification according to Metabolomics Standards Initiative [MSI]) [45]. Fifteen samples out of the over 200 signals per sample (7.5%) produced by gas chromatographic–mass spectrometry were not investigated further because they were not consistently found in other sets of samples (either too low in concentration or of poor spectral quality to be confirmed as metabolites).

4.4. Statistical Analyses

Data regarding milk and mozzarella cheese quality are expressed as mean ± SE. Differences were assessed by Student's t-test, and $p < 0.05$ value was considered significant. The chromatographic data were tabulated with one sample per row and one variable (metabolite) per column. Data pre-treatment consisted of normalizing each metabolite peak area to that of the internal standard (2-iso-propyl malic acid) followed by generalized log transformation and data scaling by autoscaling (mean-centered and divided by standard deviation of each variable). Statistical analysis of data from three biological replicates for each farm was performed by ANOVA Bonferroni correction (p-value < 0.05) by applying GraphPad PRISM software. Only metabolites significantly different between the farms were considered for further analyses. Principal component analyses (PCA) and heatmap representations were conducted by the online tool ClustVis (http://biit.cs.ut.ee/clustvis/). Unit variance scaling was applied to rows (metal/metabolite values in each farm) and single value decomposition (SVD) with imputation was used to calculate principal components. The heatmaps were generated clustering columns (Farms) by correlation distance and McQuitty linkage. Samples were classified considering geographical area of origin (North/South) and also the presence or absence of PDO trademark (Yes = Y; No = N).

Moreover, partial least square discriminant analysis (PLS-DA) [46] was performed using the statistical software package R (Foundation for Statistical Computing, Vienna, Austria). Class separation was achieved by PLS-DA, which is a supervised method that uses multivariate regression techniques to extract, via linear combinations of original variables (X), the information that can predict class membership (Y). PLS regression was performed using the plsr function included in the R pls package [47]. Classification and cross-validation were performed using the corresponding wrapper function included in the caret package [48]. A permutation test was performed to assess the significance of class discrimination. In each permutation, a PLS-DA model was built between the data (X) and the permuted class labels (Y) using the optimal number of components determined by cross validation for the model based on the original class assignment. Variable importance in projection (VIP) scores were calculated for each metabolite. The VIP score is a weighted sum of squares of the PLS loadings, taking into account the amount of explained Y-variation in each dimension.

Author Contributions: Conceptualization, G.N. and G.C.; data curation, A.S. and G.N.; formal analysis, G.M. and D.D.N.; investigation, A.S.; methodology, F.V.; resources, G.C.; software, F.V.; validation, G.M. and D.D.N.; writing—original draft, A.S.; writing—review and editing, M.D. and G.C. All authors have read and agreed to the published version of the manuscript.

Funding: This research received no external funding.

Conflicts of Interest: The authors declare no conflict of interest.

References

1. Cubero-Leon, E.; Peñalver, R.; Maquet, A. Review on metabolomics for food authentication. *Food Res. Int.* **2014**, *60*, 95–107. [CrossRef]
2. Danezis, G.P.; Tsagkaris, A.S.; Brusic, V.; Georgiou, C.A. Food authentication: State of the art and prospects. *Curr. Opin. Food Sci.* **2016**, *10*, 22–31. [CrossRef]
3. Danezis, G.P.; Tsagkaris, A.S.; Camin, F.; Brusic, V.; Georgiou, C.A. Food authentication: Techniques, trends & emerging approaches. *TrAC Trends Anal. Chem.* **2016**, *85*, 123–132.

4. Pustjens, A.; Muilwijk, M.; Weesepoel, Y.; van Ruth, S. Advances in Authenticity Testing of Geographical Origin of Food Products. In *Advances in Food Authenticity Testing*; Elsevier: Amsterdam, The Netherlands, 2016; pp. 339–367.
5. Álvarez, B.; Pascual-Alonso, M.; Rusu, A.; Bogason, S. A review on existing databases relevant for food fraud and authenticity. *Arch. Zootec.* **2013**, *62*, 73–91.
6. Tsopelas, F.; Konstantopoulos, D.; Kakoulidou, A.T. Voltammetric fingerprinting of oils and its combination with chemometrics for the detection of extra virgin olive oil adulteration. *Anal. Chim. Acta* **2018**, *1015*, 8–19. [CrossRef]
7. Merás, I.D.; Manzano, J.D.; Rodríguez, D.A.; de la Peña, A.M. Detection and quantification of extra virgin olive oil adulteration by means of autofluorescence excitation-emission profiles combined with multi-way classification. *Talanta* **2018**, *178*, 751–762. [CrossRef]
8. Agrimonti, C.; Pirondini, A.; Marmiroli, M.; Marmiroli, N. A quadruplex PCR (qxPCR) assay for adulteration in dairy products. *Food Chem.* **2015**, *187*, 58–64. [CrossRef]
9. Azad, T.; Ahmed, S. Common milk adulteration and their detection techniques. *Int. J. Food Contam.* **2016**, *3*, 22. [CrossRef]
10. Gonçalves, J.; Pereira, F.; Amorim, A.; van Asch, B. New method for the simultaneous identification of cow, sheep, goat, and water buffalo in dairy products by analysis of short species-specific mitochondrial DNA targets. *J. Agric. Food Chem.* **2012**, *60*, 10480–10485. [CrossRef] [PubMed]
11. Bonetti, E. The effectiveness of meta-brands in the typical product industry: Mozzarella cheese. *Br. Food J.* **2004**, *106*, 746–766. [CrossRef]
12. Hajdukiewicz, A. European Union agri-food quality schemes for the protection and promotion of geographical indications and traditional specialities: An economic perspective. *Folia Hortic.* **2014**, *26*, 3–17. [CrossRef]
13. Pignata, M.C.; Ferrão, S.P.B.; Oliveira, C.P.; Faleiro, A.S.; Bonomo, R.C.; Silva, W.S.; Rodrigues, L.B.; de Albuquerque Fernandes, S.A. Mechanical Parameters of the Mozzarella from Buffalo with Inclusion Levels of The Cow's Milk: Preliminary Study at the Lab Scale. *J. Bioanal. Biomed.* **2015**, *7*, 1. [CrossRef]
14. Caira, S.; Pinto, G.; Nicolai, M.A.; Chianese, L.; Addeo, F. Simultaneously tracing the geographical origin and presence of bovine milk in Italian water buffalo Mozzarella cheese using MALDI-TOF data of casein signature peptides. *Anal. Bioanal. Chem.* **2016**, *408*, 5609–5621. [CrossRef] [PubMed]
15. Russo, R.; Severino, V.; Mendez, A.; Lliberia, J.; Parente, A.; Chambery, A. Detection of buffalo mozzarella adulteration by an ultra-high-performance liquid chromatography tandem mass spectrometry methodology. *J. Mass Spectrom.* **2012**, *47*, 1407–1414. [CrossRef] [PubMed]
16. Mazzei, P.; Piccolo, A. 1H HRMAS-NMR metabolomic to assess quality and traceability of mozzarella cheese from Campania buffalo milk. *Food Chem.* **2012**, *132*, 1620–1627. [CrossRef]
17. Sassi, M.; Arena, S.; Scaloni, A. MALDI-TOF-MS platform for integrated proteomic and peptidomic profiling of milk samples allows rapid detection of food adulterations. *J. Agric. Food Chem.* **2015**, *63*, 6157–6171. [CrossRef]
18. Cevallos-Cevallos, J.M.; Reyes-De-Corcuera, J.I.; Etxeberria, E.; Danyluk, M.D.; Rodrick, G.E. Metabolomic analysis in food science: A review. *Trends Food Sci. Technol.* **2009**, *20*, 557–566. [CrossRef]
19. Gianferri, R.; Maioli, M.; Delfini, M.; Brosio, E. A low-resolution and high-resolution nuclear magnetic resonance integrated approach to investigate the physical structure and metabolic profile of Mozzarella di Bufala Campana cheese. *Int. Dairy J.* **2007**, *17*, 167–176. [CrossRef]
20. He, Q.; Kong, X.; Wu, G.; Ren, P.; Tang, H.; Hao, F.; Huang, R.; Li, T.; Tan, B.; Li, P. Metabolomic analysis of the response of growing pigs to dietary L-arginine supplementation. *Amino Acids* **2009**, *37*, 199–208. [CrossRef]
21. Nicholson, J.; Lindon, J. Metabonomics. *Nature* **2008**, *455*, 1054–1056. [CrossRef]
22. Sacco, D.; Brescia, M.; Sgaramella, A.; Casiello, G.; Buccolieri, A.; Ogrinc, N.; Sacco, A. Discrimination between Southern Italy and foreign milk samples using spectroscopic and analytical data. *Food Chem.* **2009**, *114*, 1559–1563. [CrossRef]
23. Shintu, L.; Ziarelli, F.; Caldarelli, S. Is high-resolution magic angle spinning NMR a practical speciation tool for cheese samples? Parmigiano Reggiano as a case study. *Magn. Reson. Chem.* **2004**, *42*, 396–401. [CrossRef] [PubMed]
24. Sun, H.-Z.; Wang, D.-M.; Wang, B.; Wang, J.-K.; Liu, H.-Y.; Guan, L.L.; Liu, J.-X. Metabolomics of four biofluids from dairy cows: Potential biomarkers for milk production and quality. *J. Proteome Res.* **2015**, *14*, 1287–1298. [CrossRef] [PubMed]

25. Caboni, P.; Murgia, A.; Porcu, A.; Manis, C.; Ibba, I.; Contu, M.; Scano, P. A metabolomics comparison between sheep's and goat's milk. *Food Res Int.* **2019**, *119*, 869–875. [CrossRef] [PubMed]
26. Caboni, P.; Murgia, A.; Porcu, A.; Demuru, M.; Pulina, G.; Nudda, A. Gas chromatography-mass spectrometry metabolomics of goat milk with different polymorphism at the αS1-casein genotype locus. *J. Anim. Sci.* **2016**, *99*, 6046–6051. [CrossRef]
27. Pisano, M.B.; Scano, P.; Murgia, A.; Cosentino, S.; Caboni, P. Metabolomics and microbiological profile of Italian mozzarella cheese produced with buffalo and cow milk. *Food Chem.* **2016**, *192*, 618–624. [CrossRef]
28. Scano, P.; Murgia, A.; Pirisi, F.M.; Caboni, P. A gas chromatography-mass spectrometry-based metabolomic approach for the characterization of goat milk compared with cow milk. *J Dairy Sci.* **2014**, *97*, 6057–6066. [CrossRef]
29. Chen, J.K.; Shen, C.R.; Liu, C.L. N-acetylglucosamine: Production and applications. *Mar. Drugs* **2010**, *8*, 2493–2516. [CrossRef]
30. Mendoza, M.R.; Olano, A.; Villamiel, M. Chemical indicators of heat treatment in fortified and special milks. *J. Agric. Food Chem.* **2005**, *53*, 2995–2999. [CrossRef]
31. Troyano, E.; Villamiel, M.; Olano, A.; Sanz, J.; Martínez-Castro, I. Monosaccharides andmyo-inositol in commercial milks. *J. Agric. Food Chem.* **1996**, *44*, 815–817. [CrossRef]
32. Coppola, S.; Parente, E.; Dumontet, S.; La Peccerella, A. The microflora of natural whey cultures utilized as starters in the manufacture of Mozzarella cheese from water–buffalo milk. *Le Lait* **1988**, *68*, 295–310. [CrossRef]
33. Coppola, S.; Blaiotta, G.; Ercolini, D.; Moschetti, G. Molecular evaluation of microbial diversity occurring in different types of Mozzarella cheese. *J. Appl. Microbiol.* **2001**, *90*, 414–420. [CrossRef] [PubMed]
34. Bonizzi, I.; Feligini, M.; Aleandri, R.; Enne, G. Genetic traceability of geographical origin of typical Italian water buffalo mozzarella cheese: A preliminary approach. *J. App. Microbiol.* **2006**, *102*, 667–673. [CrossRef] [PubMed]
35. Lucey, J.A.; Johnson, M.E.; Horne, D.S. Invited review: Perspectives on the basis of the rheology and texture properties of cheese. *J Dairy Sci.* **2003**, *86*, 2725–2743. [CrossRef]
36. MEGlobal. Fast Facts Diethylene Glycol. Available online: http://www.meglobal.biz/media/MEGlobal_FastFacts_DEG.pdf (accessed on 15 February 2019).
37. Kabata-Pendias, A.; Pendias, H. *Trace Elements in Soils and Plant*, 3rd ed.; CRC Press: Boca Raton, FL, USA, 2001; p. 403.
38. Rotz, C.A.; Muck, R.E. Changes in forage quality during harvest and storage. In *Forage Quality, Evaluation, and Utilization*, 1st ed.; Fahey, G.C., Collins, M., Mertens, D.R., Moser, L.E., Eds.; American Society of Agronomy, Crop Science Society of America, Soil Science Society of America: Madison, WI, USA, 1994; pp. 828–868.
39. Bontempo, L.; Paolini, M.; Franceschi, P.; Ziller, L.; García-González, D.L.; Camin, F. Characterisation and attempted differentiation of European and extra-European olive oils using stable isotope ratio analysis. *Food Chem.* **2019**, *276*, 782–789. [CrossRef]
40. Henderson, G.; Cox, F.; Ganesh, S.; Jonker, A.; Young, W.; Abecia, L.; Angarita, E.; Aravena, P.; Arenas, G.N.; Ariza, C. Rumen microbial community composition varies with diet and host, but a core microbiome is found across a wide geographical range. *Sci. Rep.* **2015**, *5*, 14567. [CrossRef]
41. Jami, E.; White, B.A.; Mizrahi, I. Potential role of the bovine rumen microbiome in modulating milk composition and feed efficiency. *PLoS ONE* **2014**, *9*, e85423. [CrossRef]
42. Lima, F.S.; Oikonomou, G.; Lima, S.F.; Bicalho, M.L.; Ganda, E.K.; de Oliveira Filho, J.C.; Lorenzo, G.; Trojacanec, P.; Bicalho, R.C. Prepartum and postpartum rumen fluid microbiomes: Characterization and correlation with production traits in dairy cows. *Appl. Environ. Microbiol.* **2015**, *81*, 1327–1337. [CrossRef]
43. Berardo, N. Prediction of the chemical composition of white clover by near-infrared reflectance spectroscopy. *Grass Forage Sci.* **1997**, *52*, 27–32. [CrossRef]
44. Troisi, J.; Sarno, L.; Martinelli, P.; Di Carlo, C.; Landolfi, A.; Scala, G.; Rinaldi, M.; D'Alessandro, P.; Ciccone, C.; Guida, M. A metabolomics-based approach for non-invasive diagnosis of chromosomal anomalies. *Metabolomics* **2017**, *13*, 140. [CrossRef]
45. Sumner, L.W.; Amberg, A.; Barrett, D.; Beale, M.H.; Beger, R.; Daykin, C.A.; Fan, T.W.-M.; Fiehn, O.; Goodacre, R.; Griffin, J.L.; et al. Proposed minimum reporting standards for chemical analysis Chemical Analysis Working Group (CAWG) Metabolomics Standards Initiative (MSI). *Metabolomics Off. J. Metabolomic Soc.* **2007**, *3*, 211–221. [CrossRef]

46. Wold, S.; Sjöström, M.; Eriksson, L. PLS-regression: A basic tool of chemometrics. *PLS Methods* **2001**, *58*, 109–130. [CrossRef]
47. Mevik, B.H.; Wehrens, R. The pls Package: Principal Component and Partial Least Squares Regression in R. *J. Stat. Softw.* **2007**, *18*, 148–152. [CrossRef]
48. Kuhn, M. Building Predictive Models in R Using the caret Package. *J. Stat. Softw.* **2008**, *28*, 1–26. [CrossRef]

Sample Availability: Samples of the compounds extract of milk and mozzarella cheese are available from the authors.

molecules

MDPI

Article

Rapid Evaluation and Optimization of Medium Components Governing Tryptophan Production by *Pediococcus acidilactici* TP-6 Isolated from Malaysian Food via Statistical Approaches

Ye Heng Lim [1], Hooi Ling Foo [1,2,*], Teck Chwen Loh [3,4,*], Rosfarizan Mohamad [1,2,5] and Raha Abdul Rahim [1,6,7]

1 Institute of Bioscience, Universiti Putra Malaysia, UPM Serdang 43400, Selangor, Malaysia; yhlim_0418@hotmail.com (Y.H.L.); farizan@upm.edu.my (R.M.); raha@utem.edu.my (R.A.R.)
2 Department of Bioprocess Technology, Faculty of Biotechnology and Biomolecular Sciences, Universiti Putra Malaysia, UPM Serdang 43400, Selangor, Malaysia
3 Department of Animal Science, Faculty of Agriculture, Universiti Putra Malaysia, UPM Serdang 43400, Selangor, Malaysia
4 Institute of Tropical Agriculture and Food Security, Universiti Putra Malaysia, UPM Serdang 43400, Selangor, Malaysia
5 Institute of Tropical Forestry and Forest Products, Universiti Putra Malaysia, UPM Serdang 43400, Selangor, Malaysia
6 Department of Cell and Molecular Biology, Faculty of Biotechnology and Biomolecular Sciences, Universiti Putra Malaysia, UPM Serdang 43400, Selangor, Malaysia
7 Office of Vice Chancellor, Universiti Teknikal Malaysia Melaka, Jalan Hang Tuah Jaya, Durian Tunggal 76100, Melaka, Malaysia
* Correspondence: hlfoo@upm.edu.my (H.L.F.); tcloh@upm.edu.my (T.C.L.); Tel.: +60-3-9769-7476 (H.L.F.); +60-3-97694814 (T.C.L.)

Academic Editors: Francesco Vinale and Maria Luisa Balestrieri
Received: 10 November 2019; Accepted: 9 December 2019; Published: 11 February 2020

Abstract: Tryptophan is one of the most extensively used amino acids in livestock industry owing to its effectiveness in enhancing the growth performance of animals. Conventionally, the production of tryptophan relies heavily on genetically modified *Escherichia coli* but its pathogenicity is a great concern. Our recent study demonstrated that a lactic acid bacterium (LAB), *Pediococcus acidilactici* TP-6 that isolated from Malaysian food was a promising tryptophan producer. However, the tryptophan production must enhance further for viable industrial application. Hence, the current study evaluated the effects of medium components and optimized the medium composition for tryptophan production by *P. acidilactici* TP-6 statistically using Plackett-Burman Design, and Central Composite Design. The optimized medium containing molasses (14.06 g/L), meat extract (23.68 g/L), urea (5.56 g/L) and $FeSO_4$ (0.024 g/L) significantly enhanced the tryptophan production by 150% as compared to the control de Man, Rogosa and Sharpe medium. The findings obtained in this study revealed that rapid evaluation and effective optimization of medium composition governing tryptophan production by *P. acidilactici* TP-6 were feasible via statistical approaches. Additionally, the current findings reveal the potential of utilizing LAB as a safer alternative tryptophan producer and provides insight for future exploitation of various amino acid productions by LAB.

Keywords: Tryptophan production; lactic acid bacteria; *Pediococcus acidilactici* TP-6; Plackett-Burman design; central composite design

1. Introduction

Fermentation medium plays an indispensable role in the industrial fermentation process due to its impact on the formation of the desired products [1]. A cost-effective medium formulation is crucial in ensuring the economic feasibility of the fermentation process. Hence, optimization of the medium composition is important in order to minimize the cost of production without compromising the production. Conventional method and statistical method are the most common methodologies employed in the optimization study. The conventional method of process optimization is also known as one-factor-at-a-time method by varying one factor while keeping the other factors unchanged until an apparent optimum condition is achieved. However, the conventional optimization method often require large number of experiments and it could be time consuming and laborious. Furthermore, this method is not suitable for multifactor optimization because it is unable to elucidate the interactions between the factors and thus incapable to detect the true optimum condition [2].

The limitation of conventional optimization method can be overcome by using statistical optimization method, which involves a collection of numerous experimental strategies, mathematical procedures and statistical inferences. Unlike conventional optimization method, the statistical optimization method is able to explain the interactions between multiple variables and determine the true optimum based on statistical approaches [3]. One of the most commonly used statistical optimization approach is response surface methodology (RSM). The first-and second-degree models are among the most frequently used approximating polynomial models in RSM. Some of the popular first-order designs that are regularly employed for optimization of the fermentation process include the Plackett-Burman Design (PBD) [4] and the Factorial Design [5]. Meanwhile, the commonly used second-order designs include Central Composite Design (CCD) [6] and Box-Behnken Design [7].

Previous optimization studies on the production of tryptophan have revolved around the conventional producer strains such as *E. coli* and *Corynebacterium glutamicum*. For instances, Faghfuri et al. [8] reported that sugar beet molasses was a good source of pyridoxal phosphate (PLP) and serine for tryptophan production by *E. coli*. Meanwhile, Cheng et al. [9] suggested that the growth and tryptophan production of *E. coli* was inhibited by acetic acid above 2 g/L and glucose concentration should be controlled at low level for tryptophan production. Moreover, Hagino and Nakayama [10] reported that molasses, casein enzymatic hydrolysate and $(NH_4)_2SO_4$ were the best carbon, organic nitrogen and inorganic nitrogen source for tryptophan production by *C. glutamicum*. However, the use of genetically engineered and pathogenic microorganism is a major concern and has urged an exploration for a safer producer. Lactic acid bacteria (LAB) have been revealed to possess the ability to produce various amino acid in several recent studies [11,12]. Apart from its versatility in amino acid production, LAB postbiotic metabolites have been extensively reported to confer various health benefits to animals and enhance their growth performance by regulating the gastrointestinal health and immune response of the animals [13–16].

Tryptophan has gained tremendous attention in recent years, particularly in medical, feed and livestock industries. Tryptophan is known as the fourth limiting amino acid in livestock feed, right after lysine, methionine and threonine amino acids [17]. It has been demonstrated to affect both growth and neurotransmitter metabolism of poultry [18]. It also affects glucose metabolism by inhibiting gluconeogenesis [19]. Moreover, L-tryptophan was found to play a crucial role in improving the growth performance, meat quality, reducing stress, regulating insulin response and protein synthesis in muscles of pigs [20], as well as improving the feed conversion and carcass yield of broilers [21]. In medical field, tryptophan is often used as sedative and antidepressant and hence it is frequently prescribed for the treatment of schizophrenia and alcoholism [22]. Tryptophan also acts as a precursor for serotonin biosynthesis, a neurotransmitter that is responsible to relieve anxiety [23].

The effects of growth medium components on amino acid productions by LAB have not been elucidated previously, despite the effects of the M-17 medium [24] and de Man, Rogosa and Sharpe (MRS) medium [11,25] being reported, for the production of amino acid by LAB. *Pediococcus acidilactici* TP-6 was previously identified as a superior producer of tryptophan in MRS medium [25]. Nevertheless, limited knowledge regarding the nutritional requirements of *P. acidilactici* TP-6 for its growth and tryptophan production was available. Thus, the objectives of this study were to evaluate the effects of medium components on the growth and tryptophan production of *P. acidilactici* TP-6 by using PBD, followed by optimization of the medium components for tryptophan production by using steepest ascent method and CCD approaches.

2. Results and Discussion

2.1. Plackett-Burman Design

The nutritional requirement of *P. acidilactici* TP-6 for tryptophan production was studied by using PBD, where each variable was represented at two levels. A dummy variable (X) that serves as an indicator for the presence of significant interactions between the variables was incorporated in the PBD. The presence of significant interactions between the variables was indicated by high effect values of the dummy variable [26]. Table 1 shows the tryptophan production and cell population of respective trial of PBD. In general, tryptophan production was not detected in most of the experimental runs, suggesting the stringent nutrient requirement of the producer strain for tryptophan production. Run 15 recorded the highest amount of tryptophan production of 22.9 mg/L of tryptophan, followed by run 18 with 21.27 mg/L of tryptophan. However, there was no significant difference ($p > 0.05$) between the net tryptophan produced achieved in run 15 and run 18 respectively. Nevertheless, the production of tryptophan by *P. acidilactici* TP-6 in the control MRS medium (29.41 mg/L) was still significantly higher ($p < 0.05$) than those achieved in the PBD. Thus, further optimization of the medium composition is mandatory to increase tryptophan yield by *P. acidilactici* TP-6.

Table 1. Plackett-Burman Design (PBD) matrixes for 22 variables with coded values and their corresponding tryptophan production and cell population of *P. acidilactici* TP-6.

Std Run	A	B	C	D	E	F	G	H	J	K	L	M	N	O	P	Q	R	S	T	U	V	W	X	Tryptophan Production (mg/L)	Cell Population (logCFU/mL)
1	1	1	1	1	1	−1	1	−1	1	1	−1	−1	1	1	−1	−1	1	−1	1	−1	−1	−1	−1	0.00 ± 0.00 ^{H}	8.47 ± 0.04 ^{I}
2	−1	1	1	1	1	1	−1	1	−1	1	1	−1	−1	1	1	−1	−1	1	−1	1	−1	−1	−1	0.00 ± 0.00 ^{H}	9.26 ± 0.01 ^{B}
3	−1	−1	1	1	1	1	1	−1	1	−1	1	1	−1	−1	1	1	−1	−1	1	−1	1	−1	−1	0.42 ± 0.05 ^{G}	8.82 ± 0.03 ^{F}
4	−1	−1	−1	1	1	1	1	1	−1	1	−1	1	1	−1	−1	1	1	−1	−1	1	−1	1	−1	0.34 ± 0.02 ^{G}	8.77 ± 0.04 ^{F}
5	−1	−1	−1	−1	1	1	1	1	1	−1	1	−1	1	1	−1	−1	1	1	−1	−1	1	−1	1	9.35 ± 0.05 ^{D}	9.10 ± 0.02 ^{D}
6	1	−1	−1	−1	−1	1	1	1	1	1	−1	1	−1	1	1	−1	−1	1	1	−1	−1	1	−1	0.00 ± 0.00 ^{H}	9.04 ± 0.01 ^{E}
7	−1	1	−1	−1	−1	−1	1	1	1	1	1	−1	1	−1	1	1	−1	−1	1	1	−1	−1	1	6.81 ± 0.07 ^{E}	8.70 ± 0.01 ^{G}
8	1	−1	1	−1	−1	−1	−1	1	1	1	1	1	−1	1	−1	1	1	−1	−1	1	1	−1	−1	16.69 ± 0.51 ^{C}	8.75 ± 0.02 ^{FG}
9	−1	1	−1	1	−1	−1	−1	−1	1	1	1	1	1	−1	1	−1	1	1	−1	−1	1	1	−1	0.00 ± 0.00 ^{H}	7.33 ± 0.01 ^{M}
10	−1	−1	1	−1	1	−1	−1	−1	−1	1	1	1	1	1	−1	1	−1	1	1	−1	−1	1	1	0.00 ± 0.00 ^{H}	7.89 ± 0.02 ^{L}
11	1	−1	−1	1	−1	1	−1	−1	−1	−1	1	1	1	1	1	−1	1	−1	1	1	−1	−1	1	0.00 ± 0.00 ^{H}	9.22 ± 0.02 ^{BC}
12	1	1	−1	−1	1	−1	1	−1	−1	−1	−1	1	1	1	1	1	−1	1	−1	1	1	−1	−1	0.00 ± 0.00 ^{H}	8.79 ± 0.02 ^{F}
13	−1	1	1	−1	−1	1	−1	1	−1	−1	−1	−1	1	1	1	1	1	−1	1	−1	1	1	−1	0.00 ± 0.00 ^{H}	9.02 ± 0.03 ^{E}
14	−1	−1	1	1	−1	−1	1	−1	1	−1	−1	−1	−1	1	1	1	1	1	−1	1	−1	1	1	4.85 ± 0.04 ^{F}	8.60 ± 0.00 ^{H}
15	1	−1	−1	1	1	−1	−1	1	−1	1	−1	−1	−1	−1	1	1	1	1	1	−1	1	−1	1	22.94 ± 0.79 ^{B}	9.19 ± 0.01 ^{C}
16	1	1	−1	−1	1	1	−1	−1	1	−1	1	−1	−1	−1	−1	1	1	1	1	1	−1	1	−1	0.00 ± 0.00 ^{H}	8.75 ± 0.03 ^{FG}
17	−1	1	1	−1	−1	1	1	−1	−1	1	−1	1	−1	−1	−1	−1	1	1	1	1	1	−1	1	0.00 ± 0.00 ^{H}	8.75 ± 0.02 ^{FG}
18	1	−1	1	1	−1	−1	1	1	−1	−1	1	−1	1	−1	−1	−1	−1	1	1	1	1	1	−1	21.27 ± 0.24 ^{B}	7.88 ± 0.02 ^{L}
19	−1	1	−1	1	1	−1	−1	1	1	−1	−1	1	−1	1	−1	−1	−1	−1	1	1	1	1	1	10.58 ± 0.25 ^{D}	8.47 ± 0.01 ^{K}
20	1	−1	1	−1	1	1	−1	−1	1	1	−1	−1	1	−1	1	−1	−1	−1	−1	1	1	1	1	0.00 ± 0.00 ^{H}	8.50 ± 0.01 ^{IJ}
21	1	1	−1	1	−1	1	1	−1	−1	1	1	−1	−1	1	−1	1	−1	−1	−1	−1	1	1	1	0.00 ± 0.00 ^{H}	9.11 ± 0.01 ^{D}
22	1	1	1	−1	1	−1	1	1	−1	−1	1	1	−1	−1	1	−1	1	−1	−1	−1	−1	1	1	0.00 ± 0.00 ^{H}	9.34 ± 0.02 ^{AB}
23	1	1	1	1	−1	1	−1	1	1	−1	−1	1	1	−1	−1	1	−1	1	−1	−1	−1	−1	1	0.35 ± 0.01 ^{G}	9.05 ± 0.01 ^{E}
24	−1	−1	−1	−1	−1	−1	−1	−1	−1	−1	−1	−1	−1	−1	−1	−1	−1	−1	−1	−1	−1	−1	−1	0.00 ± 0.00 ^{H}	7.33 ± 0.02 ^{M}
MRS	1	−1	−1	−1	−1	1	1	1	1	−1	−1	−1	−1	1	1	1	1	1	−1	−1	−1	−1	−1	29.41 ± 0.76 ^{A}	9.41 ± 0.01 ^{A}

Note: Values are mean ± SEM, n =3. Mean ± SEM within the same column that share a common superscript (A–M) are not significantly different ($p > 0.05$).

Table 2 presents the ANOVA of the PBD for the effects of medium components on tryptophan production by *P. acidilactici* TP–6. The low p-value of the model (0.0019) revealed that the model was highly significant ($p < 0.01$) and it is highly unlikely (>99% confidence) that the large F-value of the model was attributed to noise. Moreover, the model was able to elucidate 99% of variation in response, owing to its high coefficient of determination (R^2 = 0.9986). Additionally, the "predicted R^2" (0.9490) and the "adjusted R^2" (0.9919) values were close to 1 and in reasonable agreement, implying the great correlation between experimental and predicted values and hence the suggested PBD model was a good model [27]. Furthermore, the model is suitable for navigating the design space owing to its adequate signal to noise ratio, which was reflected by the high adequate precision value of 110.181 (much greater than 4).

Based on the ANOVA of tryptophan production (Table 2), majority of the studied variables contributed significantly ($p < 0.05$) to tryptophan production by *P. acidilactici* TP-6, whereby up to 17 variables showed a p-value less than 0.05 except fructose, Tween 80 and biotin, which were not significant ($p > 0.05$). Among the 17 significant variables, 16 of them were highly significant ($p < 0.01$) except glucose and $MgSO_4$, which were significant ($p < 0.05$). Additionally, the dummy variable was revealed to be highly significant ($p < 0.01$). The unexpectedly high significant effect of the dummy variable implied the presence of significant interactions between the variables [26]. Hence, a design with higher resolution is required to elucidate the interaction [28]. The net tryptophan production (Y) by *P. acidilactici* TP-6 can be expressed in terms of coded symbols (A–X) as shown in the following regression Equation (1):

$$Y = 1.21 - 0.51A - 4.25B - 0.16C + 2.23D - 2.03E - 2.56F - 1.71G + 4.97H + 1.41J + 1.98L - 1.95M - 1.51N - 1.79O - 2.34P + 0.56Q + 2.01R + 0.29S + 2.81T + 1.97V + 0.30W + 1.64X \quad (1)$$

Table 2. ANOVA of PBD for the effects of medium components on tryptophan production by *P. acidilactici* TP-6.

Source	Sum of Squares	df	Mean Square	F Value	p-Value Prob > F	
Model	2428.91	21	115.66	520.43	<0.01	significant
A-Glucose	6.13	1	6.13	27.56	0.03	significant
B-Sucrose	433.36	1	433.36	1949.90	<0.01	significant
C-Fructose	0.61	1	0.61	2.76	0.24	
D-Lactose	119.36	1	119.36	537.05	<0.01	significant
E-Molasses	98.68	1	98.68	444.03	<0.01	significant
F-Yeast extract	156.85	1	156.85	705.75	<0.01	significant
G-Peptone	70.03	1	70.03	315.12	<0.01	significant
H-Meat extract	592.62	1	592.62	2666.50	<0.01	significant
J-K_2HPO_4	47.76	1	47.76	214.88	<0.01	significant
L-Urea	93.75	1	93.75	421.83	<0.01	significant
M-NH_4NO_3	90.83	1	90.83	408.69	<0.01	significant
N-$(NH_4)_2SO_4$	54.88	1	54.88	246.92	<0.01	significant
O-$(NH_4)_2HC_6H_5O_7$	77.00	1	77.00	346.49	<0.01	significant
P-NaOAc	131.64	1	131.64	592.30	<0.01	significant
Q-$MgSO_4$	7.54	1	7.54	33.93	0.03	significant
R-$MnSO_4$	97.21	1	97.21	437.39	<0.01	significant
S-Tween 80	2.04	1	2.04	9.19	0.09	
T-$FeSO_4$	188.94	1	188.94	850.15	<0.01	significant
V-$CuSO_4$	92.90	1	92.90	418.02	<0.01	significant
W-Biotin	2.09	1	2.09	9.42	0.09	
X-dummy	64.69	1	64.69	291.09	<0.01	significant
Residual	0.44	2	0.22			
Cor Total	2429.36	23				

Note: R^2: 0.9998; Adj R^2: 0.9979; Pred R^2: 0.9737; Adeq Precision: 110.181.

Figure 1 illustrates the impact of each medium component on tryptophan production by *P. acidilactici* TP-6. Among the 22 studied variables, 10 of them, including meat extract, $FeSO_4$, lactose, $MnSO_4$, urea, $CuSO_4$, K_2HPO_4, $MgSO_4$, biotin and Tween 80, exerted a stimulatory effect on tryptophan production, whereas the remaining 12 variables demonstrated an inhibitory effect. Out of the 10 variables that exhibited positive effect, 7 of them including meat extract, $FeSO_4$, lactose, $MnSO_4$, urea, $CuSO_4$ and K_2HPO_4 were highly significant at p-value less than 0.01, whereas $MgSO_4$ was significant ($p < 0.05$) as shown in Table 2. In contrast, biotin and Tween 80 did not contribute significantly ($p > 0.05$) to the tryptophan production by *P. acidilactici* TP-6. On the other hand, 8 out of the 12 variables that exhibited a negative effect were highly significant ($p < 0.01$) and one of them was significant ($p < 0.05$). Meanwhile, the other 3 variables did not affect tryptophan production significantly ($p > 0.05$).

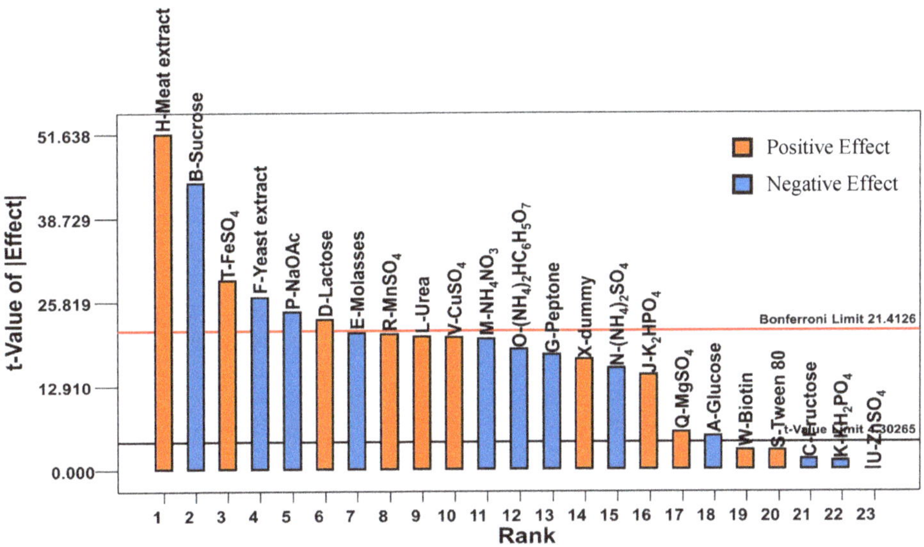

Figure 1. Effects of different medium components on tryptophan production by *P. acidilactici* TP-6.

Carbon sources play an essential role in the biosynthesis of tryptophan where the metabolism of carbon sources provides erythrose-4-phosphate and phosphoenolpyruvate, which act as precursors for tryptophan biosynthesis [29]. Among the five studied carbon sources, only lactose demonstrated a highly significant ($p < 0.01$) stimulatory effect on tryptophan production by *P. acidilactici* TP-6. Production of tryptophan by *Lactobacilli* strains in medium containing lactose as sole carbon source was also reported by Tarek and Hesham [24]. Meanwhile, the other 4 carbon sources (sucrose, molasses, glucose and fructose) exhibited a negative effect on tryptophan production by *P. acidilactici* TP-6 with sucrose, molasses and glucose being significant ($p < 0.05$) and fructose was insignificant ($p > 0.05$). This is in contrast with the findings of several other studies conducted by other researchers, where the tested carbon sources often possessed a stimulatory effect on the production of amino acid. For instance, molasses was used for the production of tryptophan by *C. glutamicum* [8]. Furthermore, molasses was reported as the most suitable carbon source for the synthesis of tryptophan by *C. glutamicum* [10]. The use of glucose for the production of various amino acid by *C. glutamicum* [30–32] and *E. coli* [33,34] was also well documented. Furthermore, sucrose was identified as the best carbon source for threonine production by *Escherichia coli* TRFC [35] and it was used as the sole carbon source for threonine production by a recombinant *E. coli* in a separate study conducted by Wang et al. [36]. Nevertheless, fructose was revealed as one of the best carbon sources for glutamate production by *Arthrobacter globiformis* [37]. The discrepancy of the effects of different carbon sources on amino acid production might be attributed to the use of different microorganisms as producer strains.

Among the three tested organic nitrogen sources, only meat extract demonstrated a highly significant stimulatory effect ($p < 0.01$), whereas yeast extract and peptone showed a highly significant inhibitory effect ($p < 0.01$). The organic nitrogen source is crucial for the biosynthesis of tryptophan, where it is responsible for supplying serine for the formation of tryptophan from (3-Indoyl)-glycerolphosphate catalyzed by the enzyme tryptophan synthase [38]. The high stimulatory effect of meat extract on tryptophan production by *P. acidilactici* TP-6 might be due to the rich serine content of meat extract [39]. Additionally, the positive effect of meat extract on tryptophan production could be attributed to its rich vitamin content, which acted as coenzymes for the activation of enzymes involved in the biosynthesis of tryptophan. For instance, meat extract contains an abundant amount of riboflavin [40], which can be converted into flavin mononucleotide, an essential coenzyme for chorismate synthase enzyme that is responsible for the synthesis of chorismate from 5-o-(1-carboxyvinyl)-3-phosphate [41]. Moreover, vitamin B_3, which is abundant in meat extract, will be metabolized into NADPH that functions as coenzyme for the action of shikimate dehydrogenase enzyme, which is accountable for the transformation of 3-dehydroshikimate into shikimate. Lim et al. [25] had demonstrated the effect of medium containing meat extract, yeast extract and peptone for the production of amino acid by LAB. In comparison, yeast extract was used for the production of tryptophan by *E. coli* in a study conducted by Faghfuri et al. [8]. In addition, Hagino and Nakayama [10] demonstrated that yeast extract was the best organic nitrogen source for tryptophan synthesis by *C. glutamicum*. Nevertheless, the present study revealed that yeast extract demonstrated an inhibitory effect on tryptophan production by *P. acidilactici* TP-6. Differences in the preference of organic nitrogen sources might be attributed to the use of different microorganism as the amino acid producer strain in the study.

On the other hand, all the inorganic nitrogen sources used in the current study contributed significantly ($p < 0.05$) to tryptophan production by *P. acidilactici* TP-6 (Table 2). However, most of them demonstrated a negative effect, except urea which exhibited a stimulatory effect. The significant positive impact of urea on the production of tryptophan might be attributed to its function to supply ammonia for the formation of anthranilate from chorismate catalyzed by the enzyme anthranilate synthase during tryptophan biosynthesis [42]. Urea was used as an inorganic nitrogen source for glutamate production by *C. glutamicum* [43] and *Brevibacterium* sp. [44], respectively, whereas NH_4NO_3 was utilized as the sole inorganic nitrogen source for glutamate production by LAB in a study conducted by Zareian et al. [45]. On the other hand, the use of $(NH_4)_2SO_4$ as inorganic nitrogen source for the production of various amino acid by *C. glutamicum* [31,32] and *E. coli* [8,33–36] were well documented.

In the meantime, 5 out of 8 mineral sources including $FeSO_4$, $MnSO_4$, $CuSO_4$, K_2HPO_4 and $MgSO_4$ exhibited positive effect on tryptophan production by *P. acidilactici* TP-6 significantly ($p < 0.05$), whereas NaOAc contributed significantly ($p < 0.05$) on tryptophan production in a negative manner. Furthermore, KH_2PO_4 and $ZnSO_4$ displayed an inhibitory effect on tryptophan production by the producer strain but the effect was insignificant ($p > 0.05$). The significant effects of minerals such as $FeSO_4$, $MnSO_4$, $CuSO_4$, K_2HPO_4 and $MgSO_4$ for the production of various amino acid have been demonstrated for *E. coli* [35,36] and *C. glutamicum* [32,46], indicating the importance of minerals for amino acid production. Many metal ions play an essential role as cofactor that is required for catalytic activity of enzymes to ensure proper functioning of biological system [47].

Findings obtained in the current study revealed that most of the divalent cations displayed a stimulatory effect on tryptophan production except Zn^{2+}. The crucial role of divalent cations such as Fe^{2+}, Mn^{2+}, Co^{2+} and Mg^{2+} on enzymes involved in biosynthesis of tryptophan has been well documented, where they often possessed a stimulatory effect. For instances, Zalkin and Kling [48] reported that the enzyme anthranilate synthase has an absolute requirement for Mg^{2+}, while Hertel et al. [49] discovered that Fe^{2+} and Co^{2+} could be used to substitute Mg^{2+} as a cofactor for anthranilate synthase. Moreover, Widholm [50] also found that Mn^{2+} or Co^{2+} could substitute Mg^{2+} for the enzyme anthranilate synthase. The negative effect of Zn^{2+} on the production of tryptophan by

P. acidilactici TP-6 in the present study might be due to its inhibitory effect on the enzyme anthranilate synthase as suggested by Hertel et al. [49] and Widholm [50].

In comparison, the biotin vitamin B employed in the current study demonstrated a stimulatory effect on the production of tryptophan by *P. acidilactici* TP-6, despite the effect being insignificant ($p > 0.05$). This is in contrast with previous reports, whereby biotin was often essential for the production of amino acid by microorganisms such as *E. coli* [33] and *Corynebacterium* [32]. One of the possible explanations for the insignificant impact of biotin on tryptophan production by *P. acidilactici* TP-6 might be attributed to different nutrient requirement between LAB and other microorganisms for amino acid production, implying that the nutrient requirement for amino acid production could be species dependent. Another possible explanation might be due to low requirement of vitamin biotin, where it is often required in minute amount. Hence, inclusion of molasses which contain high biotin content [51] in the medium formulation could provide sufficient biotin to the producer strain of *P. acidilactici* TP-6 that used in this study.

On the other hand, the non-ionic surfactant of Tween 80 that used in the present study displayed a stimulatory effect on the production of tryptophan but it was not significant ($p > 0.05$). To the best of our knowledge, there were no reports available regarding the role of Tween 80 in the production of amino acid thus far. Despite Tween 80 has been included in the medium formulation for glutamate production by *Brevibacterium* sp. in a study conducted by Nampoothiri and Pandey [44], its significance level was not elucidated. However, Tween 80 is well-known for its crucial role for the production of various LAB metabolites. For instance, the supplementation of Tween 80 was found to critically boost the bacteriocins production [52]. Tween 80 acts as biosurfactant which is responsible to modify the fluidity and permeability of the cell membrane of producer strain. This in turn facilitates the secretion of metabolites extracellularly [53].

The effects of various medium components on the growth of *P. acidilactici* TP-6 were elucidated by the 24 experimental trials of PBD. The corresponding cell population of *P. acidilactici* TP-6 in each experimental run is presented in Table 1. Among the 24 experimental runs, the highest cell population was detected in run 22 with 9.34 log CFU/mL and it was not significantly different ($p > 0.05$) as compared to control (9.41 log CFU/mL). In contrast, run 9 and run 24 showed the lowest cell population of merely 7.33 log CFU/mL. Absence of organic nitrogen source or carbon source in both runs, which was crucial for the growth of LAB [54] and could be the reason attributing to the low cell growth in both runs.

Table 3 displays the ANOVA of the PBD for the effects of medium components on the cell growth of *P. acidilactici* TP-6. The low p-value of the model (< 0.01) revealed that the model was highly significant ($p < 0.01$) and it is highly unlikely (>99% confidence) that the large F-value of the model this large was attributed to noise. Moreover, the model exhibits great predictive strength and was able to elucidate 99% of variation in response due to its high R^2 value (0.9986). Additionally, the "predicted R^2" (0.9490) and the "adjusted R^2" (0.9919) values were in reasonable agreement (difference < 0.2), implying the great correlation between experimental and predicted values and the suggested model was significant. Furthermore, the model is suitable for navigating the design space owing to its adequate signal to noise ratio, which was reflected by the high adequate precision value (44.159) that was much greater than 4.

Among the 22 studied variables, 16 variables were revealed to exhibit a significant effect ($p < 0.05$) on the growth of *P. acidilactici* TP-6. Out of the 16 significant variables, 12 of them including glucose, sucrose, molasses, yeast extract, peptone, meat extract, $(NH_4)_2SO_4$, $(NH_4)_2HC_6H_5O_7$, NaOAc, $MgSO_4$, $MnSO_4$ and biotin were highly significant ($p < 0.01$). Furthermore, the p-value of the dummy variable was less than 0.01, implying the possible presence of interactions between the variables, which have to be elucidated using a higher resolution design in the subsequent experiment [28]. The growth of

P. acidilactici TP-6 (Z) can be expressed in the term of the coded symbol as shown in the following first-order regression Equation (2):

$$Z = 8.67 + 0.17A + 0.082B + 0.023C + 0.11E + 0.28F + 0.11G + 0.21H - 0.04J - 0.025K - 0.11N + 0.14O + 0.15P + 0.11Q + 0.1R - 0.037S + 0.031U - 0.029V - 0.11W + 0.16X \quad (2)$$

Table 3. ANOVA of PBD for the effects of medium components on the growth of *P. acidilactici* TP-6.

Source	Sum of Squares	df	Mean Square	F Value	p-Value Prob > F	
Model	7.15	19	0.38	148.54	<0.01	significant
A-Glucose	0.68	1	0.68	269.27	<0.01	significant
B-Sucrose	0.16	1	0.16	63.98	<0.01	significant
C-Fructose	0.01	1	0.01	4.88	0.09	
E-Molasses	0.27	1	0.27	107.47	<0.01	significant
F-Yeast extract	1.83	1	1.83	721.80	<0.01	significant
G-Peptone	0.29	1	0.29	112.71	<0.01	significant
H-Meat extract	1.05	1	1.05	414.04	<0.01	significant
J-K_2HPO_4	0.04	1	0.04	15.23	0.02	significant
K-KH_2PO_4	0.01	1	0.01	5.82	0.07	
N-$(NH_4)_2SO_4$	0.30	1	0.30	119.34	<0.01	significant
O-$(NH_4)_2HC_6H_5O_7$	0.46	1	0.46	179.90	<0.01	significant
P-NaOAc	0.51	1	0.51	200.63	<0.01	significant
Q-$MgSO_4$	0.32	1	0.32	125.26	<0.01	significant
R-$MnSO_4$	0.25	1	0.25	100.07	<0.01	significant
S-Tween 80	0.03	1	0.03	13.04	0.02	significant
U-$ZnSO_4$	0.02	1	0.02	9.40	0.04	significant
V-$CuSO_4$	0.02	1	0.02	8.15	0.05	significant
W-Biotin	0.31	1	0.31	123.62	<0.01	significant
X-dummy	0.58	1	0.58	227.63	<0.01	significant
Residual	0.01	4	0.00			
Cor Total	7.16	23				

Note: R^2: 0.9986; Adj R^2: 0.9919; Pred R^2: 0.9490; Adeq Precision: 44.159.

The effects of medium components on the growth of *P. acidilactici* TP-6 are depicted in Figure 2. Apart from biotin, $(NH_4)_2SO_4$, K_2HPO_4, Tween 80, $CuSO_4$ and KH_2PO_4, which demonstrated an inhibitory effect, the remaining studied variables affected the cell growth of *P. acidilactici* TP-6 in a positive manner. Among the 16 positive effect variables, 11 of them were significant ($p < 0.05$), except fructose, NH_4NO_3, $FeSO_4$, lactose and urea, which were insignificant ($p > 0.05$). In addition, the present study revealed that all the tested carbon sources exerted a stimulatory effect on the cell growth of *P. acidilactici* TP-6 with glucose giving the highest stimulatory effect, followed by molasses and sucrose (Figure 2). The stimulatory effect of glucose, molasses and sucrose were significant ($p < 0.05$). Contradictorily, the stimulatory effect of fructose and lactose on the cell growth of *P. acidilactici* TP-6 was not significant ($p > 0.05$). The strong stimulatory effect of various carbon sources on the cell growth of *P. acidilactici* TP-6 implied that the carbon source was crucial for the survival and growth of producer strain of *P. acidilactici* TP-6. Furthermore, *P. acidilactici* TP-6 was able to utilize an array of carbon sources for its growth. This is in agreement with the findings reported by several studies, whereby various LAB were demonstrated to have the capability to utilize various carbon sources for their growth [55–57].

Similarly, the organic nitrogen sources used in the present study including yeast extract, meat extract and peptone also demonstrated highly significant ($p < 0.01$) stimulatory effect towards the growth of *P. acidilactici* TP-6 with yeast extract exhibited the highest stimulatory effect. The significant impact of organic nitrogen sources on the cell growth of *P. acidilactici* TP-6 could be attributed to its fastidious nutrient requirements. Typically, LAB are unable to survive on inorganic nitrogen solely. They were able to thrive in medium containing organic nitrogen such as complex proteins and peptides. A similar finding was also reported by Rodrigues et al. [58], where all the studied organic nitrogen sources contributed significantly to the growth of *Lactococcus lactis* 53 in a positive manner. The organic nitrogen with the highest stimulatory effect was also found to be yeast extract. Ooi et al. [59] also reported that yeast extract was crucial for bacteriocin production by *L. plantarum* I-UL4, whereby the bacteriocin production was only detected when yeast extract was present. Similar findings were reported by Lim et al. [60], where all the organic nitrogen sources demonstrated a significant stimulatory effect ($p < 0.05$) on the growth of *Pediococcus pentosaceus* TL-3, highlighting the crucial role of organic nitrogen sources for the growth of *P. pentosaceus* TL-3.

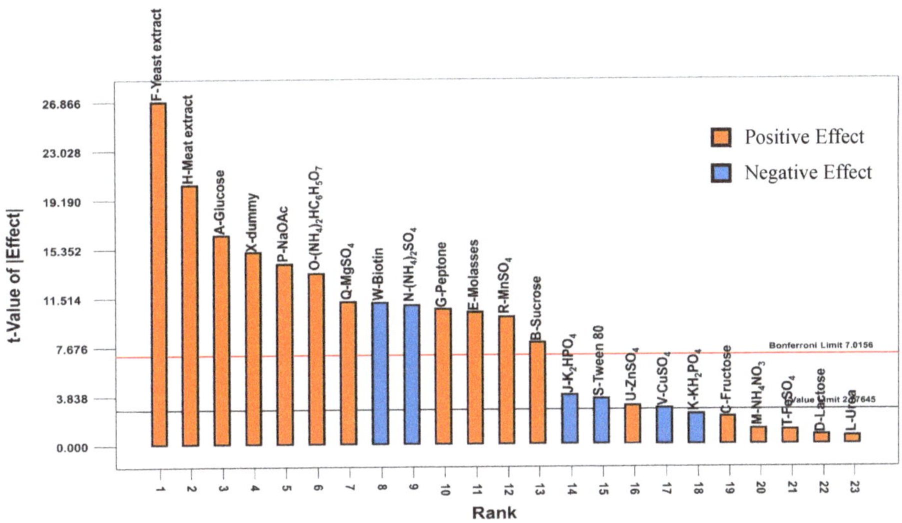

Figure 2. Effects of medium compositions on growth of *P. acidilactici* TP-6.

Unlike organic nitrogen sources, the effect of inorganic nitrogen sources on the growth of *P. acidilactici* TP-6 was less prominent, whereby only 2 out of the 4 studied inorganic nitrogen sources affected the cell growth of *P. acidilactici* TP-6 significantly ($p < 0.05$) with $(NH_4)_2HC_6H_5O_7$ demonstrated a highly significant stimulatory effect ($p < 0.01$), whereas $(NH_4)_2SO_4$ exerted a highly significant inhibitory effect ($p < 0.01$). The stimulatory effect of $(NH_4)_2HC_6H_5O_7$ on the growth of LAB was also reported by Hwang et al. [61], where $(NH_4)_2HC_6H_5O_7$ promoted the growth of *L. plantarum* Pi06. However, Rodrigues et al. [58] reported contradictory finding, where $(NH_4)_2HC_6H_5O_7$ did not exhibit significant effect on the cell growth of studied LAB. This implied that the effect of inorganic nitrogen sources on the cell growth of LAB could be strain dependent. Nevertheless, although the remaining 2 inorganic nitrogen sources (NH_4NO_3 and urea) wielded a positive effect but the effect was insignificant ($p > 0.05$). The insignificance of inorganic nitrogen sources on LAB growth could be due to inability of LAB to assimilate inorganic nitrogen sources [62]. Furthermore, de Carvalho et al. [63] also reported that urea was not significant ($p > 0.05$) for the cell growth of LAB, whereby supplementation of urea did not improve the LAB growth.

On the other hand, majority of the studied mineral sources, including NaOAc, $MgSO_4$, $MnSO_4$, $ZnSO_4$ and $FeSO_4$ displayed a stimulatory impact on the cell growth of *P. acidilactici* TP-6, whereas the remaining 3 mineral sources (K_2HPO_4, $CuSO_4$ and KH_2PO_4) displayed a negative effect. Among the 5 mineral sources with positive effect, NaOAc, $MgSO_4$ and $MnSO_4$ were highly significant at p-value less than 0.01 and $ZnSO_4$ was significant ($p < 0.05$), whereas $FeSO_4$ did not affect the growth of *P. acidilactici*

TP-6 significantly ($p > 0.05$). However, out of the 3 mineral sources that possessed a negative effect, K_2HPO_4 and $CuSO_4$ were significant ($p < 0.05$), whereas KH_2PO_4 was not significant ($p > 0.05$). Acetate ions are known for its great buffering capacity in maintaining the acidity of the medium. This might be the reason for the strong stimulatory effects of NaOAc on the growth of *P. acidilactici* TP-6. Since LAB is an acidophilic microorganism, which prefer an acidic growing environment [64], the presence of acetate ion to maintain the acidic growing environment could enhance the growth of LAB [65]. In contrast, the presence of phosphate ion could potentially lead to the elevation of pH, which in turn created an alkaline environment. As a consequence, the growth of acidophiles such as LAB could be retarded [60]. This could explain the inhibitory effect of KH_2PO_4 and K_2HPO_4 towards the growth of *P. acidilactici* TP-6. On the other hand, the positive impact of Mn^{2+} on the cell growth of *P. acidilactici* TP-6 was in agreement with the results obtained by Tomas et al. [65], whereby the supplementation of $MnSO_4$ in the culture medium had significantly enhanced the growth of *Lactobacillus salivarius* CRL 1328. Moreover, the growth promoting effect of other mineral ions such as Mg^{2+}, Fe^{2+}, Mg^{2+}, Ca^{2+}, Co^{2+} and Cu^{2+} on LAB was also well documented, where a two-fold increment in growth was noted [66]. Despite Foucaud et al. [66] reported that Cu^{2+} stimulated the growth of LAB, yet results obtained in the current study revealed that $CuSO_4$ exhibited an inhibitory effect on the growth of *P. acidilactici* TP-6. This might be due to different requirement of metal ions among different LAB strains, implying that the requirement of metal ions could be strain dependent.

Figure 2 shows that Tween 80 affected the growth of *P. acidilactici* TP-6 significantly ($p < 0.05$) in a negative manner, which was contradictory to the findings reported by Li et al. [67], whereby Tween 80 demonstrated a stimulatory effect on the growth of LAB and other microorganisms. On a separate note, the growth of *Lactobacillus casei* YIT 9018 was not affected significantly ($p > 0.05$) by Tween 80 in the study conducted by Oh et al. [68]. As a comparison, the vitamin biotin used in this study affected the growth of *P. acidilactici* TP-6 highly significantly ($p < 0.01$) in a negative manner. Similar effect was also observed for the growth of a threonine producer, *P. pentosaceus* TL-3, where biotin affected the growth of the isolate *P. pentosaceus* TL-3 significantly in a negative manner [60]. However, a contradictory finding was reported by Tripuraneni [69], where the inclusion of biotin in the growth medium of LAB enhanced the bacterial growth. The negative effect of biotin in the current study was most probably due to low biotin requirement of the producer strain. Hence, the abundant biotin content in molasses and organic nitrogen sources was sufficient to fulfil the requirement of *P. acidilactici* TP-6 [70]. Therefore, further supplementation of biotin contributed to an antagonistic effect.

A number of 17 studied variables were found to affect the production of tryptophan by *P. acidilactici* TP-6 significantly ($p < 0.05$) in the PBD study (Table 2). However, the use of all the 17 significant variables identified in the PBD for further optimization would result in large number of experimental runs. Hence, a validation test was conducted to verify the significant effects of variables identified in the PBD on tryptophan production by *P. acidilactici* TP-6 with MRS medium served as control. The medium formulation that used for validation test was described in Section 3.3. The tryptophan production, growth and serine consumption of *P. acidilactici* TP-6 that noted in different formulated media are shown in Table 4. Medium 1 that contained all the variables identified in the PBD recorded the highest tryptophan production of 26.07 mg/L, followed by Medium 5 and Medium 4 with 25.95 mg/L and 25.89 mg/L of net tryptophan produced respectively. However, there was no significant difference ($p > 0.05$) between the net tryptophan produced in Medium 1, Medium 5 and Medium 4. Nevertheless, it is noteworthy that the net tryptophan produced in the 3 media formulations was comparable to the control (26.81 mg/L), in which they were not significantly different ($p > 0.05$). This implied that Medium 1, 4 and 5 could potentially replace control MRS medium for the production of tryptophan by *P. acidilactici* TP-6.

Table 4. Growth, net tryptophan and serine produced by *P. acidilactici* TP-6 in formulated media.

Media	Cell Population (Log CFU/mL)	Tryptophan Production (mg/L)	Serine Consumption (mg/L)
1	8.95 ± 0.01 ^{B}	26.07 ± 0.86 ^{AB}	9.95 ± 0.43 ^{A}
2	7.98 ± 0.01 ^{D}	25.16 ± 0.39 ^{B}	11.75 ± 0.86 ^{A}
3	8.00 ± 0.01 ^{D}	25.00 ± 0.20 ^{B}	10.09 ± 0.73 ^{A}
4	8.06 ± 0.01 ^{C}	25.89 ± 1.05 ^{AB}	12.43 ± 0.50 ^{A}
5	8.99 ± 0.02 ^{B}	25.95 ± 0.18 ^{AB}	11.40 ± 0.45 ^{A}
MRS	9.43 ± 0.01 ^{A}	26.81 ± 4.86 ^{A}	11.47 ± 1.29 ^{A}

Note: Values are mean ± standard error of the mean (SEM), $n = 3$. Mean ± SEM within the same column that share similar superscript (A–D) are not significantly different ($p > 0.05$).

In terms of cell growth, the highest cell population of 8.99 log CFU/mL was detected in Medium 5, yet it was still significantly lower ($p < 0.05$) in comparison to the control (9.46 log CFU/mL). In contrast, the lowest cell growth was observed in Medium 2 (7.98 log CFU/mL) and Medium 3 (8 log CFU/mL), in which they were not significantly different ($p > 0.05$). Moreover, decreasing serine concentration was detected in all formulated media, implying that *P. acidilactici* TP-6 might be able to produce tryptophan via biosynthetic pathway by converting the precursor serine to tryptophan.

The results obtained in the validation test showed that Medium 1, 4 and 5 were potential medium for tryptophan production by *P. acidilactici* TP-6 since the highest tryptophan production was noted and there was no significant difference ($p > 0.05$) between the tryptophan yield. However, Medium 1 was excluded due to its multicomponent, leaving Medium 4 and 5 for the consideration as comparable growth medium. Both Medium 4 and 5 differed in their carbon source, whereby Medium 4 contained lactose and Medium 5 contained molasses as carbon source. Despite lactose exhibiting a positive effect, yet the high cost of lactose has rendered its preference as carbon source. Hence, Medium 5 was selected for further optimization, owing to its cost effectiveness as compared to Medium 4. Moreover, the use of molasses, which is an agricultural waste as sole carbon source offer additional advantage by upgrading the agricultural waste to produce value-added product of tryptophan.

2.2. Steepest Ascent Method

The vicinity of optimum concentration for each medium component in Medium 5 (Molasses, meat extract, urea and $FeSO_4$) were determined through a steepest ascent experiment consisting of 10 steps of ascension. The origin of each medium components in the steepest ascent experiment was fixed based on the high level (+1) of the PBD, whereas the direction and step length of each variable was determined based on the model of PBD (Equation (3)). According to the first-order model obtained in the PBD, meat extract, urea and $FeSO_4$ which exerted a positive effect were adjusted towards the direction of ascension since increasing the concentration of these variables would improve the production of tryptophan. In contrast, molasses which wielded a negative effect was adjusted towards the direction of dissension, as reducing the concentration of a negative variable would enhance the tryptophan production. Meanwhile, the largest coefficient which in this case, the meat extract was used as reference to compute the step length of urea and $FeSO_4$. Hence, for every 50% increment of the meat extract concentration, the concentration of urea and $FeSO_4$ were increased by 20% and 28.5% respectively. At the meantime, the concentration of molasses was reduced by 10% at each run.

The cell population, net tryptophan and serine (precursor of tryptophan) produced by *P. acidilactici* TP-6 in the steepest ascent experiment are displayed in Table 5. It was clearly evidenced that the net tryptophan produced was increasing progressively along the path of steepest ascent from the origin (27 mg/L) and achieved maximum tryptophan production in run 5 (69.05 mg/L), indicating that the net tryptophan produced was enhanced approximately 2.5 folds after the optimization by steepest ascent procedure. Nevertheless, the net tryptophan produced began to decline beyond run 5, implying that further increasing the concentration of meat extract, urea and $FeSO_4$ or reducing the concentration of molasses exerted inhibitory effect on the production of tryptophan by *P. acidilactici* TP-6. Furthermore, the net tryptophan produced in run 5 was significantly higher ($p < 0.05$) than control MRS medium, inferring that the formulated medium by steepest ascent method could be used as an alternative medium for the production of tryptophan by *P. acidilactici* TP-6.

Similar trend was observed for the cell population of *P. acidilactici* TP-6, whereby the cell growth increased from the origin (9.10 log CFU/mL) and the highest cell population was detected in run 5 (9.24 log CFU/mL). Thereafter, the cell growth remained unchanged up to run 7 (9.25 log CFU/mL) and the cell population decreased slowly as the concentration of medium components increased beyond run 7. Merely 8.77 log CFU/mL of cell population was noted at run 11. However, the cell population recorded in the control MRS medium (9.37 log CFU/mL) was still significantly higher ($p < 0.05$) than the cell population detected in different media formulations in the steepest ascent experiment. Interestingly, reduced serine content was detected in all the experimental runs, implying that *P. acidilactici* TP-6 might produce tryptophan via biosynthetic pathway, whereby serine precursor was converted to tryptophan, thereby resulted in a decreasing serine content. The concentrations of different medium components of run 5 (molasses, 15.04 g/L; meat extract, 24 g/L; urea, 5.4 g/L; $FeSO_4$, 0.022 g/L) were subsequently employed as the center point for further optimization by using CCD.

Table 5. Cell population, tryptophan production and serine consumption of *P. acidilactici* TP-6 for different media formulation in the steepest ascent experiment.

Run	Cell Population (log CFU/mL)	Tryptophan Production (mg/L)	Serine Consumption (mg/L)
1	9.10 ± 0.02 ^{D}	27.73 ± 0.04 ^{G}	8.72 ± 0.21 ^{D}
2	9.10 ± 0.01 ^{D}	35.79 ± 0.28 ^{F}	9.05 ± 0.35 ^{DE}
3	9.14 ± 0.01 ^{CD}	42.64 ± 0.40 ^{E}	9.24 ± 0.39 ^{DE}
4	9.15 ± 0.01 ^{C}	55.54 ± 0.36 ^{C}	8.93 ± 0.01 ^{DE}
5	9.24 ± 0.01 ^{B}	69.05 ± 0.55 ^{A}	9.53 ± 0.09 ^{DEF}
6	9.24 ± 0.01 ^{B}	59.84 ± 0.68 ^{B}	10.15 ± 0.12 ^{F}
7	9.25 ± 0.01 ^{B}	51.67 ± 0.92 ^{D}	9.65 ± 0.21 ^{EF}
8	9.18 ± 0.01 ^{C}	43.77 ± 0.23 ^{E}	9.00 ± 0.11 ^{DE}
9	9.16 ± 0.01 ^{C}	36.74 ± 0.91 ^{F}	7.56 ± 0.16 ^{C}
10	9.04 ± 0.01 ^{E}	21.98 ± 0.38 ^{H}	4.21 ± 0.10 ^{A}
11	8.77 ± 0.03 ^{F}	10.02 ± 0.07 ^{I}	5.11 ± 0.45 ^{B}
MRS	9.37 ± 0.01 ^{A}	27.69 ± 0.15 ^{G}	7.69 ± 0.37 ^{C}

Note: Values are mean ± standard error of the mean (SEM), $n = 3$. Mean ± SEM within the same column that share similar superscript (A–I) are not significantly different ($p > 0.05$).

2.3. *Central Composite Design*

The concentrations of molasses, meat extract, urea and $FeSO_4$ were further optimized by using CCD of Response Surface Methodology (RSM) after the steepest ascent procedure. The concentration of molasses, meat extract, urea and $FeSO_4$ were assigned to five levels—high level (+1), low level (−1), central point (0) and 2 axial points ($\pm\alpha$). Hence, a total of 30 experimental runs were suggested by the CCD and their corresponding experimental and predicted net tryptophan produced are shown in Table 6 respectively. In general, the highest net tryptophan produced was detected in runs 25–30, where all the variables were set at the center point and the respective net tryptophan produced was up to 70 mg/L, followed by run 24 (66.07 mg/L) which constituted of molasses, meat extract and urea that fixed at center point, while the $FeSO_4$ was supplemented at +2 level. The net tryptophan production recorded in runs 25–30 were significantly higher ($p < 0.05$) as compared to the other experiment runs, as well as the control MRS medium (28.18 mg/L).

Table 6. Central Composite Design (CCD) matrix with coded value and their corresponding experimental and predicted tryptophan production by *P. acidilactici* TP-6.

Std Run	A	B	C	D	Tryptophan Production (mg/L)	
					Experimental	Predicted *
1	−1	−1	−1	−1	56.24 ± 0.04 ^{JKL}	56.86
2	1	−1	−1	−1	54.33 ± 0.20 ^{N}	54.44
3	−1	1	−1	−1	53.23 ± 0.14 ^{O}	53.54
4	1	1	−1	−1	54.49 ± 0.26 ^{N}	54.72
5	−1	−1	1	−1	60.62 ± 0.17 ^{G}	60.38
6	1	−1	1	−1	57.12 ± 0.23 ^{IJK}	57.64
7	−1	1	1	−1	59.14 ± 0.32 ^{H}	59.38
8	1	1	1	−1	59.69 ± 0.26 ^{GH}	60.24
9	−1	−1	−1	1	63.25 ± 0.35 ^{E}	63.20
10	1	−1	−1	1	56.32 ± 0.44 ^{JKL}	56.78
11	−1	1	−1	1	59.25 ± 0.25 ^{H}	59.44
12	1	1	−1	1	55.89 ± 0.85 ^{LM}	56.62
13	−1	−1	1	1	64.47 ± 0.40 ^{D}	64.92
14	1	−1	1	1	58.01 ± 0.56 ^{I}	58.18
15	−1	1	1	1	63.13 ± 0.68 ^{EF}	63.48
16	1	1	1	1	60.25 ± 0.44 ^{GH}	60.34
17	−2	0	0	0	62.23 ± 0.29 ^{EF}	61.88
18	2	0	0	0	57.19 ± 0.33 ^{IJ}	56.32
19	0	−2	0	0	56.83 ± 0.12 ^{IJKL}	56.36
20	0	2	0	0	55.95 ± 0.13 ^{KLM}	55.20
21	0	0	−2	0	55.03 ± 0.22 ^{MN}	54.28
22	0	0	2	0	61.99 ± 0.42 ^{F}	61.52
23	0	0	0	−2	59.66 ± 0.27 ^{GH}	59.04
24	0	0	0	2	66.07 ± 0.05 ^{C}	65.48
25	0	0	0	0	69.33 ± 0.10 ^{AB}	69.42
26	0	0	0	0	69.55 ± 0.31 ^{AB}	69.42
27	0	0	0	0	70.22 ± 0.07 ^{A}	69.42
28	0	0	0	0	68.85 ± 0.39 ^{B}	69.42
29	0	0	0	0	69.69 ± 0.55 ^{AB}	69.42
30	0	0	0	0	68.88 ± 0.94 ^{B}	69.42
MRS	-	-	-	-	28.18 ± 0.12 ^{P}	-

Note: Values are mean ± standard error of mean (SEM), $n = 3$. Mean ± SEM within the same column that share similar superscript (A–P) are not significantly different ($p > 0.05$). * Predicted tryptophan production was calculated based on Equation (3).

The data were then analyzed with different regression models to investigate which model is best fitted to describe the relation between the variables and the production of tryptophan produced by *P. acidilactici* TP−6 as presented in Table 7. Based on the ANOVA table, it is clearly evidenced that the data were best fitted to a quadratic polynomial model. Among the 4 tested polynomial models, only the quadratic model was significant ($p < 0.05$), whereas the other polynomial models were insignificant ($p > 0.05$). Additionally, the quadratic model was highly predictive owing to its exceptionally high adjusted R^2 value (0.9837) and high predicted R^2 value (0.9586), which was not observed for other polynomial models. Furthermore, the p-value of the lack of fit test of the quadratic model (0.2196) implied that the lack of fit was not significant ($p > 0.05$) and the model can be used to explain and predict the response, which in this case the tryptophan production. This was evidenced by the good agreement between the predicted and experimental tryptophan production as presented in Table 6. Unlike the cubic polynomial model, the presence of aliased effects between the variables was not detected in the quadratic model. Hence, the production of tryptophan by *P. acidilactici* TP−6 can be best represented by the quadratic model. The following quadratic Equation (3) elucidated the effects of molasses (A), meat extract (B), urea (C) and $FeSO_4$ (D) on tryptophan production by *P. acidilactici* TP-6 (Y) in terms of coded symbols (A–D):

$$Y = 69.42 - 1.39A - 0.29B + 1.81C + 1.61D + 0.9AB - 0.084AC - 1.00AD + 0.58BC - 0.11BD - 0.45CD - 2.58A^2 - 3.41B^2 - 2.88C^2 - 1.79D^2 \quad (3)$$

Table 7. ANOVA of regression model for tryptophan production by *P. acidilactici* TP-6.

Source	Sequential *p*-Value	Lack of Fit *p*-Value	Adjusted R-Squared	Predicted R-Squared	
Linear	0.1492	<0.0001	0.1059	0.0576	
Crossproduct	0.9724	<0.0001	−0.1063	−0.3333	
Quadratic	<0.0001	0.2196	0.9837	0.9586	Suggested
Cubic	0.9875	0.0265	0.9709	0.2223	Aliased

The statistical significance of the quadratic model and the variables were evaluated by F-test and the ANOVA is presented in Table 8. The low p-value of the model (<0.01) implied that the model was highly significant ($p < 0.01$) and it was highly unlikely (>99% confidence) that the large F-value of the model this large was attributed to noise. Moreover, the model exhibited great predictive strength and was able to explain 99.2% of variation in response due to its high R^2 value (0.9916). Additionally, the "predicted R^2" (0. 9586) and the "adjusted R^2" (0.9837) values were in reasonable agreement (difference < 0.2), implying the great correlation between experimental and predicted values and the suggested model was significant. This was further supported by the insignificant lack of fit ($p > 0.05$), which was indicated by the high p-value of lack of fit test (0.22). Furthermore, the model was suitable for navigating the design space owing to its adequate signal to noise ratio which was reflected by the high adequate precision value (33.028) that was much greater than the threshold value of f4. On the other hand, the ANOVA results revealed that all the linear coefficients and quadratic coefficients affected tryptophan production significantly ($p < 0.01$) except the linear coefficient of meat extract (B). Furthermore, the interaction coefficient AB, AD, BC and BC were found to contribute significantly ($p < 0.05$) to the production of tryptophan by *P. acidilactici* TP–6.

Table 8. ANOVA for quadratic model of tryptophan production by *P. acidilactici* TP-6.

Source	Sum of Squares	df	Mean Square	F-Value	*p*-value Prob > F	
Model	814.96	14	58.21	126.07	< 0.01	significant
A	46.16	1	46.16	99.97	< 0.01	significant
B	2.08	1	2.08	4.51	0.05	
C	78.3	1	78.3	169.58	< 0.01	significant
D	61.84	1	61.84	133.92	< 0.01	significant
AB	12.86	1	12.86	27.84	< 0.01	significant
AC	0.11	1	0.11	0.24	0.63	
AD	16.07	1	16.07	34.8	< 0.01	significant
BC	5.36	1	5.36	11.62	<0.01	significant
BD	0.2	1	0.2	0.42	0.53	
CD	3.17	1	3.17	6.86	0.02	significant
A^2	182.37	1	182.37	394.96	< 0.01	significant
B^2	318.65	1	318.65	690.1	< 0.01	significant
C^2	227.15	1	227.15	491.94	< 0.01	significant
D^2	87.9	1	87.9	190.37	< 0.01	significant
Residual	6.93	15	0.46			
Lack of Fit	5.57	10	0.56	2.06	0.22	not significant
Pure Error	1.35	5	0.27			
Cor Total	821.88	29				

Note: R^2: 0.9916; Adj R^2: 0.9837; Pred R^2: 0.9586; Adeq Precision: 33.028.

The relationship between the coded variables and response was subsequently examined by constructing the three-dimensional surface plots as shown in Figures 3–8. The interaction between molasses and meat extract is depicted in Figure 3 by maintaining the concentration of urea and $FeSO_4$ at 5.4 g/L and 0.022 g/L, respectively, as the center point. Increased tryptophan production was noted with increased concentration of meat extract and molasses. The highest net tryptophan produced was detected when molasses and meat extract were both between the levels of −1 to +1. The synergistic effect of molasses and meat extract (AB) was highly significant as reflected by the p-value of less than 0.01 (Table 8). Figure 4 illustrates the response surface of tryptophan production with respect to molasses and urea by keeping the concentration of meat extract and $FeSO_4$ at the center point (24 g/L, 0.022 g/L). Increasing concentration of both molasses and urea resulted in higher net tryptophan produced. The highest net tryptophan produced was detected when molasses was in the range of −1 to 0 and urea was between 0 to +1. However, the tryptophan production was retarded when the concentration of molasses and urea was changed beyond the above-mentioned boundaries. The p-value of the interaction effect of AC (0.63) indicated that it was not significant (Table 8).

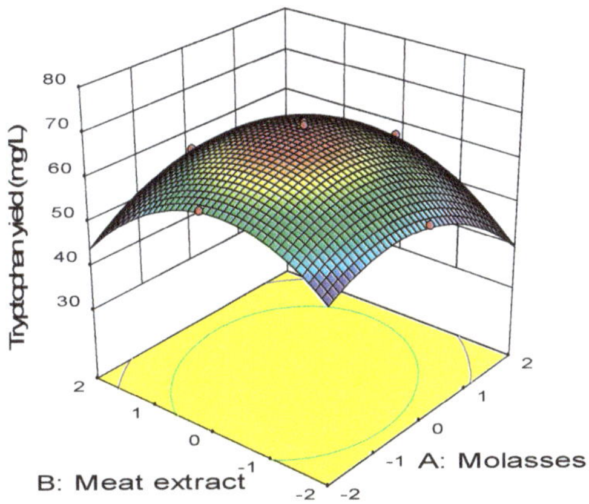

Figure 3. Response surface of tryptophan production by *P. acidilactici* TP-6 as a function of molasses and meat extract.

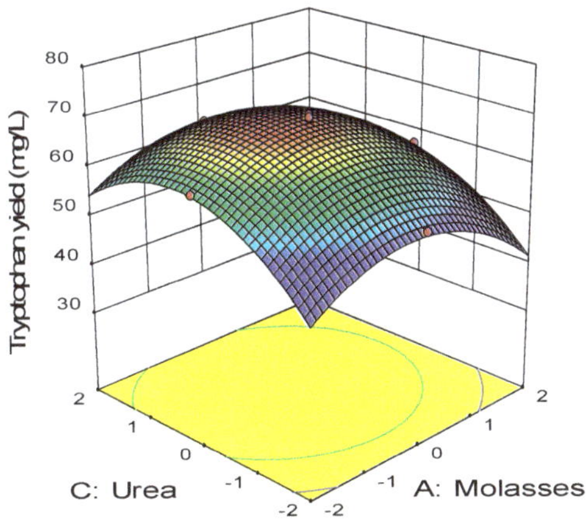

Figure 4. Response surface of tryptophan production by *P. acidilactici* TP-6 as a function of molasses and urea.

Figure 5 illustrates the combined effect between molasses and $FeSO_4$ while keeping the level of meat extract and urea at 24 g/L and 5.4 g/L, respectively, which were the center points. Similarly, increasing concentration of both molasses and $FeSO_4$ enhanced the net tryptophan produced. The highest net tryptophan production was detected when molasses was at the center point and $FeSO_4$ was between the center point (0) and high level (+1). ANOVA (Table 8) revealed that the interaction effect between molasses and $FeSO_4$ (AD) was highly significantly ($p < 0.01$) on the net tryptophan produced. Figure 6 shows the interaction between the meat extract and urea, in which the concentration of $FeSO_4$ and molasses were maintained at 0.022 g/L and 15.04 g/L, respectively, which were the center points. Increasing concentration of meat extract and urea improved the net tryptophan production by *P. acidilactici* TP-6. The highest tryptophan production was detected when both urea and meat extract were between low level (−1) and high level (+1). Increasing or decreasing the concentration of either meat extract or urea beyond these boundaries reduced the net tryptophan produced substantially. Based on Table 8, the interaction effect of meat extract and urea (BC) was highly significant ($p < 0.01$).

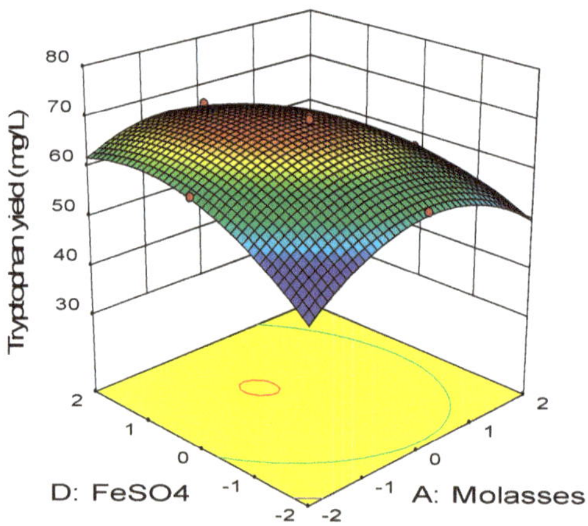

Figure 5. Response surface of tryptophan production by *P. acidilactici* TP-6 as a function of molasses and $FeSO_4$.

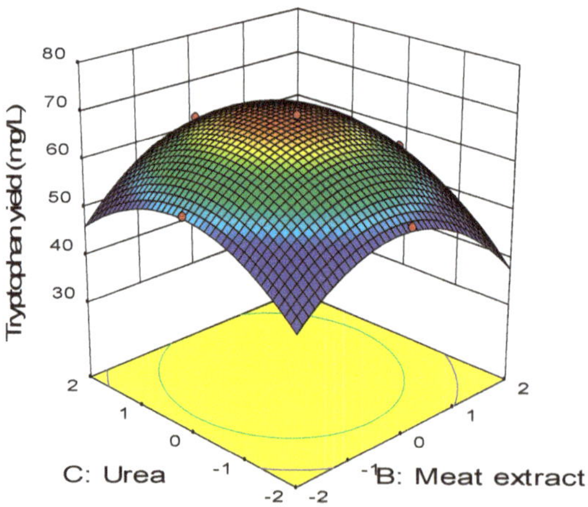

Figure 6. Response surface of tryptophan production by *P. acidilactici* TP-6 as a function of meat extract and urea.

Figure 7 depicts the three-dimensional surface plot of net tryptophan produced as a function of meat extract and $FeSO_4$ while keeping the concentration of molasses and urea at the center point (15.04 g/L, 5.4 g/L). Increasing concentration of both $FeSO_4$ and meat extract elevated the net tryptophan produced. The highest production was observed when meat extract was in the range of −1 to +1 while the $FeSO_4$ may vary between −1 to +2. The synergistic effect of meat extract and $FeSO_4$ (BD) was insignificant ($p > 0.05$) as reflected by the high p-value (0.53) in the ANOVA analysis (Table 8). On the other hand, the combined effect of urea and $FeSO_4$ is illustrated in Figure 8, in which the concentration of meat extract and molasses was kept constant at the center point (24 g/L, 15.04 g/L). Enhanced production of tryptophan was noted upon increment of the urea and $FeSO_4$ concentration. The highest net tryptophan production was observed when urea was supplemented between the range of 0 to +1 and $FeSO_4$ was in the range of −1 up to +2. The p-value of the interaction coefficient CD (0.02) revealed that it contributed significantly ($p < 0.05$) to the production of tryptophan by *P. acidilactici* TP-6.

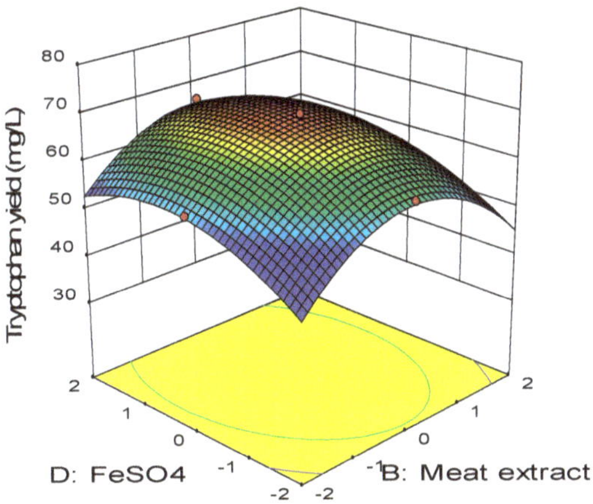

Figure 7. Response surface of tryptophan production by *P. acidilactici* TP-6 as a function of meat extract and $FeSO_4$.

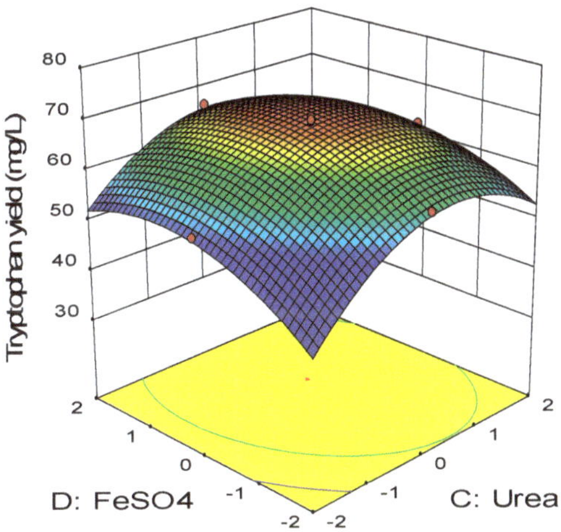

Figure 8. Response surface of tryptophan production by *P. acidilactici* TP-6 as a function of urea and $FeSO_4$.

By considering criteria such that the response is maximized and all the variables were in the range from −1 to +1, the optimum concentration of molasses (14.06 g/L), meat extract (23.68 g/L), urea (5.56 g/L) and $FeSO_4$ (0.024 g/L) were revealed with a predicted tryptophan production of 70.38 mg/L. Upon validation by cultivating *P. acidilactici* TP-6 in the optimized medium proposed by the model, up to 68.05 mg/L of tryptophan production was achieved by *P. acidilactici* TP-6 experimentally. Despite the experimental tryptophan production being slightly lower than the predicted value; however, they were not significantly different ($p > 0.05$). The tryptophan production recorded by *P. acidilactici* TP-6 in the optimized medium (68.05 mg/L) was enhanced by 150% in comparison to the control MRS medium (26.45 mg/L). Meanwhile, the tryptophan productivity by the producer strain in the optimized medium (3.40 mg/L/h) was improved by approximately 2.5 folds as compared to the control MRS medium (1.32 mg/L/h). In comparison with the production of tryptophan by *L. delbrueckii* subsp. *bulgaricus* (7.4 mg/L) in a study conducted by Simova et al. [71], *P. acidilactici* TP-6 produced a more than 9 times higher amount of tryptophan by using optimized medium as noted in the current study. Moreover, less than 2 mg/L of tryptophan production was reported for *Lactobacilli* [24], which was tremendously lower than the tryptophan production reported in the present study. Thus, the findings obtained in this study revealed that a rapid evaluation and effective optimization of medium composition governing tryptophan production by *P. acidilactici* TP-6 were feasible via statistical approaches, whereby the production of tryptophan demonstrated by *P. acidilactici* TP-6 in the optimized medium was much higher than those reported in several studies using a non-optimized medium, inferring the feasibility of utilizing *P. acidilactici* TP-6 as a food-grade alternative producer for the production of tryptophan. Additionally, the current findings revealed the potential of utilizing LAB as a safer alternative tryptophan producer, as well as providing basis for the exploitation of various amino acid productions by LAB in the future.

3. Materials and Methods

3.1. Inoculum Preparation

P. acidilactici TP-6 that was previously isolated from Malaysian fermented food, *Tempeh* [72], was employed as the producer strain for tryptophan production in this study. The strain was cultivated and preserved as described by Kareem et al. [73]. The inoculum preparation was performed as described by Lim et al. [60].

3.2. Experimental Design

PBD was first employed to elucidate the significance of medium components on the production of tryptophan by *P. acidilactici* TP-6, followed by validation of the effects of significant variables that were most crucial for tryptophan production by *P. acidilactici* TP-6. The range of optimum concentration for each significant varieble was subsequently estimated by using steepest ascent method, followed by the optimization of the concentrations of medium components for the production of tryptophan by *P. acidilactici* TP-6 via CCD of RSM.

3.3. Plackett-Burman Design

The nutritional requirement of *P. acidilactici* TP-6 for tryptophan production was evaluated by using PBD [74]. The design of experiment and statistical analysis of data were performed by using Design Expert statistical software version 9.0.6.2 (State-Ease Inc, Minneapolis, MN, USA). A number of 22 medium components which might affect amino acid production were selected based on published reports and control MRS medium composition and evaluated in the current study by assigning each variable at two distinct levels, namely low level (−1) and high level (+1) as presented in Table 9 [60].

Table 9. Coded and real values of variables selected in PBD for tryptophan production by *P. acidilactici* TP-6.

Variables	Symbol Code	Unit	Coded Values	
			−1	+1
Glucose	A	g/L	0	20
Sucrose	B	g/L	0	17.69
Fructose	C	g/L	0	19.08
Lactose	D	g/L	0	18.86
Molasses	E	g/L	0	25.08
Yeast extract	F	g/L	0	4
Peptone	G	g/L	0	10
Meat extract	H	g/L	0	8
K_2HPO_4	J	g/L	0	2
KH_2PO_4	K	g/L	0	2
Urea	L	g/L	0	3
NH_4NO_3	M	g/L	0	5
$(NH_4)_2SO_4$	N	g/L	0	5
$(NH_4)_2HC_6H_5O_7$	O	g/L	0	2
NaOAc	P	g/L	0	5
$MgSO_4$	Q	g/L	0	0.2
$MnSO_4$	R	g/L	0	0.04
Tween 80	S	mL/L	0	1
$FeSO_4$	T	g/L	0	0.01
$ZnSO_4$	U	g/L	0	0.01
$CuSO_4$	V	g/L	0	0.01
Biotin	W	g/L	0	0.06

Table 10 presents matrixes of the PBD constituting of 24 experimental runs as suggested by the software.

Table 10. PBD matrix for 22 variables with coded values for tryptophan production by *P. acidilactici* TP-6.

Std Run	A	B	C	D	E	F	G	H	J	K	L	M	N	O	P	Q	R	S	T	U	V	W	X
1	1	1	1	1	1	−1	1	−1	1	1	−1	−1	1	1	−1	−1	1	−1	1	−1	−1	−1	−1
2	−1	1	1	1	1	1	−1	1	−1	1	1	−1	−1	1	1	−1	−1	1	−1	1	−1	−1	−1
3	−1	−1	1	1	1	1	1	−1	1	−1	1	1	−1	−1	1	1	−1	−1	1	−1	1	−1	−1
4	−1	−1	−1	1	1	1	1	1	−1	1	−1	1	1	−1	−1	1	1	−1	−1	1	−1	1	−1
5	−1	−1	−1	−1	1	1	1	1	1	−1	1	−1	1	1	−1	−1	1	1	−1	−1	1	−1	1
6	1	−1	−1	−1	−1	1	1	1	1	1	−1	1	−1	1	1	−1	−1	1	1	−1	−1	1	−1
7	−1	1	−1	−1	−1	−1	1	1	1	1	1	−1	1	−1	1	1	−1	−1	1	1	−1	−1	1
8	1	−1	1	−1	−1	−1	−1	1	1	1	1	1	−1	1	−1	1	1	−1	−1	1	1	−1	−1
9	−1	1	−1	1	−1	−1	−1	−1	1	1	1	1	1	−1	1	−1	1	1	−1	−1	1	1	−1
10	−1	−1	1	−1	1	−1	−1	−1	−1	1	1	1	1	1	−1	1	−1	1	1	−1	−1	1	1
11	1	−1	−1	1	−1	1	−1	−1	−1	−1	1	1	1	1	1	−1	1	−1	1	1	−1	−1	1
12	1	1	−1	−1	1	−1	1	−1	−1	−1	−1	1	1	1	1	1	−1	1	−1	1	1	−1	−1
13	−1	1	1	−1	−1	1	−1	1	−1	−1	−1	−1	1	1	1	1	1	−1	1	−1	1	1	−1
14	−1	−1	1	1	−1	−1	1	−1	1	−1	−1	−1	−1	1	1	1	1	1	−1	1	−1	1	1
15	1	−1	−1	1	1	−1	−1	1	−1	1	−1	−1	−1	−1	1	1	1	1	1	−1	1	−1	1
16	1	1	−1	−1	1	1	−1	−1	1	−1	1	−1	−1	−1	−1	1	1	1	1	1	−1	1	−1
17	−1	1	1	−1	−1	1	1	−1	−1	1	−1	1	−1	−1	−1	−1	1	1	1	1	1	−1	1
18	1	−1	1	1	−1	−1	1	1	−1	−1	1	−1	1	−1	−1	−1	−1	1	1	1	1	1	−1
19	−1	1	−1	1	1	−1	−1	1	1	−1	−1	1	−1	1	−1	−1	−1	−1	1	1	1	1	1
20	1	−1	1	−1	1	1	−1	−1	1	1	−1	−1	1	−1	1	−1	−1	−1	−1	1	1	1	1
21	1	1	−1	1	−1	1	1	−1	−1	1	1	−1	−1	1	−1	1	−1	−1	−1	−1	1	1	1
22	1	1	1	−1	1	−1	1	1	−1	−1	1	1	−1	−1	1	−1	1	−1	−1	−1	−1	1	1
23	1	1	1	1	−1	1	−1	1	1	−1	−1	1	1	−1	−1	1	−1	1	−1	−1	−1	−1	1
24	−1	−1	−1	−1	−1	−1	−1	−1	−1	−1	−1	−1	−1	−1	−1	−1	−1	−1	−1	−1	−1	−1	−1

The following first-order model was used to express the response of the PBD:

$$Y = \beta_0 + \sum_{i=1}^{22} \beta_{i} X_i, \quad (4)$$

where:
Y = Response variable
β_0 = Interception coefficient
β_i = Coefficients of linear effects of the independent variables ($X_1 - X_{22}$)

The effects of the significant variables identified in the PBD was subsequently validated by formulating five different media as shown in Table 11.

Formulation 1 constituted of all the significant variables; Formulation 2 was made up of only the significant variables with positive effect; Formulation 3 was comprised all the variables with stimulatory effects irrespective of their significance level; Formulation 4 contained 4 main components, which represent the carbon source, organic nitrogen source, inorganic nitrogen source and mineral source with the highest positive effect; Formulation 5 constituted of the similar medium composition as Formulation 4 except lactose was replaced with molasses, which exhibited a strong stimulatory effect on the growth and the cost was the lowest among the five carbon sources.

Table 11. Media formulation to validate the effects of significant variables on tryptophan production by *P. acidilactici* TP−6.

Media Formulation	Medium Composition, g/L
Medium 1	
Glucose	20
Sucrose	17.69
Lactose	18.86
Molasses	25.08
Yeast extract	4
Peptone	10
Meat extract	8
K_2HPO_4	2
Urea	3
$(NH_4)_2SO_4$	5
$(NH_4)_2HC_6H_5O_7$	2
NaOAc	5
$MgSO_4$	0.2
$MnSO_4$	0.04
$FeSO_4$	0.01
$CuSO_4$	0.01
Medium 2	
Meat extract	8
$FeSO_4$	0.01
Lactose	18.86
$MnSO_4$	0.04
Urea	3
$CuSO_4$	0.01
K_2HPO_4	2
$MgSO_4$	0.2
Medium 3	
Meat extract	8
$FeSO_4$	0.01
Lactose	18.86
$MnSO_4$	0.04
Urea	3
$CuSO_4$	0.01
K_2HPO_4	2
$MgSO_4$	0.2
Biotin	0.06
Tween 80	1

Table 11. *Cont.*

Media Formulation	Medium Composition, g/L
Medium 4	
Lactose	18.86
Meat extract	8
Urea	3
$FeSO_4$	0.01
Medium 5	
Molasses	25.08
Meat extract	8
Urea	3
$FeSO_4$	0.01

3.4. Steepest Ascent Method

The neighborhood of the optimum concentrations of each significant variable (molasses, meat extract, urea and $FeSO_4$) was subsequently estimated by using the steepest ascent method. The direction of ascent or descent of each variable was determined by using the first-order model from PBD as a guideline, where variables bearing a positive sign was moved along the path of Steepest Ascent and vice versa. Meanwhile, the coefficient with the highest value, in this case meat extract was used as a benchmark to compute the step length of urea and $FeSO_4$. Table 12 shows the Steepest Ascent design consisting of 10 steps that governed the production of tryptophan by *P. acidilactici* TP-6. The concentration of urea and $FeSO_4$ was increased by 20% and 28.5% respectively for every 50% increment of the meat extract concentration. Meanwhile, the molasses concentration was reduced by 10% at each run.

Table 12. Steepest Ascent design for tryptophan production by *P. acidilactici* TP-6.

No.	Run	Variable level, g/L			
		Molasses (A)	Meat Extract (B)	Urea (C)	$FeSO_4$ (D)
	Δ	−2.51	4	0.6	0.003
1	Origin	25.08	8	3.0	0.010
2	Origin + Δ	22.57	12	3.6	0.013
3	Origin + 2Δ	20.06	16	4.2	0.016
4	Origin + 3Δ	17.55	20	4.8	0.019
5	Origin + 4Δ	15.04	24	5.4	0.022
6	Origin + 5Δ	12.53	28	6.0	0.025
7	Origin + 6Δ	10.02	32	6.6	0.028
8	Origin + 7Δ	7.51	36	7.2	0.031
9	Origin + 8Δ	5.00	40	7.8	0.034
10	Origin + 9Δ	2.49	44	8.4	0.037
11	Origin + 10Δ	0	48	9.0	0.040

3.5. Central Composite Design

Subsequently, the CCD was employed for the determination of optimum concentration of molasses, meat extract, urea and $FeSO_4$ that required for tryptophan production by *P. acidilactici* TP-6. Design Expert statistical software version 9.0.6.2 (State-Ease Inc, Minneapolis, MN, USA) was used for designing of experiment and statistical analysis. Table 13 presents the details of each medium component used in the CCD, where each variable was assigned to five distinct levels ($-\alpha$, −1, 0, +1, $+\alpha$). The axial distance selected in this study was two, such that the design was rotatable.

Table 14 shows a total of 30 experimental runs suggested by CCD software, comprising of 16 factorial points, 8 axial points and 6 central points. The following second-order model could be used to express the relationship of the response with each variable:

$$Y = \beta_0 + \sum \beta_j X_j + \sum \beta_{j2} X_{j2} + \sum \beta_{jk} X_j X_k \quad (5)$$

where:
Y = Response variable
β_0 = Interception coefficient
β_j = Linear coefficients
β_j^2 = Quadratic coefficients
β_{jk} = Interactive coefficients

Table 13. Coded and real values of variables selected for CCD of tryptophan production by *P. acidilactici* TP-6.

Variables	Coded Symbol	Coded Values				
		−α	−1	0	+1	+α
Molasses	A	10.02	12.53	15.04	17.55	20.06
Meat extract	B	16	20	24	28	32
Urea	C	4.2	4.8	5.4	6	6.6
$FeSO_4$	D	0.016	0.019	0.022	0.025	0.028

Table 14. CCD matrix for four variables with coded values for tryptophan production by *P. acidilactici* TP-6.

Std Run	A	B	C	D
1	−1	−1	−1	−1
2	1	−1	−1	−1
3	−1	1	−1	−1
4	1	1	−1	−1
5	−1	−1	1	−1
6	1	−1	1	−1
7	−1	1	1	−1
8	1	1	1	−1
9	−1	−1	−1	1
10	1	−1	−1	1
11	−1	1	−1	1
12	1	1	−1	1
13	−1	−1	1	1
14	1	−1	1	1
15	−1	1	1	1
16	1	1	1	1
17	−2	0	0	0
18	2	0	0	0
19	0	−2	0	0
20	0	2	0	0
21	0	0	−2	0
22	0	0	2	0
23	0	0	0	−2
24	0	0	0	2
25	0	0	0	0
26	0	0	0	0
27	0	0	0	0
28	0	0	0	0
29	0	0	0	0
30	0	0	0	0

3.6. Production of Tryptophan

Cultivation of the bacterial strain for tryptophan production was performed by inoculating 10% (*v*/*v*) of inoculum into the media, followed by incubation for 20 h at 30 °C [25].

3.7. Analytical Methods

The tryptophan content of the cultured broth was determined after separation of biomass by centrifugation for 10 min at 10,000× *g*, 4 °C. Meanwhile, the cell growth was determined by using the biomass. The determination of cell population and tryptophan content was conducted as described by Lim et al. [60].

4. Conclusions

Seventeen of the 22 studied variables were found to exhibit significant effects ($p < 0.05$) on the production of tryptophan by *P. acidilactici* TP-6, while the remaining 5 variables including fructose, KH_2PO_4, Tween 80, $ZnSO_4$ and biotin did not affect the net tryptophan produced significantly ($p > 0.05$). On the other hand, s16 variables contributed significantly ($p < 0.05$) to the growth of *P. acidilactici* TP-6, except for fructose, lactose, KH_2PO_4, urea, NH_4NO_3 and $FeSO_4$, which had no significant effect ($p > 0.05$) on the cell growth of *P. acidilactici* TP-6. A medium constituting of molasses, meat extract, urea and $FeSO_4$ was proven to be the best medium for the production of tryptophan by *P. acidilactici* TP-6, where it permitted the highest amount of tryptophan production (25.95 mg/L) with the additional advantage of cost competitiveness. Furthermore, the tryptophan production was comparable to the control MRS medium with no significant difference ($p > 0.05$). Hence, molasses, meat extract, urea and $FeSO_4$ were subsequently selected for further optimization. The application of the Steepest Ascent procedure has successfully improved the net tryptophan produced by approximately 2.5 folds, from 27 mg/L at run 1 to 69.05 mg/L at run 5. Subsequently, the optimum concentration of each medium components was determined by using CCD and it was suggested that the highest predicted tryptophan production (70.38 mg/L) could be achieved by using the combination of molasses (14.06 g/L), meat extract (23.68 g/L), urea (5.56 g/L) and $FeSO_4$ (0.024 g/L). An amount of 68.05 mg/L of tryptophan was produced by *P. acidilactici* TP-6 upon validation by cultivating the producer strain in the optimized medium. There was no significant difference ($p > 0.05$) between the tryptophan production predicted by the model with the experimental tryptophan production. Up to 150% enhancement of tryptophan production by *P. acidilactici* TP-6 was achieved by using the optimized medium, whereas the cost of the fermentation medium was reduced by 11% as compared to the control MRS medium.

Author Contributions: Conceptualization, H.L.F. and T.C.L.; Data curation, R.M. and H.L.F.; Formal analysis, Y.H.L. and H.L.F.; Funding acquisition, H.L.F., T.C.L. and R.A.R.; Investigation, Y.H.L. and H.L.F.; Methodology, Y.H.L., R.M. and H.L.F.; Project administration, H.L.F. and T.C.L.; Resources, H.L.F. and T.C.L.; Supervision, H.L.F. and R.M.; Validation, H.L.F., R.M. and R.A.R.; Writing-original draft, Y.H.L. and H.L.F.; Writing-review & editing, Y.H.L., H.L.F. and T.C.L. All authors have read and agreed to the published version of the manuscript.

Funding: This research was funded by Ministry of Education Malaysia, grant number UPM/700-1/3/LRGS.

Acknowledgments: The authors would like to thank Ministry of Education of Malaysia for funding the research grant under Long-Term Research Grant Scheme (LRGS).

Conflicts of Interest: The authors declare no conflict of interest.

Abbreviations

LAB	Lactic acid bacteria
RSM	Response surface methodology
PBD	Plackett-Burman Design
CCD	Central Composite Design
PLP	Pyridoxal phosphate
MRS	deMan, Rogosa and Sharpe
SEM	Standard error of mean
HPLC	High performance liquid chromatography
OPA	o-phthalaldehyde
FMOC	9-fluorenylmethyl chloroformate

References

1. Panda, B.P.; Ali, M.; Javed, S. Fermentation process optimization. *Res. J. Microbiol.* **2007**, *2*, 201–208.
2. Weuster-Botz, D. Experimental design for fermentation media development: Statistical design or global random search. *J. Biosci. Bioeng.* **2000**, *90*, 473–483. [CrossRef]
3. Elibol, M. Optimization of medium composition for actinorhodin production by *Streptomyces coelicolor* A3 with response surface methodology. *Process. Biochem.* **2004**, *39*, 1057–1062. [CrossRef]
4. Djekrif-Dakhmouche, S.; Gheribi-Aoulmi, Z.; Meraihi, Z.; Bennamoun, L. Application of a statistical design to the optimization of culture medium for α-amylase production by *Aspergillus niger* ATCC 16404 grown on orange waste powder. *J. Food. Eng.* **2006**, *73*, 190–197. [CrossRef]
5. Chen, H.; Xu, X.Q.; Zhu, Y. Optimization of hydroxyl radical scavenging activity of exo-polysaccharides from *Inonotus obliquus* in submerged fermentation using response surface methodology. *J. Microbiol. Biotechnol.* **2010**, *20*, 835–843.
6. Vohra, A.; Satyanarayana, T. Statistical optimization of the medium components by response surface methodology to enhance phytase production by *Pichia anomala*. *Process. Biochem.* **2002**, *37*, 999–1004. [CrossRef]
7. Montserrat, S.; Iñaki, R.; François, O.; Francesc, G.; Carles, C. Application of factorial design to the optimization of medium composition in batch cultures of *Streptomyces lividans* TK21 producing a hybrid antibiotic. *Biotechnol. Lett.* **1993**, *15*, 559–564. [CrossRef]
8. Faghfuri, E.; Fooladi, J.; Moosavi-Nejad, S.Z. L-tryptophan production by whole cells of *Escherichia coli* based on Iranian sugar beet molasses. *Jundishapur J. Microbiol.* **2013**, *6*, 1–5. [CrossRef]
9. Cheng, L.K.; Wang, J.; Xu, Q.Y.; Xie, X.X.; Zhang, Y.J.; Zhao, C.G.; Chen, N. Effect of feeding strategy on L-tryptophan production by recombinant *Escherichia coli*. *Ann. Microbiol.* **2012**, *62*, 1625–1634. [CrossRef]
10. Hagino, H.; Nakayama, K. L-tryptophan production by analog-resistant mutants derived from a phenylalanine and tyrosine double auxotroph of *Corynebacterium glutamicum*. *Agric. Biol. Chem.* **1975**, *39*, 343–349.
11. Toe, C.J.; Foo, H.L.; Loh, T.C.; Rosfarizan, M.; Raha, A.R.; Zulkifli, I. Extracellular proteolytic activity and amino acid production by lactic acid bacteria Isolated from Malaysian foods. *Int. J. Mol. Sci.* **2019**, *20*, 1777. [CrossRef] [PubMed]
12. Norfarina, M.N.; Mohd Shamzi, M.; Loh, T.C.; Foo, H.L.; Raha, A.R.; Tan, J.S.; Rosfarizan, M. Comparative analyses on medium optimization using one-factor-at-a-time, response surface methodology and artificial neural network for lysine–methionine biosynthesis by *Pediococcus pentosaceus* RF-1. *Biotechnol. Biotechnol. Equip.* **2017**, *31*, 935–947.
13. Izuddin, W.I.; Loh, T.C.; Foo, H.L.; Samsudin, A.A.; Humam, A.M. Postbiotic, *L. Plantarum* RG14 improves ruminal epithelium growth, immune status and upregulates the intestinal barrier function in post-weaning lambs. *Sci. Rep.* **2019**, *9*, 9938. [CrossRef] [PubMed]
14. Abdulla, N.R.; Mohd Zamri, A.N.; Sabow, A.B.; Kareem, K.Y.; Nurhazirah, S.; Foo, H.L.; Awis, Q.S.; Loh, T.C. Physico-Chemical properties of breast muscle in broiler chickens fed probiotics, antibiotics or antibiotic–probiotic mix. *J. Appl. Anim. Res.* **2017**, *45*, 64–70. [CrossRef]
15. Kareem, K.Y.; Loh, T.C.; Foo, H.L.; Asmara, S.A.; Akit, H. Influence of postbiotic RG14 and inulin combination on cecal microbiota, organic acid concentration and cytokine expression in broiler chickens. *Poult. Sci.* **2016**, *96*, 966–975. [CrossRef]
16. Kareem, K.Y.; Loh, T.C.; Foo, H.L.; Asmara, S.A.; Akit, H.; Abdulla, N.R.; Ooi, M.F. Carcass, meat and bone quality of broiler chickens fed with postbiotic and prebiotic combinations. *Int. J. Probiotzics Prebiotics* **2015**, *10*, 23.
17. Leuchtenberger, W.; Huthmacher, K.; Drauz, K. Biotechnological production of amino acids and derivatives: Current status and prospects. *Appl. Microbiol. Biot.* **2005**, *69*, 1–8. [CrossRef]
18. Rosebrough, R.W. Crude protein and supplemental dietary tryptophan effects on growth and tissue neurotransmitter levels in the broiler chicken. *Br. J. Nutr.* **1996**, *76*, 87–96. [CrossRef]
19. Moshirfar, A.; Kamara, K.; Castonguay, T.W. Intragastrically administered tryptophan blocks gluconeogenesis in 48-hr starved rats. *J. Nutr. Biochem.* **1996**, *7*, 567–570. [CrossRef]
20. Iwuji, T.C.; Akinmutimi, A.H.; Ogbuewu, I.P.; Etuk, I.F.; Odoemelam, V.U. Roles of tryptophan in monogastric nutrition: A review. *Adv. Agric. Sci. Eng. Res.* **2014**, *4*, 1544–1556.

21. Duarte, K.F.; Junqueira, O.M.; Filardi, R.D.S.; Siqueira, J.C.D.; Puzotti, M.M.; Garcia, E.A.; Molino, A.D.B.; Laurentiz, A.C.D. Digestible tryptophan requirements for broilers from 22 to 42 days old. *R. Bras. Zootec.* **2013**, *42*, 728–733. [CrossRef]
22. Mateus, D.M.R.; Alves, S.S.; Da Fonseca, M.M.R. Kinetics of L-tryptophan production from indole and L-serine catalyzed by whole cells with tryptophanase activity. *J. Biosci. Bioeng.* **2004**, *97*, 289–293. [CrossRef]
23. Turner, E.H.; Loftis, J.M.; Blackwell, A.D. Serotonin a la carte: Supplementation with the serotonin precursor 5-hydroxytryptophan. *Pharmacol. Ther.* **2006**, *109*, 325–338. [CrossRef] [PubMed]
24. Tarek, M.; Hesham, H.M. Screening of potential infants' *Lactobacilli* isolates for amino acids production. *Afr. J. Microbiol. Res.* **2010**, *4*, 226–232.
25. Lim, Y.H.; Foo, H.L.; Loh, T.C.; Rosfarizan, M.; Norhani, A. Comparative studies of versatile extracellular proteolytic activities of lactic acid bacteria and their potential for extracellular amino acid productions as feed supplements. *J. Anim. Sci. Biotechnol.* **2019**, *10*, 1–13. [CrossRef]
26. Miller, J.N. Experimental design and optimisation (4): Plackett–Burman designs. *Anal. Methods* **2013**, *5*, 1901–1903.
27. Li, Y.; Liu, Z.; Zhao, H.; Xu, Y.; Cui, F. Statistical optimization of xylanase production from new isolated *Penicillium oxalicum* ZH-30 in submerged fermentation. *Biochem. Eng. J.* **2007**, *34*, 82–86. [CrossRef]
28. Pirie, P.; Naeimpoor, F.; Hejazi, P. A microcosm study on P-Nitrophenol biodegradation in soil slurry by *Alcaligenes faecalis*: Plackett-Burman design. *Iran. J. Chem. Eng.* **2011**, *8*, 57–68.
29. Voet, D.; Voet, J.G. *Biochemistry*, 3rd ed.; John Wiley and Sons, Inc.: Hoboken, NJ, USA, 2004.
30. Tavakkoli, M.; Hamidi-Esfahani, Z.; Azizi, M.H. Optimization of *Corynebacterium glutamicum* glutamic acid production by response surface methodology. *Food Bioproc. Tech.* **2012**, *5*, 92–99. [CrossRef]
31. Li, J.; Ma, C.; Ma, Y.; Li, Y.; Zhou, W.; Xu, P. Medium optimization by combination of response surface methodology and desirability function: An application in glutamine production. *Appl. Microbiol. Biotechnol.* **2007**, *74*, 563–571. [CrossRef] [PubMed]
32. Kiefer, P.; Heinzle, E.; Wittmann, C. Influence of glucose, fructose and sucrose as carbon sources on kinetics and stoichiometry of lysine production by *Corynebacterium glutamicum*. *J. Ind. Microbiol. Biotechnol.* **2002**, *28*, 338–343. [CrossRef] [PubMed]
33. Lee, M.H.; Lee, H.W.; Park, J.H.; Ahn, J.O.; Jung, J.K.; Hwang, Y.I. Improved L-threonine production of *Escherichia coli* mutant by optimization of culture conditions. *J. Biosci. Bioeng.* **2006**, *101*, 127–130. [CrossRef] [PubMed]
34. Okamoto, K.; Ikeda, M. Development of an industrially stable process for L-threonine fermentation by an L-methionine-auxotrophic mutant of *Escherichia coli*. *J. Biosci. Bioeng.* **2000**, *89*, 87–89. [CrossRef]
35. Chen, N.; Huang, J.; Feng, Z.B.; Yu, L.; Xu, Q.Y.; Wen, T.Y. Optimization of fermentation conditions for the biosynthesis of L-threonine by *Escherichia coli*. *Appl. Biochem. Biotechnol.* **2009**, *158*, 595–604. [CrossRef]
36. Wang, J.; Cheng, L.K.; Chen, N. High-level production of L-threonine by recombinant *Escherichia coli* with combined feeding strategies. *Biotechnol. Biotechnol. Equip.* **2014**, *28*, 495–501. [CrossRef]
37. Roy, D.K.; Chatterjee, S.P. Production of glutamic acid by *Arthrobacter globiformis*: Influence of cultural conditions. *Folia Microbiol.* **1989**, *34*, 11–24. [CrossRef]
38. Coruzzi, G.; Last, R.; Dudareva, N.; Amrhein, N. Amino Acids. In *Biochemistry and Molecular Biology of Plants*; Buchanan, B.B., Gruissem, W., Jones, R.L., Eds.; John Wiley and Sons, Inc.: Hoboken, NJ, USA, 2015; pp. 289–336.
39. Alexander, J.C.; Beckneh, C.; Elvehjem, C.A. The alanine, cystine, glycine and serine content of meat. *J. Nutr.* **1953**, *51*, 319–328. [CrossRef]
40. Powers, H.J. Riboflavin (vitamin B-2) and health. *Am. J. Clin. Nutr.* **2003**, *77*, 1352–1360. [CrossRef]
41. Osborne, A.; Thorneley, R.N.; Abell, C.; Bornemann, S. Studies with substrate and cofactor analogues provide evidence for a radical mechanism in the chorismate synthase reaction. *J. Biol. Chem.* **2000**, *275*, 35825–35830. [CrossRef]
42. Lin, X.; Xu, S.; Yang, Y.; Wu, J.; Wang, H.; Shen, H.; Wang, H. Purification and characterization of anthranilate synthase component I (TrpE) from *Mycobacterium tuberculosis* H37Rv. *Protein Expr. Purif.* **2009**, *64*, 8–15. [CrossRef]
43. Davati, N.; Hamidi Esfahani, Z.; Shojaosadati, S.A. Optimization of medium composition for microbial production of glutamic acid from Date fruit wastes using fractional factorial method. *Iran. J. Food Sci. Technol.* **2010**, *7*, 61–67.

44. Nampoothiri, K.M.; Pandey, A. Solid state fermentation for L-glutamic acid production using *Brevibacterium* sp. *Biotechnol. Lett.* **1996**, *18*, 199–204. [CrossRef]
45. Zareian, M.; Ebrahimpour, A.; Bakar, F.A.; Mohamed, A.K.S.; Forghani, B.; Ab-Kadir, M.S.B.; Saari, N. A glutamic acid-producing lactic acid bacteria isolated from Malaysian fermented foods. *Int. J. Mol. Sci.* **2012**, *13*, 5482–5497. [CrossRef]
46. Ikeda, M.; Katsumata, R. Metabolic engineering to produce tyrosine or phenylalanine in a tryptophan-producing *Corynebacterium glutamicum* strain. *Appl. Environ. Microbiol.* **1992**, *58*, 781–785. [CrossRef]
47. Rodwell, V.; Bender, D.; Botham, K.M.; Kennelly, P.J.; Weil, P.A. *Harper's Illustrated Biochemistry*, 30th ed.; McGraw-Hill Education: New York, NY, USA, 2015.
48. Zalkin, H.; Kling, D. Anthranilate synthetase. Purification and properties of component I from *Salmonella typhimurium*. *Biochemistry* **1968**, *7*, 3566–3573. [CrossRef] [PubMed]
49. Hertel, S.C.; Hieke, M.; Gröger, D. Anthranilate synthase from *Ruta graveolens*: Partial purification and properties. *Biochem. Physiol. Pflanz.* **1991**, *187*, 121–129. [CrossRef]
50. Widholm, J. Anthranilate synthetase from 5-methyltryptophan-susceptible and-resistant cultured Daucus carota cells. *Biochim. Biophys. Acta.* **1972**, *279*, 48–57. [CrossRef]
51. Momose, H.; Takagi, T. Glutamic acid production in biotin-rich media by temperature-sensitive mutants of *Brevibacterium lactofermentum*, a novel fermentation process. *Agric. Biol. Chem.* **1978**, *42*, 1911–1917. [CrossRef]
52. Todorov, S.D.; Dicks, L.M. Bacteriocin production by *Lactobacillus pentosus* ST712BZ isolated from boza. *Braz. J. Microbiol.* **2007**, *38*, 166–172. [CrossRef]
53. Saraniya, A.; Jeevaratnam, K. Optimization of nutritional and non-nutritional factors involved for production of antimicrobial compounds from *Lactobacillus pentosus* SJ65 using response surface methodology. *Braz. J. Microbiol.* **2014**, *45*, 81–88. [CrossRef] [PubMed]
54. Saeed, A.H.; Salam, A.I. Current limitations and challenges with lactic acid bacteria: A review. *Food Nutr. Sci.* **2013**, *4*, 73–87.
55. de Carvalho, A.A.T.; Mantovani, H.C.; Paiva, A.D.; De Melo, M.R. The effect of carbon and nitrogen sources on bovicin HC5 production by *Streptococcus bovis* HC5. *J. Appl. Microbiol.* **2009**, *107*, 339–347. [CrossRef] [PubMed]
56. Oh, H.; Wee, Y.J.; Yun, J.S.; Han, S.H.; Jung, S.; Ryu, H.W. Lactic acid production from agricultural resources as cheap raw materials. *Bioresour. Technol.* **2005**, *96*, 1492–1498. [CrossRef] [PubMed]
57. Mohamed, I.A.; Loh, T.C.; Foo, H.L.; Lau, W.H.; Awis, Q.S. Biodegradation of palm kernel cake by cellulolytic and hemicellulolytic bacterial cultures through solid state fermentation. *Sci. World J.* **2014**, *2014*, 1–8. [CrossRef] [PubMed]
58. Rodrigues, L.; Teixeira, J.; Oliveira, R.; Van Der Mei, H.C. Response surface optimization of the medium components for the production of biosurfactants by probiotic bacteria. *Process. Biochem.* **2006**, *41*, 1–10. [CrossRef]
59. Ooi, M.F.; Nurzafirah, M.; Foo, H.L.; Loh, T.C.; Rosfarizan, M.; Raha, A.R.; Arbakariya, A. Effects of carbon and nitrogen sources on bacteriocin-inhibitory activity of postbiotic metabolites produced by *Lactobacillus plantarum* I-UL4. *Malays. J. Microbiol.* **2015**, *11*, 176–184.
60. Lim, Y.H.; Foo, H.L.; Loh, T.C.; Rosfarizan, M.; Raha, A.R.; Zulkifli, I. Optimized medium via statistical approach enhanced threonine production by *Pediococcus pentosaceus* TL-3 isolated from Malaysian food. *Microb. Cell Fact.* **2019**, *18*, 1–19. [CrossRef]
61. Hwang, C.F.; Chang, J.H.; Houng, J.Y.; Tsai, C.C.; Lin, C.K.; Tsen, H.Y. Optimization of medium composition for improving biomass production of *Lactobacillus plantarum* Pi06 using the Taguchi array design and the Box-Behnken method. *Biotechnol. Bioprocess. Eng.* **2012**, *17*, 827–834. [CrossRef]
62. Hutkins, R.W. *Microbiology and Technology of Fermented Foods*; Blackwell Publishing: Ames, IA, USA, 2008; Volume 22.
63. de Carvalho, I.P.C.; Detmann, E.; Mantovani, H.C.; Paulino, M.F.; Valadares Filho, S.D.C.; Costa, V.A.C.; Gomes, D.I. Growth and antimicrobial activity of lactic acid bacteria from rumen fluid according to energy or nitrogen source. *Rev. Bras. Zootec.* **2011**, *40*, 1260–1265. [CrossRef]
64. Thu, T.V.; Foo, H.L.; Loh, T.C.; Bejo, M.H. Inhibitory activity and organic acid concentrations of metabolite combinations produced by various strains of *Lactobacillus plantarum*. *Afr. J. Biotechnol.* **2011**, *10*, 1359–1363.

65. Tomas, J.M.S.; Bru, E.; Nader-Macia, M.E. Different combinations of salts affect the growth and bacteriocin production by *Lactobacillus salivarius* CRL 1328. *J. Chem. Technol. Biotechnol.* **2010**, *85*, 91–99. [CrossRef]
66. Foucaud, C.; Francois, A.; Richard, J. Development of a chemically defined medium for the growth of *Leuconostoc mesenteroides*. *Appl. Environ. Microbiol.* **1997**, *63*, 301–304. [CrossRef] [PubMed]
67. Li, J.Y.; Zhang, L.W.; Du, M.; Han, X.; Yi, H.X.; Guo, C.F.; Zhang, Y.C.; Luo, X.; Zhang, Y.H.; Shan, Y.J.; et al. Effect of tween series on growth and cis-9, trans-11 conjugated linoleic acid production of *Lactobacillus acidophilus* F0221 in the presence of bile salts. *Int. J. Mol. Sci.* **2011**, *12*, 9138–9154. [CrossRef]
68. Oh, S.; Rheem, S.; Sim, J.; Kim, S.; Baek, Y. Optimizing conditions for the growth of *Lactobacillus casei* YIT 9018 in tryptone-yeast extract-glucose medium by using response surface methodology. *Appl. Environ. Microbiol.* **1995**, *61*, 3809–3814. [CrossRef] [PubMed]
69. Tripuraneni, S. Effect of Nutrient Supplements on Cucumber Fermentation by Lactic Acid Bacteria. Master's Thesis, University of Arkansas, Little Rock, AR, USA, 2011.
70. Rosen, K. Production of Baker's Yeast. In *Yeast Biotechnology*; Berry, D.R., Russell, I., Steward, G.C., Eds.; Unwin Hyman Ltd.: London, UK, 2012; pp. 471–500.
71. Simova, E.; Simov, Z.; Beshkova, D.; Frengova, G.; Dimitrov, Z.; Spasov, Z. Amino acid profiles of lactic acid bacteria, isolated from kefir grains and kefir starter made from them. *Int. J. Food. Microbiol.* **2006**, *107*, 112–123. [CrossRef]
72. Lim, Y.S. Isolation of Bacteriocinogenic Lactic Acid Bacteria and Purification of Selected Bacteriocins from Traditional Fermented Foods. Master's Thesis, Universiti Putra Malaysia, Serdang, Malaysia, 2003.
73. Kareem, K.Y.; Foo, H.L.; Loh, T.C.; Ooi, M.F.; Asmara, S.A. Inhibitory activity of postbiotic produced by strains of *Lactobacillus plantarum* using reconstituted media supplemented with inulin. *Gut. Pathog.* **2014**, *6*, 1–7. [CrossRef]
74. Azin, A.; Rosfarizan, M.; Raha, A.R.; Rosli, M.I.; Farideh, N.; Tan, J.S.; Sahar, A. Cyclodextrin glycosyltransferase biosynthesis improvement by recombinant *Lactococcus lactis* NZ: NSP: CGT: Medium formulation and culture condition optimization. *Biotechnol. Biotechnol. Equip.* **2015**, *29*, 555–563.

Sample Availability: Samples of the compounds are not available from the authors.

molecules

MDPI

Article

Enhancement of Versatile Extracellular Cellulolytic and Hemicellulolytic Enzyme Productions by *Lactobacillus plantarum* RI 11 Isolated from Malaysian Food Using Renewable Natural Polymers

Nursyafiqah A. Mohamad Zabidi [1], Hooi Ling Foo [1,2,*], Teck Chwen Loh [3,4,*], Rosfarizan Mohamad [1,2,5] and Raha Abdul Rahim [2,6,7]

1 Department of Bioprocess Technology, Faculty of Biotechnology and Biomolecular Sciences, Universiti Putra Malaysia, UPM Serdang 43400, Selangor, Malaysia; syafiqah_zabidi92@yahoo.com (N.A.M.Z.); farizan@upm.edu.my (R.M.)
2 Institute of Bioscience, Universiti Putra Malaysia, UPM Serdang 43400, Selangor, Malaysia; raha@upm.edu.my
3 Department of Animal Science, Faculty of Agriculture, Universiti Putra Malaysia, UPM Serdang 43400, Selangor, Malaysia
4 Institute of Tropical Agriculture and Food Security, Universiti Putra Malaysia, UPM Serdang 43400, Selangor, Malaysia
5 Institute of Tropical Forestry and Forest Products, Universiti Putra Malaysia, UPM Serdang 43400, Selangor, Malaysia
6 Department of Cell and Molecular Biology, Faculty of Biotechnology and Biomolecular Sciences, Universiti Putra Malaysia, UPM Serdang 43400, Selangor, Malaysia
7 Office of Vice Chancellor, Universiti Teknikal Malaysia Melaka, Jalan Hang Tuah Jaya, Durian Tunggal 76100, Melaka, Malaysia
* Correspondence: hlfoo@upm.edu.my (H.L.F.); tcloh@upm.edu.my (T.C.L.); Tel.: +60-3-9769-7476 (H.L.F.); +60-3-97694814 (T.C.L.)

Received: 15 December 2019; Accepted: 17 March 2020; Published: 3 June 2020

Abstract: *Lactobacillus plantarum* RI 11 was reported recently to be a potential lignocellulosic biomass degrader since it has the capability of producing versatile extracellular cellulolytic and hemicellulolytic enzymes. Thus, this study was conducted to evaluate further the effects of various renewable natural polymers on the growth and production of extracellular cellulolytic and hemicellulolytic enzymes by this novel isolate. Basal medium supplemented with molasses and yeast extract produced the highest cell biomass (log 10.51 CFU/mL) and extracellular endoglucanase (11.70 μg/min/mg), exoglucanase (9.99 μg/min/mg), β-glucosidase (10.43 nmol/min/mg), and mannanase (8.03 μg/min/mg), respectively. Subsequently, a statistical optimization approach was employed for the enhancement of cell biomass, and cellulolytic and hemicellulolytic enzyme productions. Basal medium that supplemented with glucose, molasses and soybean pulp (F5 medium) or with rice straw, yeast extract and soybean pulp (F6 medium) produced the highest cell population of log 11.76 CFU/mL, respectively. However, formulated F12 medium supplemented with glucose, molasses and palm kernel cake enhanced extracellular endoglucanase (4 folds), exoglucanase (2.6 folds) and mannanase (2.6 folds) specific activities significantly, indicating that the F12 medium could induce the highest production of extracellular cellulolytic and hemicellulolytic enzymes concomitantly. In conclusion, *L. plantarum* RI 11 is a promising and versatile bio-transformation agent for lignocellulolytic biomass.

Keywords: lactic acid bacteria; probiotic; *Lactobacillus plantarum*; media enhancement; cellulolytic enzyme; hemicellulolytic enzyme; lignocellulosic biomass

1. Introduction

Bacteria utilized nutrients found in the environment for its survival and cellular biosynthesis [1]. The nutritional requirements of bacteria can be designed specifically for each strain to optimize its growth in the laboratory [2,3]. As for some heterotrophs, organic carbon such as sugars, amino acids and fats are crucial for their growth [4]. The presence of free amino acids or glucose alone can generate energy essential for bacterial growth [5,6]. Enzyme-producing microbes depend on glucose as their main carbon source, while nitrogen sources of tryptone and peptone support well the bacteria growth [7]. The statistical optimization method can be employed to overcome the limitation of conventional optimization method of growth medium, whereby the statistical optimization method is able to describe the interactions between multiple variables and determine the true optimum [8–10]. Fractional Factorial Design (FFD) [11] is a popular first-order design of a statistical optimization approach, which is regularly employed for optimization of a culture condition.

Copious amounts of lignocellulosic biomass are expelled worldwide annually and hence they are the most abundant renewable natural polymers [12–14] that has vast potential to be used as nutrients for bacteria cultivation and production of biochemicals. Lignocellulose comprises cellulose, hemicellulose and lignin as complex polymers [15–19]. Lignin is the most recalcitrant substance in lignocellulose biomass and contributes to the physical strength of the plant cell wall, whereas cellulose and hemicellulose play an important role in cell wall construction [12]. The intra- and inter-molecular hydrogen bond of cellulose contribute to its intricate structure with a high degree of polymerization [20]. Effective degradation of lignocellulosic biomass requires the presence of both cellulolytic and hemicellulolytic enzymes.

Cellulolytic enzymes are produced by a myriad of bacteria and fungi, aerobes or anaerobes, mesophiles or thermophiles when they grow on lignocellulosic materials [21], which can be employed as a biotransformation agent in the degradation of agricultural biomasses. Two cellulolytic enzyme systems of microorganisms have been reported, which are free cellulolytic and hemicellulolytic enzymes and a protein scaffold that is known as cellulosome [22,23]. The production of cellulolytic enzymes by the producer microorganism can be induced by specific substrates [24], such as lignocellulosic biomass [25]. Despite extensive reports on the production of lignocellulosic enzymes by certain fungal and bacterial species, the efficiency of biotransformation of agro-industrial waste remains low [25], whereby phenolic compounds generated during pre-treatment of biomass have been reported to hinder cellulase and hemicellulase activities [26,27]. Moreover, a consortium of bacterial or fungal cultures are unable to achieve complete biotransformation of agro-wastes due to antagonistic interactions among the consortium members [28]. Hence, continuous efforts of screening and isolation of novel extracellular cellulase and hemicellulase producers are important to overcome these challenges.

The safety status of enzyme producers is a great concern [17] in medical, food and animal feed industries. LAB are perceived as Generally Regarded as Safe (GRAS) microorganisms by the Food and Drug Administration [29] of the United States of America since they are generally non-toxicogenic and non-pathogenic bacteria [29–33], which have been present in various fermented foods for decades and do not have any deleterious effects on human. In addition, LAB have a long history in industrial processes as food starters and biocontrol agents, and as producers of high-value compounds [34,35]. *Lactobacillus plantarum*, one of the major and important species of LAB [36], are notable as probiotics since they enhance the gastrointestinal health and performances in animals by promoting the gut immune response [35–40]. Furthermore, the extracellular metabolites of *L. plantarum* that are known as postbiotics have been proven extensively to be a potential alternative for in-feed antibiotic growth promoters to improve meat quality and growth performance of broiler chicken, laying hens and piglets [38–42]. Recently, several strains of *L. plantarum* have been reported to be versatile extracellular proteolytic enzyme producers and capable of producing an array of essential amino acids [39,43].

It is widely known that LAB are fastidious microorganisms, which required complex media compositions for their growth [44,45] and hence they could not ferment inexpensive lignocellulose feedstocks directly, whereby fermentation of lignocellulose hydrolysates by LAB has been frequently reported and is the most developed technology [29,46]. In addition, LAB are known to have the capability to degrade pentoses and hexoses via the phosphoketolase pathway [42,44].

Cellulolytic and hemicellulolytic enzymes are important industrial enzymes that are widely used in the production of food, animal feed, pulp, paper and textile. Prior to 2019, LAB have not been reported elsewhere as efficient non-treated complex biomass bioconversion agents that have the capability of producing extracellular cellulolytic and hemicellulolytic enzymes concomitantly. Thus, we have made an attempt to explore the potential of locally isolated LAB as an efficient non-treated complex biomass bioconversion agent using palm kernel cake (PKC) as a model for agricultural lignocellulosic biomass, and we have reported very recently that *Lactobacillus plantarum* RI 11, a probiotic isolated from Malaysian food, is a potential lignocellulosic biomass degrader, whereby versatile extracellular cellulolytic and hemicellulolytic enzymes were produced simultaneously when it was grown on commercially available de Man, Rogosa and Sharpe (MRS) medium [46]. However, limited knowledge regarding the effects of various non-treated complex renewable natural polymers on the growth, cellulolytic, and hemicellulolytic enzyme productions was available for LAB. Therefore, this study was conducted to investigate further the effects of various combinations of renewable natural polymers derived from agricultural wastes on the growth and productions of extracellular cellulolytic and hemicellulolytic enzymes by *L. plantarum* RI 11. Subsequently, the statistical optimization method of FFD was employed for further enhancement of cellulolytic and hemicellulolytic enzyme productions by *L. plantarum* RI 11.

2. Results and Discussion

2.1. Growth Profile of *L. plantarum* RI 11 in Media Supplemented with Renewable Natural Polymers

In this study, the effects of various renewable natural polymers of agricultural biomasses, such as molasses, rice straw, PKC and soybean pulp on the viability of *L. plantarum* RI 11 were determined for 7 days of incubation. Agricultural residues are the major source of lignocellulosic biomass, which is renewable, highly available and inexpensive [15,16] as rich resources of fermentable sugars [17] that can act as a precursor or intermediates in the production of biofuels and bio-based products [18]. It is a prevalent concern to the countries that have profited from agricultural industry as the disposal of biomass waste is a great challenge in the management of biomass waste [17], whereby the bioconversion of agricultural lignocellulosic biomasses remain economically unfeasible due to unavailable of efficient biocatalysts or bioagents.

In this experiment, glucose and yeast extract were used as a control for carbon and nitrogen sources, respectively [3]. Moreover, the composition of commercially available selective MRS medium was used as a reference to formulate growth medium comprising various renewable natural polymers. Figure 1a–g show the cell population of *L. plantarum* RI 11 grown in different medium formulations collected at different incubation times, whereby different renewable natural polymers that affected the growth of *L. plantarum* RI 11 substantially attributed to different nutrients were provided by different renewable natural polymers [1–3]. The initial cell population of approximately log 8 CFU/mL was observed for all formulated media as shown in Figure 1a–g.

Figure 1a shows the growth profile of *L. plantarum* RI 11 in control MRS medium. The maximum cell population of approximately log 11.5 CFU/mL was detected at 18 h of incubation. It was the highest cell population of *L. plantarum* RI 11 that noted in this experiment, indicating that *L. plantarum* RI 11 grown well in MRS medium since it is a selective medium for *L. plantarum*. However, the cell population decreased gradually after 1-day of incubation to log 8.5 CFU/mL at 5 days of incubation and maintained its cell population for approximately log 8.7 CFU/mL at the end of day 7 of incubation.

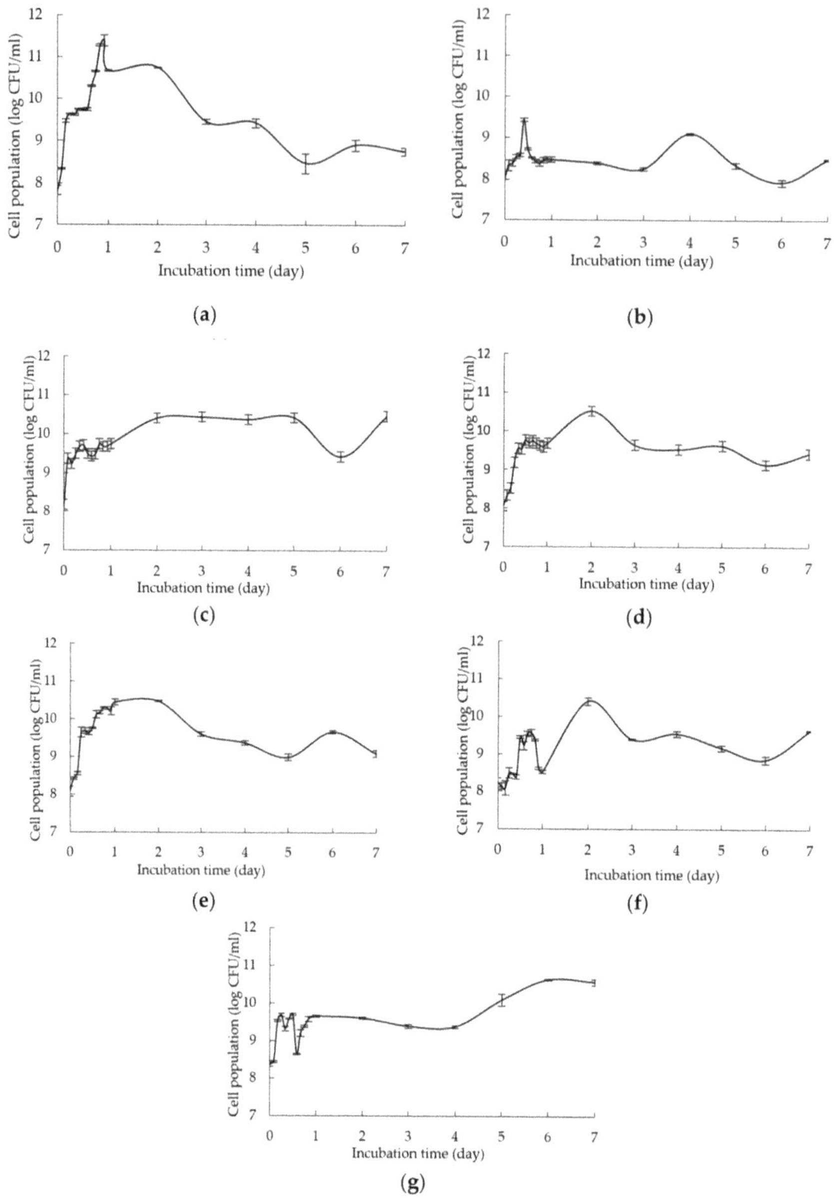

Figure 1. Growth profile of *Lactobacillus plantarum* RI 11 in: (**a**) control MRS medium (**b**) M1; rice straw and yeast extract (**c**) M2; rice straw and soybean pulp (**d**) M3; molasses and yeast extract (**e**) M4; molasses and soybean pulp (**f**) M5; PKC and yeast extract and (**g**) M6; PKC and soybean pulp. The error bars represent the standard deviation of cell population of triplicate samples ($n = 3$).

Figure 1b,d,g show similar growth profiles of *L. plantarum* RI 11 in media supplemented with rice straw and yeast extract (M1), molasses and yeast extract (M3), and PKC and soybean pulp (M6), respectively. The cell population of *L. plantarum* RI 11 was increased slightly from log 7.98 ± 0.12 to log 8.72 ± 0.03 CFU/mL at 12 h of incubation, log 8.04 ± 0.09 to log 9.76 ± 0.01 at 16-h of incubation and log

8.36 ± 0.05 to log 9.70 ± 0.02 CFU/mL at 12-h of incubation when grown in M1, M3 and M6 media, respectively, suggesting that the highest cell population was obtained from 12-h to 16-h of incubation for M1, M3 and M6 media, respectively. The cell population was maintained at log 8.3 CFU/mL, log 9.5 CFU/mL and log 9.4 CFU/mL for M1, M3 and M6 media, respectively, from 18-h to 22-h of incubation. In comparison, the steepest log phase of *L. plantarum* RI 11 was observed for M3 medium (Figure 1d), followed by M6 (Figure 1g) and M1 (Figure 1b) medium, respectively. M3 medium provided readily used nutrients of molasses and yeast extract as soluble carbon and nitrogen sources for the growth of *L. plantarum* RI 11. Hence, the cell population of *L. plantarum* RI 11 in M3 was higher than M1 and M6 media. Molasses is a by-product of the sugar processing industry, which is rich in reducing sugars, nitrogen, trace elements and vitamins [47–51] that can promote the growth of bacteria [52]. Refined sugars, such as glucose or sucrose have been used more commonly as a carbon source in comparison to rice straw and PKC [47,53], whereas molasses is commonly used fermentation substrates since it is cheap, widely available and comprised mainly sugars that act as nutrients for all microorganisms.

M6 medium that consisted of PKC and soybean pulp as complex carbon and nitrogen sources produced a steeper log phase as compared to M1 medium that consisted of rice straw and yeast extract as the complex carbon and soluble nitrogen sources. Soybean pulp is a cheap and good dietary fibre [47,48], which has been reported to encourage the growth of beneficial bacteria, such as *Lactobacillus* spp. [54]. Nevertheless, rice straw is highly heterogenous as compared to the combination of PKC and soybean pulp to support the growth of *L. plantarum* RI 11 [55]. Due to the abundance of glucose, sucrose and fructose [49–53], the cell population of *L. plantarum* RI 11 in M3 medium was the highest as reported by a previous study of Zajšek et al. [56].

In comparison, Figure 1c illustrates a shorter exponential phase of *L. plantarum* RI 11 when grown in M2 medium that contained complex carbon and nitrogen sources of rice straw and soybean pulp. The initial cell population of log 8.17 ± 0.04 was increased to log 10.41 ± 0.05 CFU/mL at day-2 of incubation, whereby the cell population was then maintained throughout 7-days of incubation. Despite both M2 and M6 media containing the same nitrogen source of soybean pulp, they contained heterogenous rice straw and PKC as carbon source, respectively, indicating that limited nutrients were available from the degradation of rice straw for the growth of *L. plantarum* RI 11 in comparison to PKC and thus produced shorter and steeper log phase as shown in Figure 1c.

Figure 1e shows the growth curve of *L. plantarum* RI 11 in M4 medium, which has a similar log phase to control MRS medium. The cell population was increased from initial log 8.06 CFU/mL to log 10.44 CFU/mL at day 1 of incubation. The cell population was then maintained for a day before it decreased to log 9 CFU/mL. M4 medium consisted of molasses and soybean pulp as carbon and nitrogen sources, respectively. The cell population of control MRS medium supported slightly higher cell population (log 10.67 ± 0.02) of *L. plantarum* RI 11 at 24 h of incubation as compared to M4 medium with log 10.44 ± 0.08 CFU/mL, suggesting that M4 medium has contributed a comparable effect to the growth of *L. plantarum* RI 11, whereby both molasses and soybean pulp promoted the growth of *L. plantarum* RI 11 due to their high contents of sugar and protein [47–53].

Figure 1f, in contrast, shows the longest and gradual slope of exponential growth of *L. plantarum* RI 11 supported by M5 medium that contained PKC and yeast extract as carbon and nitrogen sources, respectively. The initial cell population of log 8.4 CFU/mL was increased to log 9.54 CFU/mL that occurred at 16-h of incubation. Interestingly, the stationary phase occurred at 48 h of incubation, indicating that PKC was more complex substrate that could support longer growth duration of *L. plantarum* RI 11. Similar observation was also noted for medium M6 (Figure 1g) that comprised PKC and soybean pulp as the complex carbon and nitrogen sources.

Interestingly, when any carbon source combined with soybean pulp, a better cell population and shorter log phase of *L. plantarum* RI 11 was noted. This could be attributed to soybean pulp that contains a substantial amount of galactose, mannose, arabinose and uronic acid, which served as the growth promoter and enzyme inducers [47,48] for *L. plantarum* RI 11. Although soybean pulp contains mainly dietary fibre (49.85%) and lignin (29.50%), it also contains 19.79% of hemicellulose and low amount of cellulose (0.56%). The proximate analysis (results not shown) showed that the total carbon content of soybean pulp that used in this study was 71.53%. Generally, the results of Figure 1 implied that molasses was the most favorable renewable natural polymer as carbon source, followed by PKC and rice straw for the growth of *L. plantarum* RI 11. In addition, the combination of molasses and soybean pulp as soluble carbon and complex nitrogen sources promoted and sustained the growth of *L. plantarum* RI 11 that was comparable to the control MRS medium. Thus, the results obtained in this study revealed the versatility of *L. plantarum* RI 11 that possessed the ability to alter its physiological response and entered the stationary phase [57,58] accordingly when it was present in growth medium supplemented with different combinations of carbon and nitrogen sources.

2.2. Extracellular Cellulolytic and Hemicellulolytic Enzyme Activities of L. plantarum *RI 11*

Extracellular cellulolytic (endoglucanase, exoglucanase and, β-glucosidase) and hemicellulolytic (mannanase) enzyme activities of *L. plantarum* R1 11 that grown in different media mixtures were subsequently determined in this study. The cellulolytic and hemicellulolytic enzymes are essential to be present concomitantly to degrade the complex agriculture lignocellulosic biomass effectively [59], whereby endoglucanase, exoglucanase and β-glucosidase hydrolyze the glycosidic bonds [15] that are present in the cellulose polymer synergistically [19]. Endoglucanase will imitate the hydrolysis of β-1,4-glycosidic bonds of cellulose polymer randomly to produce reducing and non-reducing ends [59–61] of shorter cellulose polymers. Then, exoglucanase will degrade both reducing and non-reducing ends of the shorter cellulose polymer [59–61] to release glucose monomer. β-glucosidase will degrade cellobiose that produced by endo- and exoglucanase to 2 glucose monomers [61,62]. Endoglucanase is also known as CMCase due to its ability to hydrolyze carboxymethylcellulose (CMC), which was used as a substrate in the current study, whereas exoglucanase activity was determined by using avicel as a substrate and hence it is also known as avicelase, whereby avicel was degraded into cellobiose by exoglucanase. As for β-glucosidase activity determination, 4-nitrophenyl-β-D-glucopyranoside (PNPG) was used as a substrate in this study, whereas locust bean gum (LBG) was used as a substrate for mannanase activity determination [63–65]. The reducing sugar that released by both cellulolytic and hemicellulolytic enzymes was detected spectrophotometrically by using a dinitrosalicyclic acid (DNS) reagent [64,65].

Figure 2a–c demonstrate the specific endoglucanase, exoglucanase and β-glucosidase enzyme activities detected at pH 5, 6.5 and 8, respectively, that grown in the MRS control medium. It was clearly shown that *L. plantarum* RI 11 was able to produce versatile extracellular cellulolytic and hemicellulolytic enzymes that active from pH 5 to 8. Interestingly, the hemicellulolytic (mannanase) activity was only detected at pH 8 (Figure 2c) in comparison to the cellulolytic enzyme activities.

Endoglucanase-CMCase, exoglucanase-avicelase and β-glucosidase activities were detected at a broad pH range of pH 5, 6.5 and 8, respectively, as shown in Figure 2a–c, respectively, whereby three isozymes of endoglucanase were detected at pH 5 between 4–14 h of incubation (Figure 2a), one isozyme of endoglucanase was detected at pH 6.5 between 8–12 h of incubation (Figure 2b) and pH 8 at 12–14 h of incubation (Figure 2c). In comparison, two isozymes of exoglucanase were detected at pH 5 (Figure 2a) and pH 6.5 (Figure 2b) at 6–14 h and 6–12 h of incubation, respectively. Only one isozyme of exoglucanase was detected at pH 8 between 6–10 h of incubation (Figure 2c). As for β-glucosidase activity, only one isozyme was detected at pH 5, 6.5 and 8 at 12 h of incubation. However, β-glucosidase activity at pH 8 was approximately half of the activity that obtained at pH 5 and 6.5, respectively. The concomitant presence of versatile extracellular cellulolytic (endoglucanase-CMCase, exoglucanase-avicelase and β-glucosidase) and hemicellulolytic (mannanase) enzyme activities of *L. plantarum* that active at a broad pH range (from pH 5 to pH 8) corresponded well to the finding of Lee et al. [46], who has reported that LAB strains that isolated from Malaysian foods produced multi extracellular cellulolytic and hemicellulolytic enzymes simultaneously, which are active under broad pH conditions when grown in selective MRS medium. Similar observations were also reported by Nidetzky et al. [66], Eriksson [67] and Lin et al. [68]. Furthermore, Lee et al. [46] reported that β-glucosidase specific enzyme activity was the lowest among the detected cellulolytic enzymes, in which the same results were noted in this study. β-glucosidase gene is commonly harbored by LAB, especially *L. plantarum* species [69]. Hence, β-glucosidase specific activity was detected even in the control MRS medium [70]. Throughout this experiment, β-glucosidase was observed to be produced later as compared to the other two cellulolytic enzymes, as reported by Lee et al. [46].

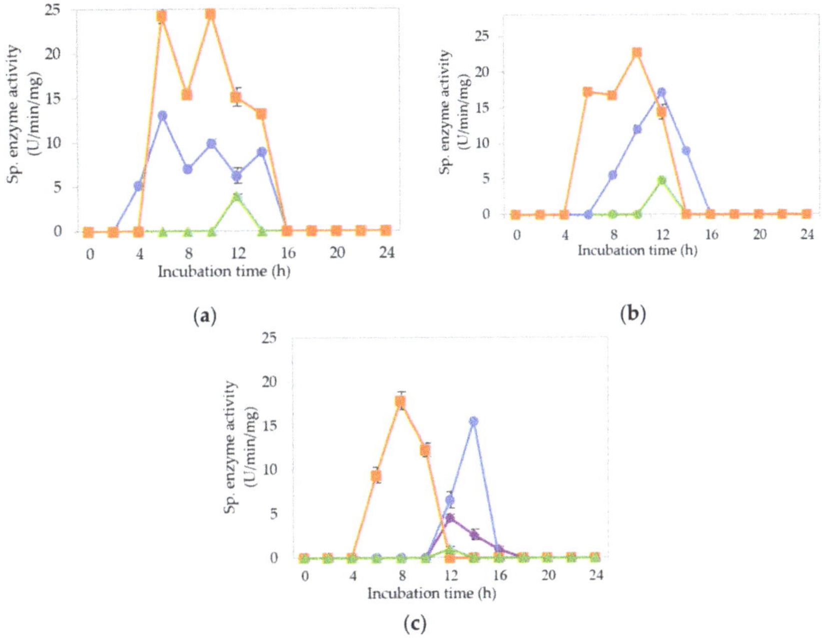

Figure 2. Specific extracellular cellulolytic and hemicellulolytic activities; (•) endoglucanase; (■) exoglucanase; (▲) β-glucosidase and (♦) mannanase of *Lactobacillus plantarum* RI 11 grown in MRS control medium for 24 h. The enzyme activities were detected at three different pH levels: (**a**) pH 5; (**b**) pH 6.5; (**c**) pH 8. The error bars represent the standard deviation of specific enzyme activity of triplicate samples (*n* = 3).

Figure 3a–c illustrate the specific extracellular cellulolytic and hemicellulolytic enzyme activities at pH 5, 6.5 and 8, respectively, when *L. plantarum* R1 11 was grown in basal medium supplemented with rice straw and yeast extract as carbon and nitrogen sources (M1).

Surprisingly, significant ($p < 0.05$) levels of endoglucanase, exoglucanase and β-glucosidase activities were detected at pH 5, 6.5 and 8, respectively, as found in MRS control medium. One isozyme of endoglucanase was recorded at pH 5 and 6.5, respectively, both at 10-h of incubation with 4.07 μg/min/mg and 1.68 μg/min/mg of specific activities. As for exoglucanase, one isozyme was detected at pH 5 between 8–12 h of incubation, whereas two isozymes were detected at pH 6.5 between 4 to 10-h of incubation. For β-glucosidase activity, only one isozyme was detected at pH 5, 6.5 and 8 after 12-h of incubation, whereby the specific activity was reduced to half as compared to the β-glucosidase activity that obtained in MRS control medium. The specific enzyme activities recorded for β-glucosidase were 2.54 nmol/min/mg and 2.10 nmol/min/mg at pH 5 and 6.5, respectively. As for pH 8, a significant ($p<0.05$) level of β-glucosidase was detected, which was 6.05 nmol/min/mg. No mannanase activity was detected when *L. plantarum* R1 11 was grown in M1 medium. The cellulolytic and hemicellulolytic enzyme levels were comparably low in M1 medium. Hence, it can be concluded that though yeast extract was able to promote the growth of *L. plantarum* R1 11, but, due to the heterogeneity of the rice straw [55], a low level of cellulolytic and hemicellulolytic enzymes was secreted by the *L. plantarum* R1 11.

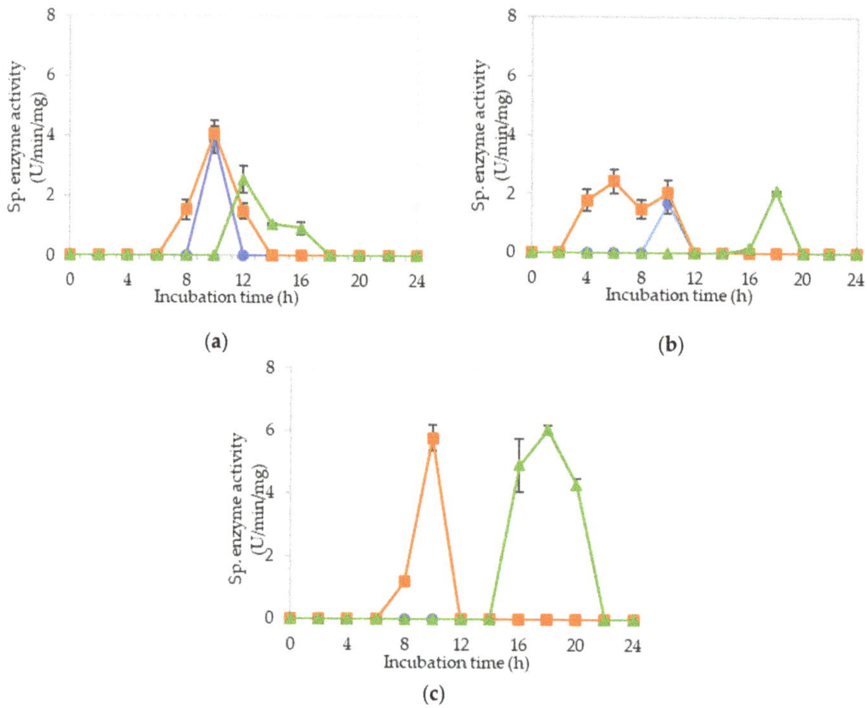

Figure 3. Specific extracellular cellulolytic and hemicellulolytic activities; (•) endoglucanase; (■) exoglucanase; (▲) β-glucosidase and (♦) mannanase of Lactobacillus plantarum RI 11 grown in M1 (rice straw and yeast extract) for 24 h. The enzyme activities were detected at three different pH levels: (**a**) pH 5; (**b**) pH 6.5; (**c**) pH 8. The error bars represent the standard deviation of specific enzyme activity of triplicate samples (n = 3).

Figure 4a–c demonstrate the significantly (p<0.05) lower specific extracellular endoglucanase, exoglucanase and β-glucosidase activities at pH 5, 6.5 and 8, respectively, when *L. plantarum* R1 11 was cultivated in basal medium supplemented with rice straw and soybean pulp (M2) medium. Interestingly, only one isozyme of extracellular endoglucanase was detected at pH 5, 6.5 and 8, respectively, between 4–6 h of incubation with significantly (p<0.05) higher activity of 3.1 μg/min/mg detected at pH 5. However, this specific activity was significantly lower (p < 0.05) than those specific activity observed for other media. Similar levels of enzymatic activities were observed in both Figures 3 and 4. This was most probably due to the same carbon source used in both media. Carbon source is very crucial in the production of enzyme [71]; hence, it can be concluded that rice straw could not promote the production of cellulolytic and hemicellulolytic enzymes. As for the exoglucanase specific activity, only one isozyme was detected at pH 5 and 6.5 between 4–6 h and between 8–10 h of incubation respectively. As for β-glucosidase, it was detected at pH 5 and 6.5. However, significantly higher (p<0.05) specific activity of 4.1 nmol/min/mg at 16 h of incubation was detected at pH 5. In comparison, much lower mannanase activity was recorded between 1.2 to 1.5 μg/min/mg at 14-16 h of incubation at pH 8. Generally, insignificant (p>0.05) level of cellulolytic and hemicellulolytic enzyme specific activities were obtained when *L. plantarum* R1 11 was grown in basal medium supplemented with rice straw and soybean pulp, attributing to the recalcitrant nature of rice straw [55].

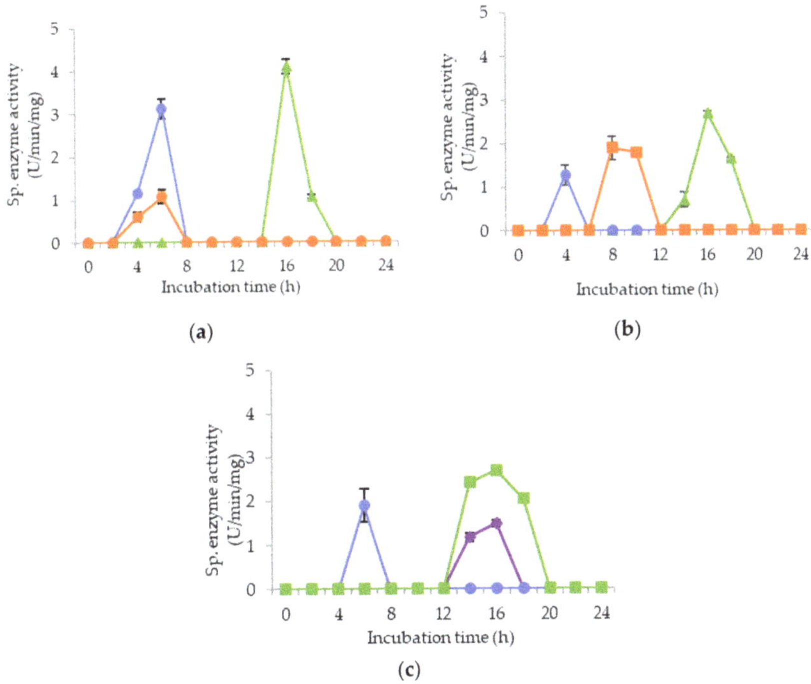

Figure 4. Specific extracellular cellulolytic and hemicellulolytic activities; (•) endoglucanase; (■) exoglucanase; (▲) β-glucosidase and (♦) mannanase of Lactobacillus plantarum RI 11 grown in M2 (rice straw and soybean pulp) for 24 h. The enzyme activities were detected at three different pH levels: (**a**) pH 5; (**b**) pH 6.5; (**c**) pH 8. The error bars represent the standard deviation of specific enzyme activity of triplicate samples (n = 3).

Figure 5a–c show the specific extracellular endoglucanase, exoglucanase and β-glucosidase activities detected at pH 5, 6.5 and 8, respectively, when *L. plantarum* RI 11 was grown in basal medium supplemented with molasses and yeast extract (M3). At pH 5, two isozymes of endoglucanase were detected between 4–12 h of incubation. However, only one isozyme of endoglucanase was detected at pH 6.5 and 8, respectively, between 4–10 h of incubation. Simultaneously, exoglucanase specific activity was also recorded with endoglucanase activity from pH 5 to 8, despite only one isozyme was detected at pH 5 between 2–6 h of incubation. Nevertheless, two isozymes of exoglucanase were detected at pH 6.5 and 8, respectively, between 4–16 h of incubation. In comparison, the highest specific activity of β-glucosidase was only produced after endoglucanase and exoglucanase were excreted by *L. plantarum* RI 11, which occurred between 10–16 h of incubation. Interestingly, three significant ($p<0.05$) enzymatic activities of mannanase were detected at pH 8 between 6–16 h of incubation (Figure 5c).

The finding in this experiment was contradicted with a study reported by Kim et al. [72], whereby the present of mannan-like polysaccharide was essential to induce the production of mannanase. However, mannan-like polysaccharide was not present in the M3 medium that contained molasses as carbon source. Nevertheless, El-Sharouny et al. [73] reported that the production of mannanase was affected by glucose. Molasses, as reported by Hamouda et al. [53], consists of sucrose, glucose, fructose, xylose and maltose, hence it was proven that molasses can induce the production of mannanase. Furthermore, the mannanase enzyme that purified by El-Sharouny et al. [73] was active in extreme alkaline conditions, which was similar to the observation of this experiment.

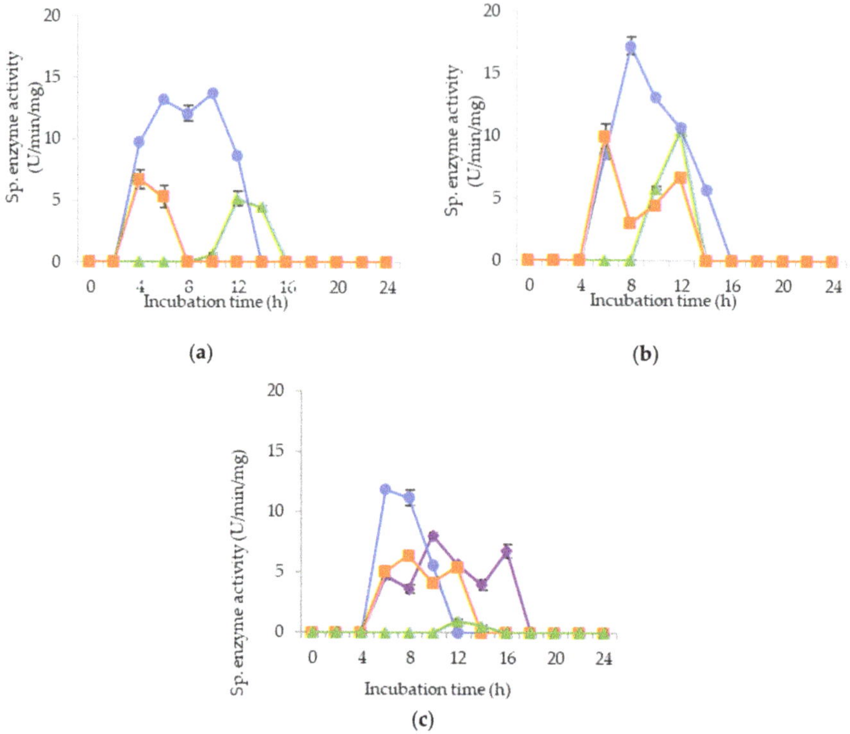

Figure 5. Specific extracellular cellulolytic and hemicellulolytic activities; (•) endoglucanase; (■) exoglucanase; (▲) β-glucosidase and (♦) mannanase of *Lactobacillus plantarum* RI 11 grown in M3 (molasses and yeast extract) for 24 h. The enzyme activities were detected at three different pH levels: (**a**) pH 5; (**b**) pH 6.5; (**c**) pH 8. The error bars represent the standard deviation of specific enzyme activity of triplicate samples ($n = 3$).

Molasses is acidic in nature, in which it provides a favorable environment for LAB to grow and to produce useful metabolite [49–53]. Molasses is also cellulosic [69], in which it will act an inducer for the production of cellulolytic enzymes by cellulolytic microorganisms [45] before 20-h of incubation [74] as shown in this experiment, whereby M3 medium was able to induce the production of three cellulolytic enzymes over a broad pH range by *L. plantarum* R1 11 with the highest and significant endoglucanase was noted from pH 5 to 6.5 in M3 media. In addition, M3 medium is a more economical medium that contained readily fermented sugars [50–52] to produce comparable extracellular cellulolytic and hemicellulolytic enzymes in comparison to the MRS control medium

Figure 6a–c demonstrate the specific endoglucanase, exoglucanase and β-glucosidase obtained at pH 5, 6.5 and 8, respectively, when *L. plantarum* R1 11 was grown in basal medium supplemented with molasses and soybean pulp (M4). Significantly ($p < 0.05$) higher endoglucanase specific activities that obtained approximately at 8 h of incubation was detected at pH 5 in comparison to pH 6.5 and pH 8, respectively, whereby only one isozyme of endoglucanase was produced in M4 formulated medium. The exoglucanase, on the other hand, recorded a higher specific activity of approximately 10–20 µg/min/mg at pH 5 and 8, respectively. Though the highest endoglucanase was detected in M4 medium, it was only active at pH 5. In contrast, M3 medium that consists of molasses and yeast extract produced more stable endoglucanase enzymes throughout the broad pH range. As for the β-glucosidase activity, low specific activity of approximately 3 nmol/min/mg was obtained between 4–14 h of incubation from pH 5 to pH 8, despite of insignificant ($p > 0.05$) activity, was noted. Interestingly, mannanase activity was detected at pH 6.5 and pH 8 between 12–18 h and 8–12 h of incubation, respectively. However, the highest specific activity of 8.05 μg/min/mg was noted at pH 6.5. This finding was similar to the study of Kim et al. [72], in which mannan-like polysaccharides from soybean pulp induced mannanase productions. Interestingly, mannanase was produced concomitantly with the production of endoglucanase, as reported by Sachslesner et al. [75]. Hence, it was generally observed that, throughout this experiment, mannanase enzyme was detected together with endoglucanase activity. However, not all endoglucanase enzyme productions were accompanied by the production of mannanase enzyme.

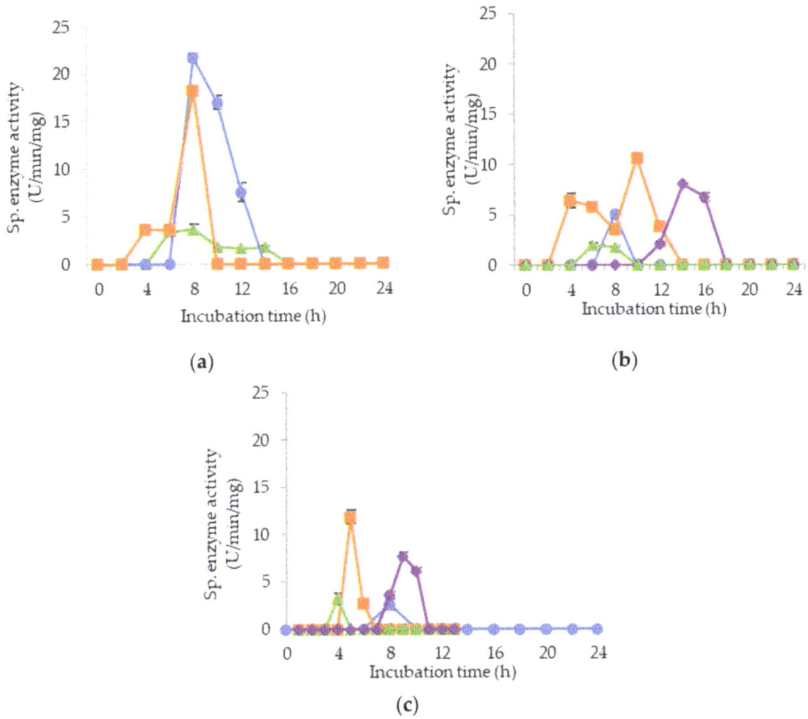

Figure 6. Specific extracellular cellulolytic and hemicellulolytic activities; (•) endoglucanase; (■) exoglucanase; (▲) β-glucosidase and (♦) mannanase of *Lactobacillus plantarum* RI 11 grown in M4 (molasses and soybean pulp) for 24 h. The enzyme activities were detected at three different pH levels: (**a**) pH 5; (**b**) pH 6.5; (**c**) pH 8. The error bars represent the standard deviation of specific enzyme activity of triplicate samples ($n = 3$).

Figure 7a–c show the specific endoglucanase, exoglucanase and β-glucosidase activities which were detected at pH 5, 6.5 and 8, respectively, when *L. plantarum* RI 11 was grown in basal medium supplemented with PKC and yeast extract (M5). For this formulation, one isozyme of endoglucanase was present at pH 6.5 and 8 respectively with significantly lowest ($p < 0.05$) activities of 1–2.7 μg/min/mg detected. As for exoglucanase activity, although it was detected from pH 5–8, an insignificant ($p > 0.05$) level of activity was recorded. Interestingly, significant specific activities of β-glucosidase were detected between 4–8 h of incubation, having 7.55 μg/min/mg at pH 8 as the highest activity. Surprisingly, no mannanase was detected from M5 medium. This finding was in contrast with the study reported by Kim et al. [72], whereby PKC that comprised of mannan polysaccharides did not induce the mannanase production in this study. Nevertheless, the results of this experiment were in close agreement with the finding of El-Sharouny et al. [73], whereby the production of mannanase was induced by the presence of simple sugars such as glucose.

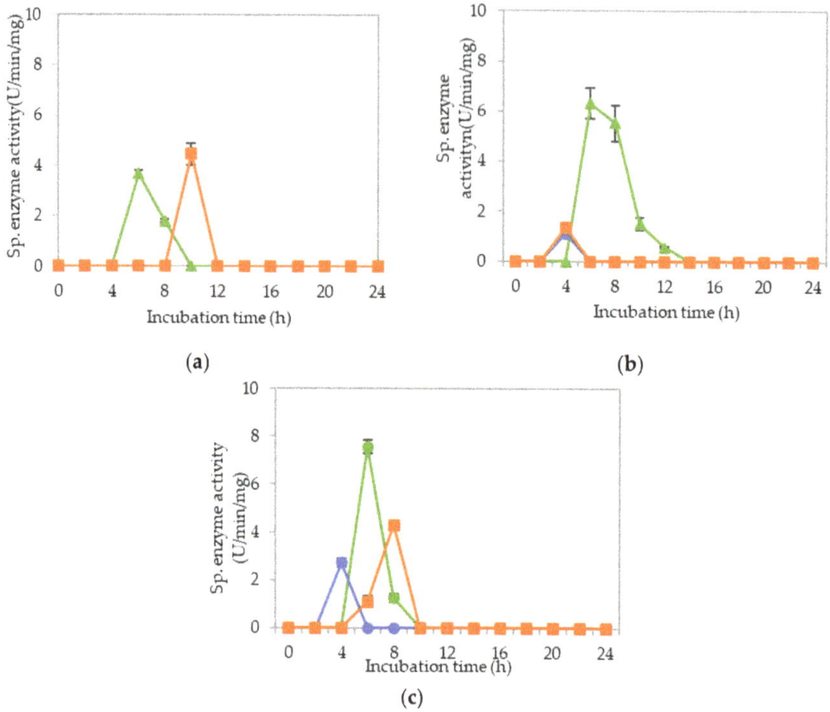

Figure 7. The specific extracellular cellulolytic and hemicellulolytic activities; (•) endoglucanase; (■) exoglucanase; (▲) β-glucosidase and (♦) mannanase of *Lactobacillus plantarum* RI 11 grown in M5 (PKC and yeast extract) for 24 h. The enzyme activities were detected at three different pH levels: (**a**) pH 5; (**b**) pH 6.5; (**c**) pH 8. The error bars represent the standard deviation of specific enzyme activity of triplicate samples (n = 3).

Figure 8a–c illustrate the specific endoglucanase, exoglucanase and β-glucosidase activities recorded at pH 5, 6.5 and 8, respectively, when *L. plantarum* R1 11 was grown in basal medium supplemented with PKC and soybean pulp (M6). Surprisingly, extracellular endoglucanase activity was not detected at all pH and only one isozyme of exoglucanase was detected at pH 5 with an insignificant ($p<0.05$) activity of 2.56 μg/min/mg was recorded. Similarly, although β-glucosidase was noted from pH 5 to 8, the specific enzyme activity was diminutive. Mannanase, as usual, was detected only at pH 8 with low level of specific enzyme activity at a later incubation time of 16–18 h. The basal medium supplemented with PKC and soybean pulp did not induce substantial production of extracellular cellulolytic and hemicellulolytic enzymes by *L. plantarum* RI 11 in comparison to the other renewable natural polymers such as molasses and rise straw as shown clearly in Figures 3, 5 and 6, respectively, indicating that the combination of PKC and soybean pulp was not a favorable medium for the production of extracellular cellulolytic and hemicellulolytic enzymes by *L. plantarum* RI 11. However, mannanase was detected later throughout the incubation, though the enzymatic level was very low. This finding coincided with the study of El-Sharouny et al. [73], as well as noted in Figure 7 of this study.

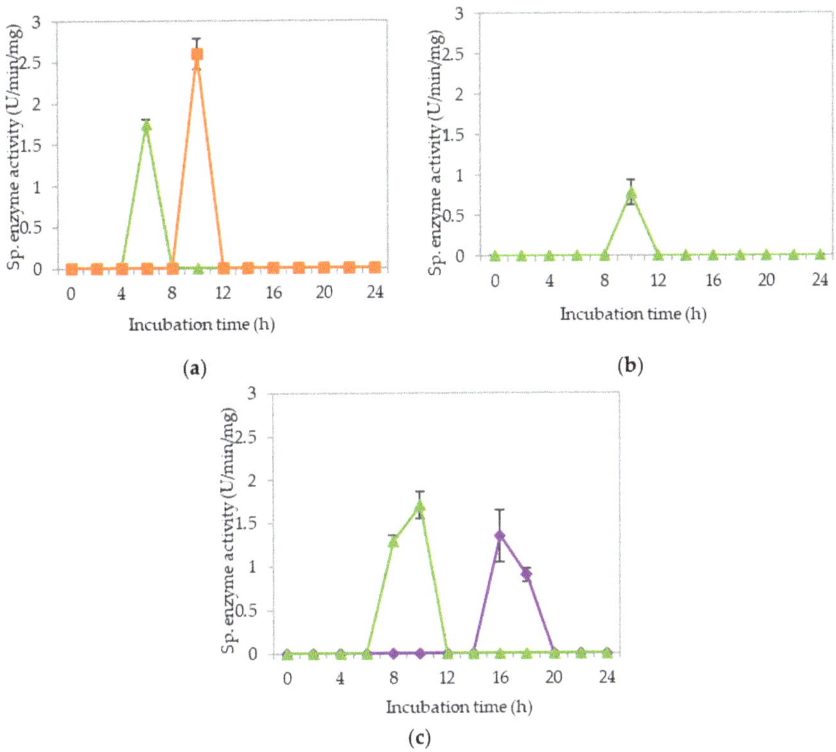

Figure 8. Specific extracellular cellulolytic and hemicellulolytic activities; (•) endoglucanase; (■) exoglucanase; (▲) β-glucosidase and (♦) mannanase of Lactobacillus plantarum RI 11 grown in M6 (PKC and soybean pulp) for 24 h. The enzyme activities were detected at three different pH levels: (**a**) pH 5; (**b**) pH 6.5; (**c**) pH 8. The error bars represent the standard deviation of specific enzyme activity of triplicate samples (*n* = 3).

Generally, the overall results of this experiment showed that the basal medium supplemented with molasses and yeast extract (M3) recorded as the best basal medium that was comparable with the MRS control medium to produce extracellular cellulolytic and hemicellulolytic enzymes, respectively. Cellulolytic enzyme is a complex enzyme comprising endoglucanase, exoglucanase and β-glucosidase [75,76], whereas hemicellulolytic enzyme encompasses xylanase and mannanase [77]. The extracellular cellulolytic and hemicellulolytic enzyme activities of *L. plantarum* RI 11 were detected within 24 h of incubation, and hence prolong incubation time resulting in no further extracellular enzyme production despite the cell viability that could maintain up to seven days of incubation at 30 °C as shown in Figure 1. This could be due to carbon starvation [57] and led to a halt of the protein and DNA synthesis and hence the enzyme production. Interestingly, β-glucosidase specific activity was only detected after the production of both endoglucanase and exoglucanase specific activities, supporting the synergistic enzymatic degradation mechanism of cellulosic polymer, whereby β-glucosidase cleaved the cellobiose molecules that released from the enzymatic reaction of both endoglucanase and exoglucanase [15,17] into monomer glucose [23,60]. Generally, β-glucosidase activity is the rate limiting factor among the cellulolytic enzymes, which is very sensitive to glucose inhibition [70,77]. However, the glucose and xylose contents of molasses are good inducers that could stimulate bacteria producer to synthesize β-glucosidase. Nonetheless, among the cellulolytic enzymes, the specific enzyme activity of β-glucosidase was the lowest, as reported by Lee et al. [46]. Moreover,

Spano et al. [69] reported that β-glucosidase gene isolated from wine *L. plantarum* was regulated by abiotic stresses.

The results obtained in this study also indicated that the extracellular hemicellulolytic enzyme of mannanase was significantly ($p > 0.05$) produced by *L. plantarum* RI 11 between 6–16 h of incubation (3.96 to 8.09 μg/min/mg) that active at alkaline pH 8, as reported by Rasul et al. [74]. Furthermore, mannanase and endoglucanase were usually produced concomitantly and production of mannanase was always accompanied by the production of endoglucanase [75]. Mannanase of *L. plantarum* RI 11 was most likely to be induced by the presence of mannose or mannan-like polysaccharides [72] or sugar [73] in the formulated media. Mannanase is essential to cleave mannan to mannose that could serve as energy source for the survival of bacteria producer cells [69,70]. However, since lower specific enzyme activities of β-glucosidase and mannanase were secreted by the *L. plantarum* RI 11, it was possible that the cellulose degradation by endoglucanase and exoglucanase was inhibited by the presence of cellobiose [23]. Hence, both β-glucosidase and mannanase enzymes were essential to warrant complete degradation of renewable natural polymers [71–74].

2.3. Fractional Factorial Design for Growth Enhancement of *L. plantarum RI 11*

Factorial Design is a popular statistical optimization method to determine important factors that influence the yield of experiments and the interactions that might exist between the factors [78–80]. Full factorial design is more suitable for factors less than 4. Fractional Factorial Design (FFD) is more appropriate when the resources are limited [79] and there are more than five investigated factors [79,80]. It is important to know the effect of medium components on the growth of *L. plantarum* RI 11. Therefore, the medium formulation that gave the highest viable cell count was optimized subsequently by FFD. Altogether, 16 media formulations were suggested by FFD based on carbon and nitrogen sources used in this study. Both control MRS medium and reconstituted control MRS (CRMRS) medium were used as control media in this experiment, with the former being commercial and the latter was reconstituted. The renewable natural polymers that used as carbon sources in this experiment were rice straw, molasses and PKC, whereas soybean pulp and yeast extract were employed as nitrogen sources in this experiment. In each formulated medium, other minerals and salts such as Tween 80, sodium acetate, potassium hydrogen phosphate, diammonium hydrogen citrate, magnesium sulphate heptahydrate and manganese sulphate tetrahydrate were added according to the composition of control MRS medium. Tween 80 and potassium hydrogen phosphate were reported to have a positive effect on the growth of microorganisms [48] since both elements can maintain pH value to prolong a fermentation process.

Table 1 shows the viable cell count of *L. plantarum* RI 11 grown in 16 formulated media, and both control MRS medium and CRMRS medium for seven days of incubation. The viable cell count of *L. plantarum* RI 11 decreased significantly ($p < 0.05$) for the first five days of incubation when grown in F1 medium (basal media with no carbon and nitrogen source) and no growth was detected after day 5 of incubation. Under unfavorable conditions, all bacteria will develop starvation-survival strategy to sustain their growth [81,82] and bacterial cell will enter a dormant state [83–85] when grown in basal medium without carbon and nitrogen sources. F4 medium (glucose, molasses and soybean pulp) and F5 medium (rice straw, yeast extract and soybean pulp) recorded the highest viable cell count of log 11 CFU/mL at 4-days of incubation, which was also noted at Section 2.1, whereby nitrogen source of soybean pulp [86] was sufficient to promote the growth of *L. plantarum* RI 11 [48,49] and too much of nitrogen source was reported to cause cell death [87]. Soybean pulp has been reported to increase the cell population of *Lactobacillus* sp. [86,88], whereas rice straw can also act as the nitrogen source [55]. All nitrogen sources used in this study were organic and especially yeast extract [87] in formulated F5 medium supported the rapid growth and high viable count of *L. plantarum* RI 11. Yeast extract that consists of amino acids, peptides, vitamins and carbohydrates [87] is a good nitrogen source to stimulate metabolic stress [88]. In addition, yeast extract also contains unidentifiable growth factors that can stimulate the growth of microorganisms [83].

Table 1. Viable cell count of *Lactobacillus plantarum* RI 11 grown in media designed by FFD.

Day	Viable Cell Density(Log CFU/mL)								
	F1	**F2**	**F3**	**F4**	**F5**	**F6**	**F7**	**F8**	**F9**
0	7.99 ± 0.04^{Ad}	8.53 ± 0.00^{Db}	9.43 ± 0.21^{Ba}	9.25 ± 0.29^{Ea}	9.40 ± 0.19^{Da}	9.25 ± 0.15^{Fa}	8.48 ± 0.04^{Cbc}	8.28 ± 0.03^{Dbcd}	7.91 ± 0.10^{Dd}
1	7.85 ± 0.30^{Ae}	9.62 ± 0.26^{Ee}	9.70 ± 0.04^{Ba}	9.74 ± 0.03^{Da}	9.63 ± 0.11^{Dab}	9.67 ± 0.04^{DEa}	8.43 ± 0.02^{Cd}	9.50 ± 0.04^{Aab}	9.56 ± 0.09^{Cab}
2	6.56 ± 0.14^{Bj}	9.73 ± 0.03^{Bd}	9.73 ± 0.02^{Bd}	9.73 ± 0.02^{Dd}	9.64 ± 0.04^{Dd}	9.76 ± 0.01^{Dd}	7.57 ± 0.08^{Di}	9.22 ± 0.04^{Be}	10.50 ± 0.01^{Ac}
3	6.40 ± 0.10^{Bc}	10.05 ± 0.07^{Bc}	10.14 ± 0.09^{Aab}	10.45 ± 0.03^{BCab}	10.46 ± 0.05^{Cab}	10.73 ± 0.01^{Ba}	9.05 ± 0.19^{Babc}	8.13 ± 0.14^{DEabc}	9.41 ± 0.12^{Cab}
4	5.92 ± 0.09^{Ck}	10.61 ± 0.23^{Ac}	9.20 ± 0.01^{Cef}	11.76 ± 0.01^{Aa}	11.76 ± 0.01^{Aa}	10.99 ± 0.05^{Bb}	8.68 ± 0.02^{Chi}	8.74 ± 0.10^{Cghi}	9.32 ± 0.07^{Ce}
5	5.30 ± 0.00^{Dj}	9.25 ± 0.18^{Cg}	10.42 ± 0.09^{Ad}	10.74 ± 0.01^{Bc}	11.69 ± 0.04^{Aa}	11.43 ± 0.12^{Ab}	9.49 ± 0.09^{Agf}	9.43 ± 0.04^{ABgf}	10.54 ± 0.07^{Acd}
6	0 ± 0^{Ej}	9.77 ± 0.08^{Bc}	9.26 ± 0.16^{Cde}	10.34 ± 0.06^{BCb}	11.10 ± 0.07^{Da}	9.43 ± 0.13^{EFd}	7.46 ± 0.06^{Di}	7.94 ± 0.11^{Eh}	10.42 ± 0.06^{Ab}
7	0 ± 0^{Eh}	10.20 ± 0.17^{ABb}	10.19 ± 0.16^{Ab}	10.14 ± 0.16^{Dbc}	11.29 ± 0.07^{Ba}	10.05 ± 0.16^{Cbc}	7.28 ± 0.17^{Dg}	7.90 ± 0.10^{Ef}	10.01 ± 0.11^{Bbc}
	F10	**F11**	**F12**	**F13**	**F14**	**F15**	**F16**	**MRS**	**CRMRS**
0	8.08 ± 0.17^{Dcd}	8.28 ± 0.04^{BCbcd}	8.12 ± 0.14^{CDbcd}	8.20 ± 0.01^{Dbcd}	8.13 ± 0.12^{Cbcd}	8.29 ± 0.16^{Ebcd}	8.19 ± 0.05^{Ebcd}	9.13 ± 0.03^{Ea}	9.01 ± 0.19^{Ca}
1	9.67 ± 0.08^{Aa}	9.29 ± 0.08^{Abc}	7.89 ± 0.08^{De}	9.04 ± 0.09^{Ac}	8.23 ± 0.10^{Cd}	9.69 ± 0.04^{Ca}	9.73 ± 0.02^{Ca}	9.65 ± 0.00^{Dab}	9.62 ± 0.04^{Bab}
2	9.68 ± 0.06^{Ad}	9.14 ± 0.02^{Ae}	7.87 ± 0.17^{Dh}	9.00 ± 0.02^{Af}	8.26 ± 0.08^{Cg}	10.76 ± 0.01^{Aa}	10.56 ± 0.02^{Ac}	10.73 ± 0.01^{Aab}	9.70 ± 0.02^{ABd}
3	8.45 ± 0.11^{Cabc}	8.07 ± 0.03^{Cabc}	7.86 ± 0.05^{Dbc}	8.44 ± 0.07^{Cabc}	8.24 ± 0.02^{Cabc}	10.76 ± 0.01^{Aa}	8.42 ± 0.05^{Eabc}	10.10 ± 0.08^{Cab}	9.99 ± 0.16^{Aab}
4	9.15 ± 0.02^{Bef}	8.21 ± 0.08^{BCj}	8.52 ± 0.16^{ABi}	8.60 ± 0.05^{BCi}	8.59 ± 0.14^{Bi}	9.99 ± 0.08^{Bd}	8.92 ± 0.18^{Dfgh}	10.36 ± 0.05^{Bc}	8.99 ± 0.07^{Cfg}
5	9.40 ± 0.08^{Bgf}	8.36 ± 0.09^{Bi}	8.73 ± 0.02^{Ah}	8.77 ± 0.01^{Bh}	8.32 ± 0.04^{Ci}	10.16 ± 0.08^{Be}	9.97 ± 0.05^{BCe}	10.62 ± 0.08^{Acd}	9.50 ± 0.08^{Bf}
6	9.31 ± 0.06^{Bde}	9.11 ± 0.08^{Aef}	8.40 ± 0.10^{ABCg}	8.20 ± 0.06^{Dgh}	8.95 ± 0.15^{Af}	9.31 ± 0.07^{Dde}	10.16 ± 0.11^{Bb}	9.43 ± 0.05^{Dd}	8.46 ± 0.06^{Dg}
7	9.30 ± 0.01^{Bd}	7.66 ± 0.14^{Df}	8.35 ± 0.07^{BCe}	7.89 ± 0.10^{Ef}	9.08 ± 0.11^{Ad}	9.36 ± 0.06^{Dd}	9.78 ± 0.08^{Cc}	9.04 ± 0.20^{Ed}	8.54 ± 0.04^{De}

Note: Mean ± SEM. The same superscript within the same column (lowercase) indicates not significantly different ($p < 0.05$). Mean ± SEM. The same superscript within the same row (uppercase) indicates not significantly different ($p < 0.05$).

The growth of *L. plantarum* RI 11 increased substantially when soluble sugars such as glucose or molasses was combined with soybean pulp as the organic nitrogen source [83,85]. Amino acids were made available upon the degradation of nitrogen source [83]. In addition, media containing glucose and disaccharides of glucose would decrease the pH of the media [83], which was more favorable for the growth of *L. plantarum* RI 11. Different carbon sources will enter the metabolic network at different entry points of glycolysis [89]. Nerveless, the bacteria will usually use preferred substrates or substrates that give a higher growth rate [90,91]. For an example, F4 medium that contained both glucose and molasses that function as carbon source, *L. plantarum* RI 11 has to choose either glucose or molasses as its sole carbon source. Selection of the carbon sources is made at the level of carbohydrate-specific induction [83] known as carbon catabolite repression [90,91]. For instance, the presence of sucrose in molasses will promote the high biomass yield due to the presence of intermediates in the metabolism of fructose. Too many carbon and nitrogen sources as found in F15 and F16 media would increase the cell density until a maximum point in which it would decrease subsequently, due to inhibitory effects of the substrates [90,92]. Between soluble and complex carbon sources, cellulose-producing microorganisms will usually select complex carbon sources, as reported by Dou et al. [93].

A similar level of cell densities was detected for both F4 and F5 media, although both media sources have different compositions. It could be due to both yeast extract and soybean pulp, in high concentration, acting as carbon sources as well as nitrogen sources [90]. F4 and F5 media contained less substrates when compared to F15 and F16 media, which contributed to different cell populations obtained. Therefore, the results of this experiment demonstrated that the medium compositions of F4 (glucose, molasses and soybean pulp) and F5 (rice straw, soybean pulp and yeast extract) contributed comparable cell populations with control MRS medium.

2.4. Fractional Factorial Design for Enhancement of Cellulolytic and Hemicellulolytic Enzyme Productions

The corresponding extracellular cellulolytic (endoglucanase-CMCase, exoglucanase-avicelase and, β-glucosidase) and hemicellulolytic (mannanase) enzyme activities of *L. plantarum* R1 11 that grown in 16 different media sources formulated by FFD were determined subsequently. Figure 9 illustrated that *L. plantarum* R1 11 produced different levels of extracellular cellulolytic and hemicellulolytic enzyme activities in various formulated media sources. Surprisingly, the medium that produced high viable cell density of *L. plantarum* R1 11 in Section 2.3 did not correspond to the medium that generated high cellulolytic and hemicellulolytic specific enzyme activities in this experiment.

There was no production of cellulolytic and hemicellulolytic enzymes when *L. plantarum* R1 11 was grown in F1 medium (basal medium without any carbon and nitrogen sources). *L. plantarum* R1

11 was most likely to be in a dormant stage, repressing its metabolic activity [82] until the favorable growth condition was available [87]. Both carbon and nitrogen sources were highly important for the growth and production of metabolites by microorganisms [94,95].

Generally, less enzymes with low level of enzymatic activities were noted when *L. plantarum* R1 11 was grown in F10, F15 and F16 formulated media. F10 was a basal medium supplemented with both soluble (glucose) and heterogeneous renewable natural polymers of PKC, yeast extract and soybean waste, indicating that the supplementation of many carbon sources in the growth medium would not necessarily enhance the production of cellulolytic and hemicellulolytic enzymes by *L. plantarum* R1 11. Glucose is a simple carbohydrate [44], which acts as a good energy source [83]. However, glucose would repress the cellulolytic and hemicellulolytic enzyme activities [83,85,93]. Nevertheless, a low endoglucanase activity (0.90 μg/min/mg) was detected at pH 5. The repression effect of PKC was greater and hence only a low amount of activity was detected, although *L. plantarum* RI 11 was able to grow on PKC [93]. F15 and F16 media that consisted PKC demonstrated low level of mannanase activity. However, a higher level of mannanase activity was detected in F16 medium. PKC has been reported to be a good substrate for mannanase production [94]. PKC, which comprises of mannan, will induce the production of mannanase [72]. F16 also contained glucose, rice straw, molasses, soybean pulp and yeast extract. Hence, PKC was a better substrate to produce mannanase by *L. plantarum* RI 11 [72,84].

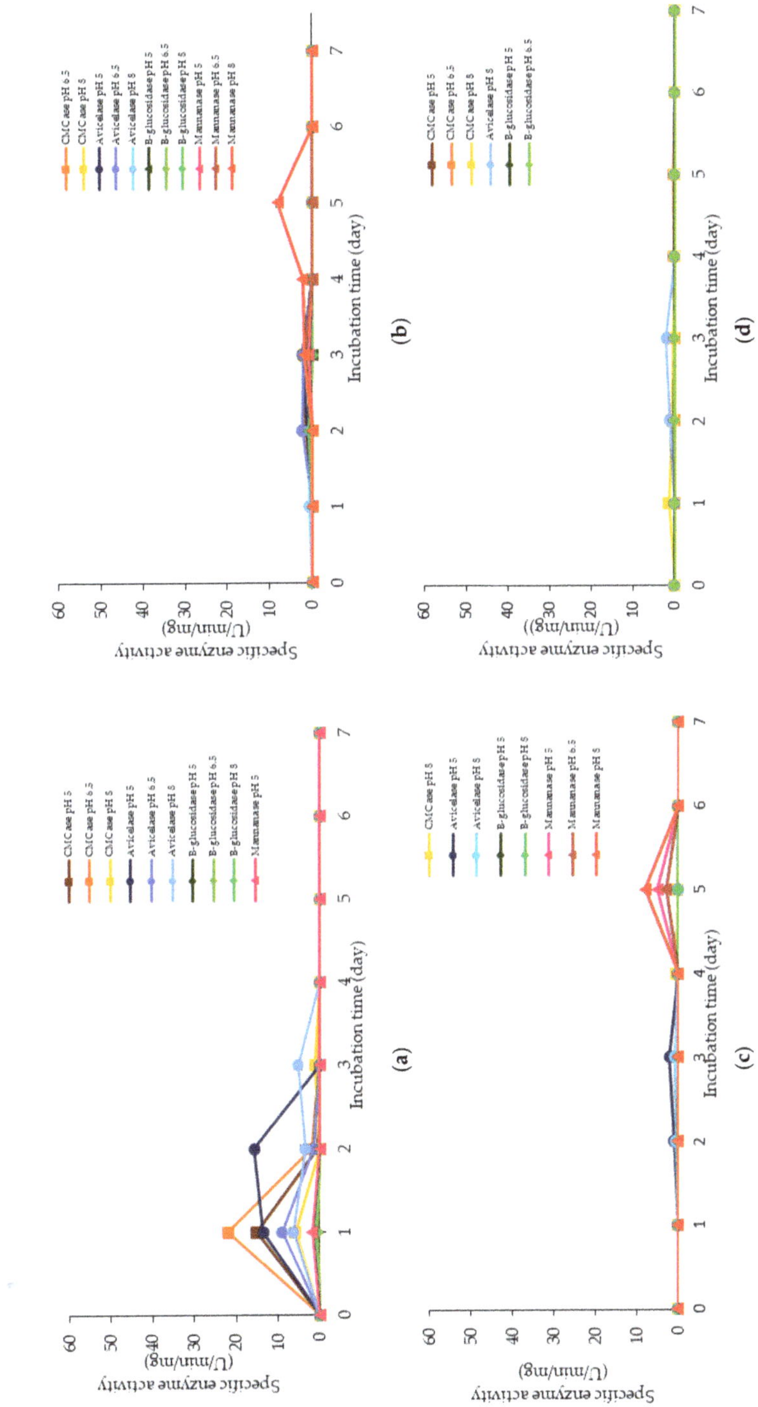

Figure 9. *Cont.*

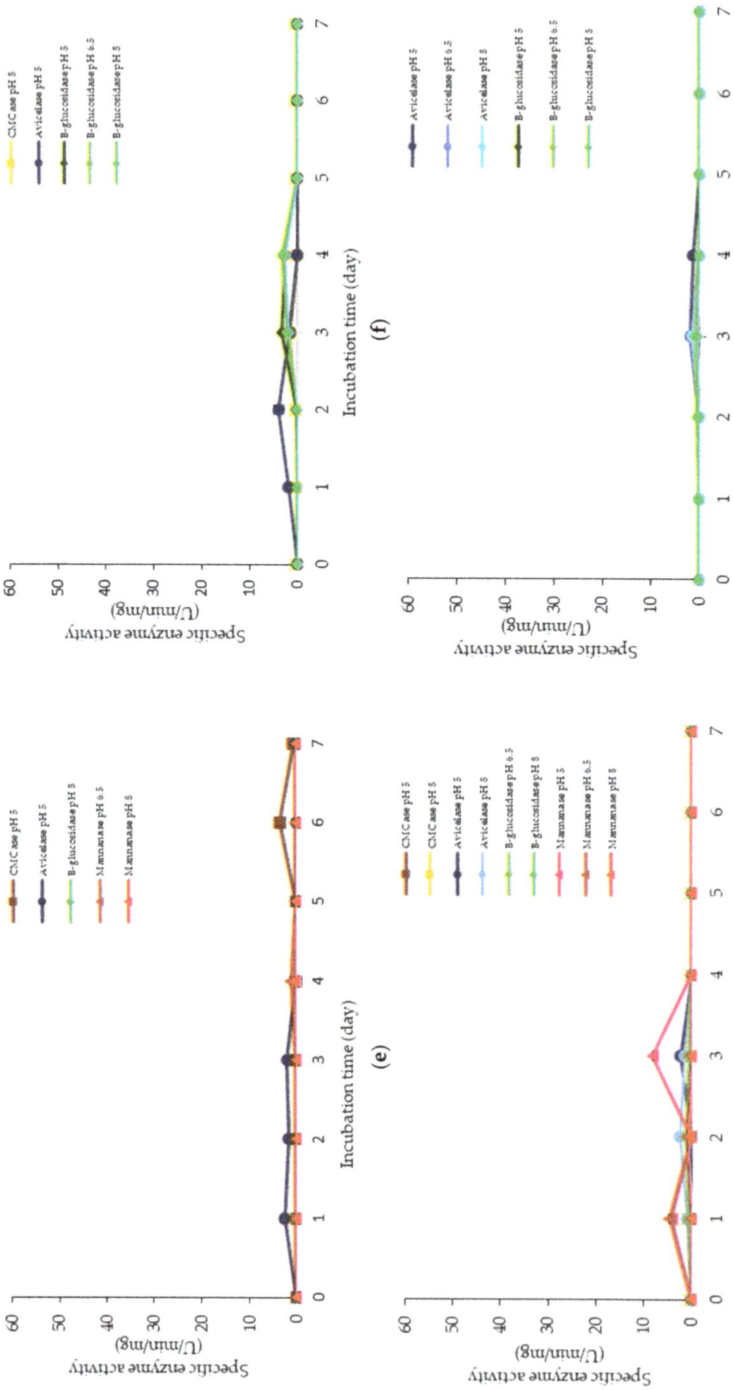

Figure 9. *Cont.*

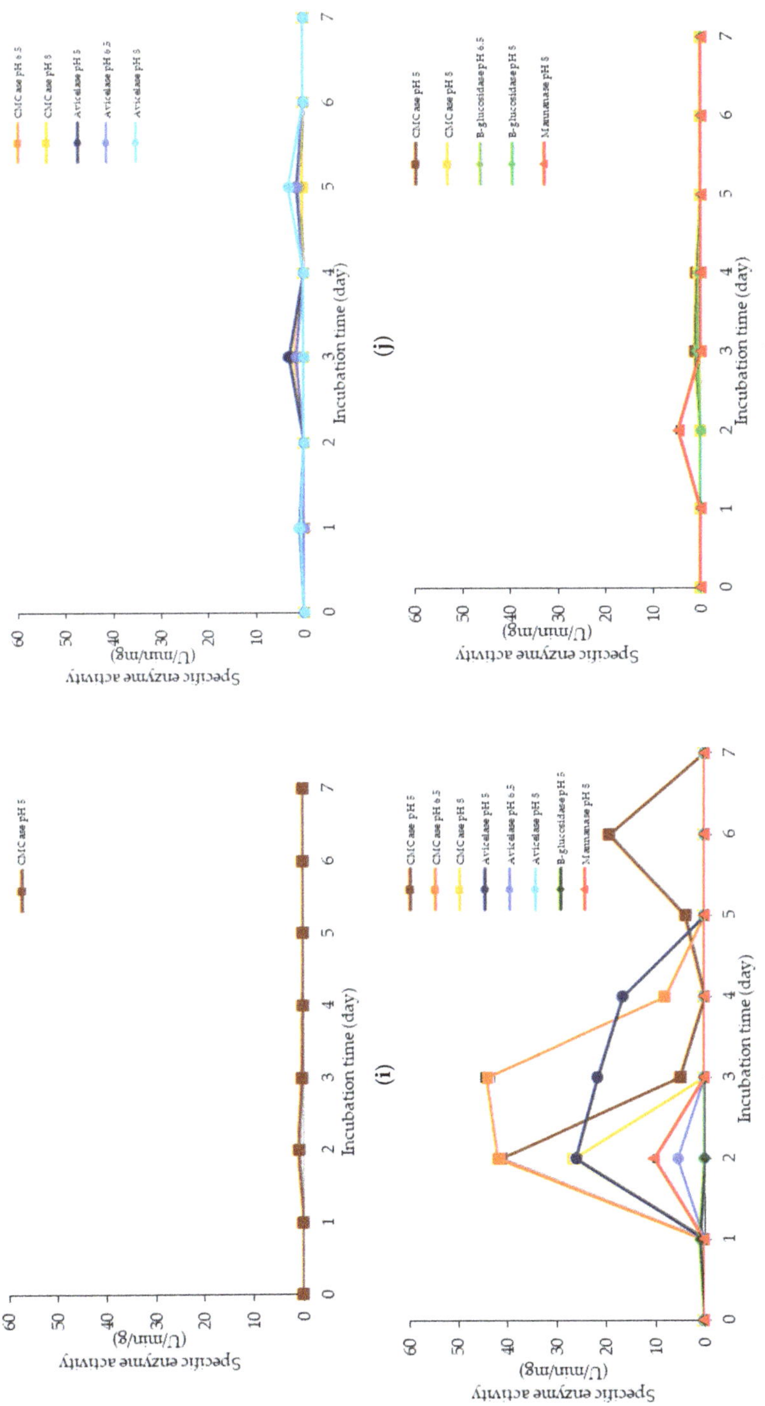

Figure 9. *Cont.*

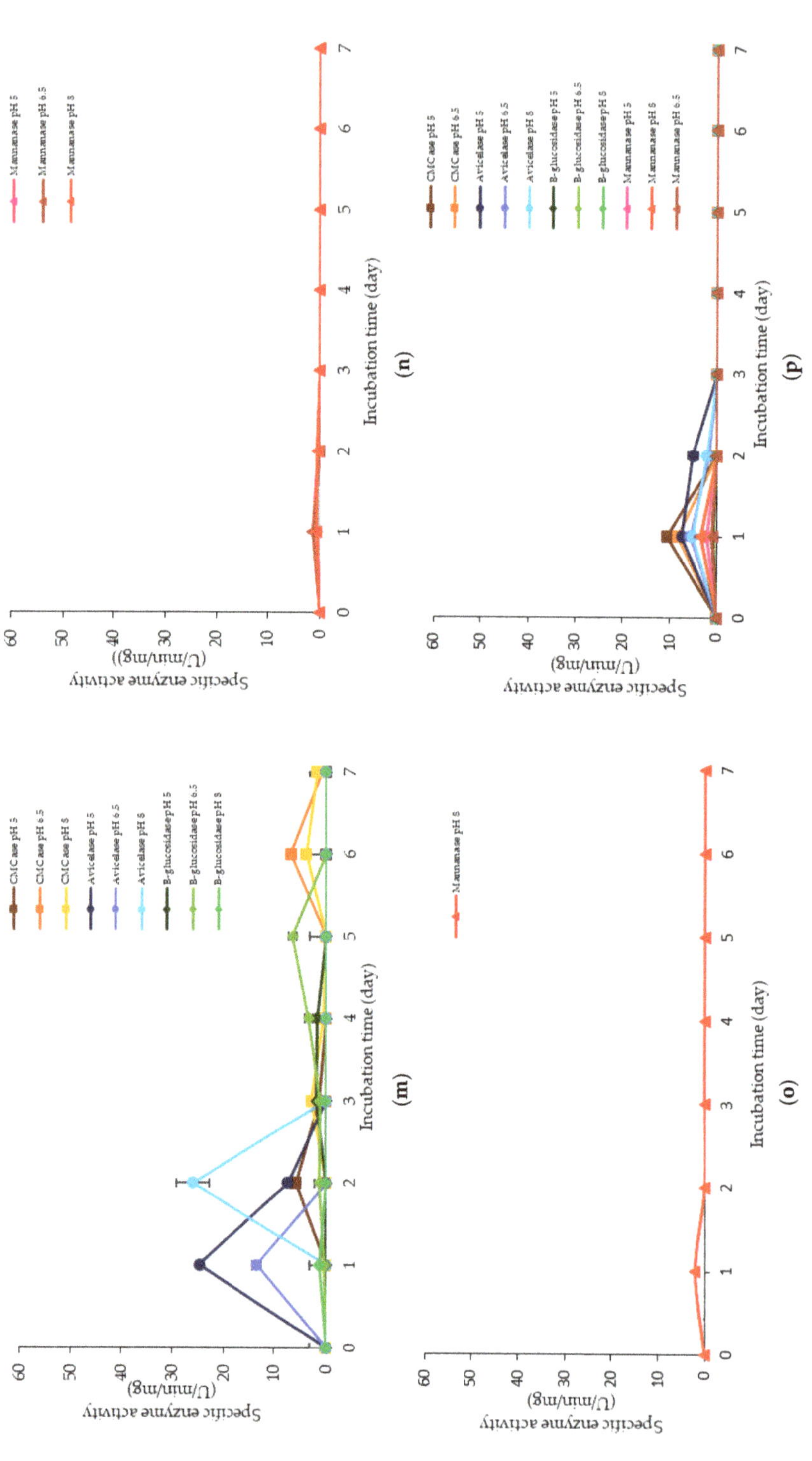

Figure 9. *Cont.*

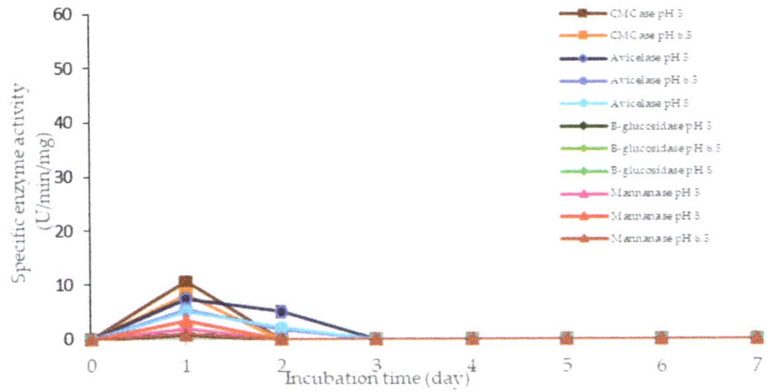

(q)

Figure 9. Specific extracellular cellulolytic and hemicellulolytic activities of Lactobacillus plantarum RI 11 grown in (**a**) F2; (**b**) F3; (**c**) F4; (**d**) F5; (**e**) F6; (**f**) F7; (**g**) F8; (**h**) F9; (**i**) F10; (**j**) F11; (**k**) F12; (**l**) F13; (**m**) F14; (**n**) F15; (**o**) F16; (**p**) MRS; (**q**) CRMRS during seven days of incubation. The enzyme activities were detected at three different pH levels. The error bars represent the standard deviation of specific enzyme activity of triplicate samples ($n = 3$). The medium composition was listed in Table 4.

Similarly, a low level of enzyme activity was detected for F5, F6, F7, F9, F11, and F13 media. Different enzyme productions were detected, but the levels of enzymatic activity were insignificant ($p>0.05$). F5 medium was a basal media supplemented with rice straw, yeast extract and soybean pulp. The positive effects of yeast extract and soybean pulp as a nitrogen source [95] have been demonstrated on the viable cell density of the microorganisms [95]. In this experiment, rice straw was acted as both carbon and nitrogen sources [84]. Hence, most of the medium components contributed to the growth of the cell, resulting in microbial biomass yield. Rice straw, which consisted of a similar proportion of cellulose and hemicellulose [96,97], induced the production of cellulolytic enzymes instead of hemicellulolytic enzymes by *L. plantarum* RI 11. Figure 9d shows similar levels of endo-, exoglucanase and β-glucosidase at different pH that produced at different incubation times. Endoglucanase specific activity that active from pH 5–8 was detected from day-1 to day-5 of incubation, whereas exoglucanase was detected at day-2 and day-3 of incubation. Moreover, β-glucosidase, on the other hand, was detected at day-2 to day-5 of incubation, which suggested strongly the synergistic cellulose degradation by *L. plantarum* RI 11 [76,96]. β-glucosidase enzyme production, which happened later than the incubation time, supported a previous report by Lee et al. [46].

F6 medium supplemented with glucose, rice straw and soybean pulp has recorded the highest viable cell population with corresponding extracellular cellulolytic and hemicellulolytic enzyme productions in comparison to F5 medium that induced only cellulolytic enzymes. This was most probably due to nutrient content of both rice straw and soybean pulp that consisted both cellulose and hemicellulose fractions [98,99]. However, soybean pulp is rich in mono- and oligosaccharides [86], which might contribute to the inhibition of cellulolytic and hemicellulolytic enzyme productions. Similar observation was noted in Figure 9h, whereby F9 medium was consisted of PKC and soybean pulp. Despite PKC being a good substrate for various hydrolytic enzyme productions [94], the presence of soybean pulp would inhibit the production of various hydrolytic enzymes [88]. Increased specific activity of all cellulolytic enzymes was observed in F7 medium, which was supplemented with molasses and rice straw. Therefore, the increased specific activity of all cellulolytic enzymes could be attributed to molasses and rice straw supplementations, which were highly heterogeneous in nature. Molasses is rich in mono- and oligosaccharides [46] and oligosaccharides also can be generated by incomplete digestion of endo- and exoglucanase enzymes [99,100]. However, since *L. plantarum* RI 11 was able to produce β-glucosidase enzymes [70], it was most likely that the oligosaccharides were converted to glucose. Hence, no apparent effect of enzyme inhibition was noted in this experiment.

Both F11 and F13 media contained PKC and yeast extract, despite a difference with the addition of molasses in the former medium, and rice straw for the latter medium. Surprisingly, endoglucanase enzyme activity was detected in both media, probably due to the presence of PKC [91]. In addition, exoglucanase was also detected in F11 medium. However, β-glucosidase was detected in F13 medium, implying that molasses promoted exoglucanase enzyme production and rice straw promoted β-glucosidase enzyme production by *L. plantarum* RI 11.

F3, F4 and F8 media was observed to promote the production of all cellulolytic and hemicellulolytic enzymes simultaneously with mannanase being recorded as the highest activity of approximately 8 U/min/mg. F3 and F4 media comprised molasses and soybean pulp with the addition of yeast extract for F3 medium and addition of glucose for F4 medium, respectively. F8 medium, on the other hand, comprised of glucose, molasses, rice straw and yeast extract. The enzyme production for both F3 and F4 media occurred at later incubation time as the enzyme production of F8 medium was occurred at early stage of the incubation. For F3 and F4 media, cellulolytic enzyme activity was detected at the early stage of incubation, then followed by the mannanase production at the later of incubation. Molasses, which is consisted mainly mono-and oligosaccharides [55,60–62], induced the production of endo- and exoglucanase in this experiment. In comparison, *L. plantarum* RI 11 required a longer time to fully utilize soybean pulp and hence the mannanase was produced at the later stage of incubation. A similar observation was noted for F8 medium. However, mannanase was detected at day-1 and day-3 of incubation along with other cellulolytic enzyme productions.

F12 medium recorded the highest specific activity of endo-, exoglucanase and mannanase as compared to the other formulated media. F14 medium, on the other hand, recorded the highest β-glucosidase enzyme production. Both F12 and F14 media were the best in producing the cellulolytic and hemicellulolytic enzymes concomitantly. Approximately 41 μg/min/mg of endoglucanase specific activity was detected at pH 5 and 6.5, respectively. However, the endoglucanase specific activity was reduced to half at pH 8. Similar levels of endoglucanase at pH 8 was detected for exoglucanase at pH 5. Approximately 26 μg/min/mg was recorded for exoglucanase at day-2 of incubation. F12 medium, which comprised of glucose, molasses and PKC, was a good substrate to produce these enzymes at day-2 of incubation. PKC and molasses induced the production of endoglucanase, exoglucanase and mannanase enzymes in this experiment. The mono- and oligosaccharides of molasses and PKC were favorable medium components to produce both cellulolytic and hemicellulolytic enzymes, rather than inhibiting as shown by soybean pulp [89,90]. The results obtained in this study have clearly shown that both endo- and exoglucanase activities were highly correlated with each other [98,101] and reacted synergistically for efficient agriculture biomasses degradation. Approximately, 10 μg/min/mg of mannanase was detected only at pH 8. It could be deduced that the mannanase production by *L. plantarum* RI 11 was active at pH 8, which was similar to the mannanase production by *Bacillus* sp. that was also active at pH 8 [102].

As for F14 medium, exoglucanase specific activity was reported to be similar to those produced by F12 medium that were active at a broad pH range. F14 medium that contained glucose, rice straw and PKC was a good substrate for exoglucanase, but not for endoglucanase production. A lower level of endoglucanase was detected in F14 medium, supporting a previous notion that molasses supported endoglucanase production. However, the highest β-glucosidase was detected in F14 medium at pH 5, which was in agreement with the study of carbon and nitrogen sources' influences on the growth and sporulation of *Bacillus thuringiensis* for biopesticide production [103]. Despite sugar or monosaccharides possibly inhibiting enzyme production, it only affected other cellulolytic enzymes but not β-glucosidase [103]. Hence, higher β-glucosidase specific activity was resulted, although glucose was present in the media.

3. Materials and Methods

3.1. Microorganisms and Maintenance

L. plantarum RI 11 isolated from Malaysian fermented food (*ikan rebus*) was obtained from the Laboratory of Industrial Biotechnology, Department of Bioprocess Technology, Faculty of Biotechnology and Biomolecular Sciences, Universiti Putra Malaysia (UPM) [35]. The stock culture of *L. plantarum* RI 11 was maintained at −20 °C in MRS broth (Merck, Darmstadt, Germany) containing 20% (*v*/*v*) of glycerol (Merck) [46]. The reviving procedure of *L. plantarum* RI 11 was performed according to the method described by Foo et al. [35].

3.2. Media Formulation for *L. plantarum* RI 11

The active *L. plantarum* strain was washed once with sterile 0.85% (*w*/*v*) NaCl (Merck) solution and adjusted to 10^9 CFU/mL to be used as inoculum. The adjusted inoculum of *L. plantarum* RI 11 was then grown in basal media containing 1 g L^{-1} Tween 80 (Merck), 5 g L^{-1} sodium acetate (Merck, Darmstadt, Germany), 2 g L^{-1} potassium hydrogen phosphate (Merck), 2 g L^{-1} diammonium hydrogen citrate (Merck), 0.2 g L^{-1} magnesium sulphate heptahydrate (Merck) and 0.04 g L^{-1} manganese sulphate tetrahydrate (Merck), which was supplemented with different combination of carbon and nitrogen sources. The renewable natural polymers that were used as carbon sources were rice straw (15.46 g/L), molasses (25.09 g/L) and PKC (11.86 g/L), whereas soybean pulp (51.54 g/L) and yeast extract (28.34 g/L) were used as nitrogen sources in this experiment.

Table 2 shows six media combinations of carbon and nitrogen sources supplemented in basal medium were designed in this experiment, whereas Table 3 shows the carbon and nitrogen content of each renewable natural polymers that were used for the formulation of growth media in this study. Glucose (Merck), PKC, rice straw and molasses were employed as carbon sources, whereas yeast extract and soybean pulp were used as nitrogen sources. The amount of each carbon and nitrogen source supplemented in the basal medium was determined with the reference to the carbon and nitrogen contents of commercially available MRS broth that were employed as a control medium in this study. Yeast extract was purchased from Ohly (Hamburg, Germany). Both glucose and yeast extract were employed as the control for carbon and nitrogen sources in this experiment.

Table 2. Carbon and nitrogen sources supplemented to each formulated medium.

Medium	C Source	Concentration (g/L)	N Source	Concentration (g/L)
M1	Rice straw	15.46	Yeast extract	28.34
M2	Rice straw	15.46	Soybean pulp	51.54
M3	PKC	11.86	Yeast extract	28.34
M4	PKC	11.86	Soybean pulp	51.54
M5	Molasses	25.09	Yeast extract	28.34
M6	Molasses	25.09	Soybean pulp	51.54
Control	Glucose	20.00	Yeast extract	28.34

Table 3. Carbon and nitrogen contents for renewable natural polymers employed in this study.

	Component	Carbon Content (%)	Nitrogen Content (%)
Renewable natural polymers	PKC	31.00	-
	Molasses	14.65	-
	Rice straw	23.77	-
	Soybean pulp	-	35.77
Reconstituted MRS medium	Glucose	18.38	-
	Yeast extract	-	65.05

3.3. *Extracellular Cellulolytic and Hemicellulolytic Enzymes Activities of* L. plantarum *RI 11*

Ten (10) % (*v*/*v*) of the active *L. plantarum* RI 11 culture (log 9 CFU/mL) was transferred into 9 mL of the basal medium supplemented with renewable natural polymers derived from agro wastes and incubated for 7 days at 30 °C. Samples were harvested every 2-h intervals for 7 days of incubation. The cell biomass of *L. plantarum* RI 11 was collected by centrifugation at 10,000× *g* for 15 min (Hitachi Himac CR22611 High Speed Refrigerated Centrifuge, Hitachi Koki Co Ltd., Tokyo, Japan) [46]. The cell population of *L. plantarum* RI 11 of each sampling interval was freshly analysed to determine the cell viability, whereas the supernatant was filtered by 0.2 µm cellulose acetate hydrophilic syringe filter (Minisart, Sartorius Stedim Biotech GmbH, Göttingen, Germany) to obtain cell-free supernatant (CFS). CFS was then used for the determinations of protein concentration, extracellular cellulolytic and hemicellulolytic enzymatic activities in triplicate analyses.

3.4. *Protein Concentration Determination*

The protein concentration was determined in triplicate analyses by using the Bradford reagent (Sigma, St. Louis, MO, USA) method [104] and bovine serum albumin (BSA) (Sigma) was used as a reference.

3.5. *Cell Viability Determination*

Standard plate count method [43] was conducted to determine the cell population of *L. plantarum* RI 11. The cell pellet was washed once with 0.85% (*w*/*v*) sodium chloride (NaCl) (Merck) solution. Tenfold serial dilution (10^0 to 10^{-9}) was conducted for colony forming unit/mL (CFU/mL) determination using 0.85% (*w*/*v*) NaCl solution as the diluent. An aliquot of 50 µL from the respective diluted cell population was spread evenly on the MRS agar and incubated at 30 °C for 48 h. The cell viability determination was conducted in triplicate.

3.6. Quantification of Extracellular Cellulolytic and Hemicellulolytic Enzyme Activities

Extracellular cellulolytic and hemicellulolytic enzyme assays were determined in triplicate with three types of assay buffers: 0.1 M sodium acetate (Merck) pH 5 buffer, 0.1 M sodium phosphate (Merck) pH 6.5 buffer and 0.1 M Tris-HCl (Merck) pH 8 buffer as described by Lee et al. [46]. The substrates used were 2.0% (*w*/*v*) of carboxymethylcellulose (Sigma), 0.5% (*w*/*v*) of avicel pH10 (Fluka, Tokyo, Japan), 5 mM 4-nitrophenyl- β-D-glucopyranoside (PNPG) (Sigma) and 1% (*w*/*v*) of locust bean gum from ceratonia silique seed (LBG) (Sigma) for endo-β-1,4-glucanase (CMCase), exo-β-1,4-glucanase (avicelase), β-glucosidase and mannanase activity, respectively [63–65,105]. CMCase, avicelase and β-glucosidase assays represented cellulolytic enzyme activities, while mannanase assay denoted hemicellulolytic enzyme activity. CFS collected from different medium formulation and incubation times were incubated with respective substrates prepared in three different buffers and incubated at 37 °C for 1 h. The reducing sugar was determined by using dinitrosalicyclic acid (DNS, Sigma) assay method for CMCase, avicelase and mannanase assays [63–65,105–107]. The absorbance of each sample was determined at 540 nm by Cary 50 Probe UV-visible spectrophotometer (Agilent Technologies, Santa Clara, CA, USA) [63–65,105].

As for β-glucosidase assay, the CFS was incubated with respective substrate prepared in three different buffers as described above at 37 °C for 30 min, followed by adding 500 μL of cold 0.5 M sodium carbonate (Sigma) to terminate the enzyme activity and the absorbance of assay mixture was determined at 373 nm [63] by Cary 50 Probe UV-visible spectrophotometer (Agilent, Santa Clara, CA, USA). The positive controls of all enzyme assays comprised CFS and respective buffers, while negative control comprised respective substrates and buffers. Glucose was utilized as the CMCase and avicelase reference. As for β-glucosidase and mannanase, p-nitrophenol and mannose were used as the references, respectively. The specific enzyme activity (U/min/mg) for CMCase, avicelase and mannanase was defined as the amount of reducing sugar produced (μg) per minute of incubation time in 1 mg of protein under experimental conditions. The enzyme activity (U/min/mg) of β-glucosidase was defined as the amount of enzyme that released 1 nmol of p-nitrophenol per minute of incubation time in 1 mg of protein under experimental conditions.

3.7. Fractional Factorial Design for Growth Enhancement and L. plantarum RI 11

Three replicates for each of the basal media supplemented with renewable natural polymers were prepared according to Table 4. The renewable natural polymers that used as carbon sources were rice straw (15.46 g/L), molasses (25.09 g/L) and PKC (11.86 g/L), whereas soybean pulp (51.54 g/L) and yeast extract (28.34 g/L) were used as nitrogen sources in this experiment. In each formulated medium, other minerals and salts such as Tween 80 (1.00 g/L), sodium acetate (5.00 g/L), potassium hydrogen phosphate (2.00 g/L), diammonium hydrogen citrate (2.00 g/L), magnesium sulphate heptahydrate (0.20 g/L) and manganese sulphate tetrahydrate (0.05 g/L) were added as the additional nutrients. The amount of the minerals and salts followed the compositions of control MRS medium.

Table 4. FFD matrix for six variables with coded values for the growth enhancement, cellulolytic and hemicellulolytic enzyme productions of *Lactobacillus plantarum* RI 11.

Run	Factor					
	A	B	C	D	E	F
F1	−1	−1	−1	−1	−1	−1
F2	1	−1	−1	−1	1	−1
F3	−1	1	−1	−1	1	1
F4	1	1	−1	−1	−1	1
F5	−1	−1	1	−1	1	1
F6	1	−1	1	−1	−1	1
F7	−1	1	1	−1	−1	−1
F8	1	1	1	−1	1	−1
F9	−1	−1	−1	1	−1	1
F10	1	−1	−1	1	1	1
F11	−1	1	−1	1	1	−1
F12	1	1	−1	1	−1	−1
F13	−1	−1	1	1	1	−1
F14	1	−1	1	1	−1	−1
F15	−1	1	1	1	−1	1
F16	1	1	1	1	1	1

MRS and CRMRS was used as a control. Factor A, Glucose (20.00 g/L); B, Molasses (25.09 g/L); C, Rice straw (15.46 g/L); D, PKC (11.86 g/L); E, Yeast extract (28.34 g/L); F, Soybean pulp (51.54 g/L).

3.8. Statistical Analysis

Analysis of variance (ANOVA) was used to determine the significant difference between results. The significance of the differences between the mean was compared by using Duncan's Multiple Range Test System at *p*-value less than 0.05. Statistical analyses were performed using SAS statistical software (version 9.4). The results of statistical analysis were presented as mean ± standard error of the mean (SEM).

4. Conclusions

The results obtained in current study suggested that renewable natural polymers such as rice straw, molasses, PKC and soybean pulp were able to induce *L. plantarum* RI11 to produce versatile extracellular cellulolytic and hemicellulolytic enzymes that active from acidic to alkaline pH conditions. Diverse levels of a few isozymes of extracellular cellulolytic and hemicellulolytic enzymes that produced within the 24 h of incubation warrant extensive research effort in the mechanism elucidation of these inducible enzymes by the renewable natural polymers. Different natural biomass substrates elicited different effects on *L. plantarum* RI11 variability and cell population, as well as the production of various extracellular cellulolytic and hemicellulolytic enzymes. The versatile extracellular cellulolytic and hemicellulolytic enzymes production by *L. plantarum* RI11 that active under broad pH range possessed vast application potential in various industries. Amongst the formulated media, F12 medium that consisted of glucose, molasses and PKC promoted the growth of *L. plantarum* RI 11 that comparable to commercially available control MRS medium together with the concomitant production of the highest activities of extracellular endoglucanase, exoglucanase, β-glucosidase and mannanase enzymes that active under broad pH conditions. Hence, *L. plantarum* RI 11 isolated from Malaysian food is a versatile cellulolytic and hemicellulolytic enzymes' producer and as a potential bioagent for biotransformation of renewable natural polymers. This new finding warrants future explorations in the production of extracellular cellulolytic and hemicellulolytic enzymes by the non-toxigenic and non-pathogenic LAB.

Author Contributions: Conceptualization, H.L.F. and T.C.L.; Data curation, R.M. and H.L.F.; Formal analysis, N.A.M.Z.; Funding acquisition, H.L.F., T.C.L. and R.A.R.; Investigation, N.A.M.Z.; Methodology, N.A.M.Z., R.M. and H.L.F.; Project administration, H.L.F. and T.C.L.; Resources, H.L.F. and T.C.L.; Supervision, H.L.F. and R.M.; Validation, H.L.F., R.M. and R.A.R.; Writing—Original draft, N.A.M.Z. and H.L.F.; Writing—Review and editing, N.A.M.Z., H.L.F. and T.C.L. All authors have read and agreed to the published version of the manuscript.

Funding: This project was supported by the High Impact Putra Research Grant of Universiti Putra Malaysia (Grant No. UPM/700-2/1/GPB/2018/9598900).

Conflicts of Interest: The authors declare no conflict of interest.

Abbreviations

GRAS	Generally recognized as safe
LAB	Lactic acid bacteria
L. plantarum	*Lactobacillus plantarum*
MRS	deMan, Rogosa and Sharpe
PKC	Palm kernel cake
CFU	Colony forming unit
FFD	Fractional factorial design
CRMRS	Reconstituted MRS
CFS	Cell free supernatant
NaCl	Sodium chloride
DNS	3,5-dinitrosalicyclic acid
ANOVA	Analysis of variance
SEM	Standard error of mean

References

1. Singh, R.M.M. Role of Carbon and Nitrogen Sources in Bacterial Growth and Sporulation. *Appl. Microbiol.* **1971**, *22*, 131–132. [CrossRef]
2. Lochhead, A.G.; Burton, M.O. An Essential Bacterial Growth Factor Produced By Microbial Synthesis. *Can. J. Bot.* **1953**, *31*, 7–22. [CrossRef]
3. De Man, J.C.; Rogosa, M.; Sharpe, M.E. A Medium For The Cultivation Of Lactobacilli. *J. Appl. Bacteriol.* **1960**, *23*, 130–135. [CrossRef]
4. Hayek, S.A. Current Limitations and Challenges with Lactic Acid Bacteria: A Review. *Food Nutr. Sci.* **2013**, *4*, 73–87. [CrossRef]
5. Kirchman, D.L. The contribution of monomers and other low molecular weight compounds to the flux of DOM in aquatic ecosystems. In *Aquatic Ecosystems – Dissolved Organic Matter*; Findlay, S., Sinsabaugh, R.L., Eds.; Academic Press: New York, NY, USA, 2003; pp. 217–241.
6. Mikkelsen, D.; Flanagan, B.; Dykes, G.A.; Gidley, M. Influence of different carbon sources on bacterial cellulose production byGluconacetobacter xylinusstrain ATCC 53524. *J. Appl. Microbiol.* **2009**, *107*, 576–583. [CrossRef] [PubMed]
7. Ravindran, R.; Jaiswal, A.K. Microbial Enzyme Production Using Lignocellulosic Food Industry Wastes as Feedstock: A Review. *Bioengineering* **2016**, *3*, 30. [CrossRef]
8. Lim, Y.H.; Foo, H.L.; Loh, T.C.; Mohamad, R.; Rahim, R.A. Rapid Evaluation and Optimization of Medium Components Governing Tryptophan Production by Pediococcus acidilactici TP-6 Isolated from Malaysian Food via Statistical Approaches. *Molecules* **2020**, *25*, 779. [CrossRef]
9. Lim, Y.H.; Foo, H.L.; Loh, T.C.; Mohamad, R.; Rahim, R.A.; Idrus, Z. Optimized medium via statistical approach enhanced threonine production by Pediococcus pentosaceus TL-3 isolated from Malaysian food. *Microb. Cell Factories* **2019**, *18*, 125. [CrossRef]
10. Djekrif-Dakhmouche, S.; Gheribi-Aoulmi, Z.; Meraihi, Z.; Bennamoun, L. Application of a statistical design to the optimization of culture medium for α-amylase production by Aspergillus niger ATCC 16404 grown on orange waste powder. *J. Food Eng.* **2006**, *73*, 190–197. [CrossRef]

11. Guo, F.; Li, X.; Zhao, J.; Li, G.; Gao, P.; Han, X. Optimizing Culture Conditions by Statistical Approach to Enhance Production of Pectinase from Bacillus sp. Y1. *BioMed Res. Int.* **2019**, *2019*, 8146948. [CrossRef]
12. Abdullah, N.; Sulaim, F. The Oil Palm Wastes in Malaysia. In *Biomass Now - Sustainable Growth and Use*; IntechOpen: London, UK, 2013.
13. Hengniran, P. Future Potential of Forest and Agriculture Residues for the Energy Production in Thailand-Strategies for a Better Utilization. Ph.D. Thesis, University of Hamburg, Hamburg, Germany, 2010.
14. Pariza, M.W.; Johnson, E.A. Evaluating the Safety of Microbial Enzyme Preparations Used in Food Processing: Update for a New Century. *Regul. Toxicol. Pharmacol.* **2001**, *33*, 173–186. [CrossRef] [PubMed]
15. Maki, M. The prospects of cellulase-producing bacteria for the bioconversion of lignocellulosic biomass. *Int. J. Boil. Sci.* **2009**, *5*, 500–516. [CrossRef] [PubMed]
16. Somerville, C.; Youngs, H.; Taylor, C.; Davis, S.C.; Long, S. Feedstocks for Lignocellulosic Biofuels. *Science* **2010**, *329*, 790–792. [CrossRef] [PubMed]
17. Brodeur, G.; Yau, E.; Badal, K.; Collier, J.; Ramachandran, K.B.; Ramakrishnan, S. Chemical and Physicochemical Pretreatment of Lignocellulosic Biomass: A Review. *Enzym. Res.* **2011**, *2011*, 1–17. [CrossRef]
18. Luterbacher, J.S.; Rand, J.M.; Alonso, D.M.; Han, J.; Youngquist, J.T.; Maravelias, C.T.; Pfleger, B.F.; Dumesic, J.A. Nonenzymatic Sugar Production from Biomass Using Biomass-Derived -Valerolactone. *Science* **2014**, *343*, 277–280. [CrossRef] [PubMed]
19. Isikgor, F.H.; Becer, C.R. Lignocellulosic biomass: a sustainable platform for the production of bio-based chemicals and polymers. *Polym. Chem.* **2015**, *6*, 4497–4559. [CrossRef]
20. Pu, Y.; Hu, F.; Huang, F.; Davison, B.H.; Ragauskas, A. Assessing the molecular structure basis for biomass recalcitrance during dilute acid and hydrothermal pretreatments. *Biotechnol. Biofuels* **2013**, *6*, 15. [CrossRef]
21. Ponnambalam, A.S.; Deepthi, R.S.; Ghosh, A.R. Qualitative display and measurement of enzyme activity of isolated cellulolytic bacteria. *Biotechnol. Bioinf. Bioeng.* **2011**, *1*, 33–37.
22. Martinez, D.; Berka, R.M.; Henrissat, B.; Saloheimo, M.; Arvas, M.; Baker, S.E.; Chapman, J.; Chertkov, O.; Coutinho, P.M.; Cullen, D.; et al. Genome sequencing and analysis of the biomass-degrading fungus Trichoderma reesei (syn. Hypocrea jecorina). *Nat. Biotechnol.* **2008**, *26*, 553–560. [CrossRef]
23. Peng, X.; Qiao, W.; Mi, S.; Jia, X.; Su, H.; Han, Y. Characterization of hemicellulase and cellulase from the extremely thermophilic bacterium Caldicellulosiruptor owensensis and their potential application for bioconversion of lignocellulosic biomass without pretreatment. *Biotechnol. Biofuels* **2015**, *8*, 131. [CrossRef]
24. Han, S.O.; Cho, H.-Y.; Yukawa, H.; Inui, M.; Doi, R.H. Regulation of Expression of Cellulosomes and Noncellulosomal (Hemi)Cellulolytic Enzymes in Clostridium cellulovorans during Growth on Different Carbon Sources. *J. Bacteriol.* **2004**, *186*, 4218–4227. [CrossRef] [PubMed]
25. Sánchez, C. Lignocellulosic residues: Biodegradation and bioconversion by fungi. *Biotechnol. Adv.* **2009**, *27*, 185–194. [CrossRef] [PubMed]
26. Tejirian, A.; Xu, F. Inhibition of enzymatic cellulolysis by phenolic compounds. *Enzym. Microb. Technol.* **2011**, *48*, 239–247. [CrossRef] [PubMed]
27. Dashtban, M. Fungal Bioconversion of Lignocellulosic Residues; Opportunities & Perspectives. *Int. J. Boil. Sci.* **2009**, *5*, 578–595.
28. Tshikantwa, T.S.; Ullah, M.W.; He, F.; Yang, G. Current Trends and Potential Applications of Microbial Interactions for Human Welfare. *Front. Microbiol.* **2018**, *9*, 1156. [CrossRef] [PubMed]
29. Generally Regarded As Safe (GRAS). Available online: https://www.fda.gov/food/food-ingredients-packaging/generally-recognized-safe-gras (accessed on 22 February 2020).
30. Sewalt, V.; Shanahan, D.; Gregg, L.; La Marta, J.; Carrillo, R. The Generally Recognized as Safe (GRAS) Process for Industrial Microbial Enzymes. *Ind. Biotechnol.* **2016**, *12*, 295–302. [CrossRef]
31. Åvall-Jääskeläinen, S.; Palva, A. Lactobacillus surface layers and their applications. *FEMS Microbiol. Rev.* **2005**, *29*, 511–529. [CrossRef]
32. Choi, S.S.; Kang, B.Y.; Chung, M.J.; Kim, S.D.; Park, S.H.; Kim, J.S.; Kang, C.Y.; Ha, N.J. Safety assessment of potential lactic acid bacteria Bifidobacterium longum SPM1205 isolated from healthy Koreans. *J. Microbiol.* **2005**, *43*, 493–498.
33. Gueimonde, M.; Frias, R.; Ouwehand, A. Assuring the continued safety of lactic acid bacteria used as probiotics. *Boilogia* **2006**, *61*, 755–760. [CrossRef]

34. Hill, C.; Guarner, F.; Reid, G.; Gibson, G.R.; Merenstein, D.J.; Pot, B.; Morelli, L.; Canani, R.B.; Flint, H.J.; Salminen, S.; et al. The International Scientific Association for Probiotics and Prebiotics consensus statement on the scope and appropriate use of the term probiotic. *Nat. Rev. Gastroenterol. Hepatol.* **2014**, *11*, 506–514. [CrossRef]
35. Foo, H.L.; Loh, T.C.; Law, F.L.; Lim, Y.S.; Kufli, C.N.; Rusul, G. Effect of feeding Lactobacillus plantarum I-UL4 isolated from Malaysian Tempeh on growth performance, faecal flora and lactic acid bacteria and plasma cholesterol concentrations in postweaning rats. *Food Sci. Biotechnol.* **2003**, *12*, 403–408.
36. Moghadam, M.S.; Foo, H.L.; Leow, T.C.; Rahim, R.A.; Loh, T.C. Novel bacteriocinogenic Lactobacillus plantarum strains and their differentiation by sequence analysis of 16S rDNA, 16S-5S intergenic spacer region and randomly amplied polymorphic DNA analysis. *Food Technol. Biotechnol.* **2010**, *48*, 476–483.
37. Kareem, K.Y.; Foo, H.L.; Loh, T.C.; Foong, O.M.; Samsudin, A.A. Inhibitory activity of postbiotic produced by strains of Lactobacillus plantarum using reconstituted media supplemented with inulin. *Gut Pathogens* **2014**, *6*, 23. [CrossRef] [PubMed]
38. Loh, T.C.; Thanh, N.; Foo, H.L.; Hair-Bejo, M. Effects of feeding metabolite combinations from lactobacillus plantarum on plasma and breast meat lipids in Broiler Chickens. *Braz. J. Poultry Sci.* **2013**, *15*, 307–316. [CrossRef]
39. Lim, Y.H.; Foo, H.L.; Loh, T.C.; Mohamad, R.; Abdullah, N. Comparative studies of versatile extracellular proteolytic activities of lactic acid bacteria and their potential for extracellular amino acid productions as feed supplements. *J. Anim. Sci. Biotechnol.* **2019**, *10*, 15. [CrossRef]
40. Kareem, K.Y.; Loh, T.C.; Foo, H.L.; Akit, H.; Samsudin, A.A. Effects of dietary postbiotic and inulin on growth performance, IGF1 and GHR mRNA expression, faecal microbiota and volatile fatty acids in broilers. *BMC Veter- Res.* **2016**, *12*, 163. [CrossRef]
41. Kareem, K.Y.; Loh, T.C.; Foo, H.L.; Asmara, A.S.; Akit, H.; Abdulla, N.R.; Ooi, M.F. Carcass, meat and bone quality of broiler chickens fed with postbiotic and prebiotic combinations. *Int. J. Probiotics Prebiotics.* **2015**, *10*, 23.
42. Tarraran, L.; Mazzoli, R. Alternative strategies for lignocellulose fermentation through lactic acid bacteria: the state of the art and perspectives. *FEMS Microbiol. Lett.* **2018**, *365*. [CrossRef]
43. Toe, C.J.; Foo, H.L.; Loh, T.C.; Mohamad, R.; Rahim, R.A.; Zulkifli, I. Extracellular Proteolytic Activity and Amino Acid Production by Lactic Acid Bacteria Isolated from Malaysian Foods. *Int. J. Mol. Sci.* **2019**, *20*, 1777. [CrossRef]
44. Heenan, C.; Adams, M.; Hosken, R.; Fleet, G. Growth Medium for Culturing Probiotic Bacteria for Applications in Vegetarian Food Products. *LWT* **2002**, *35*, 171–176. [CrossRef]
45. Douillard, F.P.; De Vos, W.M. Functional genomics of lactic acid bacteria: from food to health. In *Microbial Cell Factories*; BioMed Central: London, UK, 2014; Volume 13, p. S8.
46. Boguta, A.; Bringel, F.; Martinussen, J.; Jensen, P.R. Screening of lactic acid bacteria for their potential as microbial cell factories for bioconversion of lignocellulosic feedstocks. *Microb. Cell Factories* **2014**, *13*, 97. [CrossRef] [PubMed]
47. Taskila, S.; Ojamo, H. The Current Status and Future Expectations in Industrial Production of Lactic Acid by Lactic Acid Bacteria. In *Lactic Acid Bacteria-R & D for Food, Health and Livestock Purposes*; IntechOpen: London, UK, 2013.
48. Lee, F.; Wan, S.; Foo, H.L.; Loh, T.C.; Mohamad, R.; Rahim, R.A.; Zulkifli, I. Comparative Study of Extracellular Proteolytic, Cellulolytic, and Hemicellulolytic Enzyme Activities and Biotransformation of Palm Kernel Cake Biomass by Lactic Acid Bacteria Isolated from Malaysian Foods. *Int. J. Mol. Sci.* **2019**, *20*, 4979. [CrossRef] [PubMed]
49. Cuenca, A.R.; Suárez, M.J.V.; Mateos-Aparicio, I. Soybean seeds and its by-product okara as sources of dietary fibre. Measurement by AOAC and Englyst methods. *Food Chem.* **2008**, *108*, 1099–1105. [CrossRef]
50. Lu, F.; Liu, Y.; Li, B. Okara dietary fiber and hypoglycemic effect of okara foods. *Bioact. Carbohydr. Diet. Fibre* **2013**, *2*, 126–132. [CrossRef]
51. Shikha; Sharan, A.; Darmwal, N. Improved production of alkaline protease from a mutant of alkalophilic Bacillus pantotheneticus using molasses as a substrate. *Bioresour. Technol.* **2007**, *98*, 881–885. [CrossRef] [PubMed]

52. Coelho, L.F.; De Lima, C.J.B.; Rodovalho, C.D.M.; Bernardo, M.P.; Contiero, J. Lactic acid production by new Lactobacillus plantarum LMISM6 grown in molasses: optimization of medium composition. *Braz. J. Chem. Eng.* **2011**, *28*, 27–36. [CrossRef]
53. Razmovski, R.; Vučurović, V. Ethanol production from sugar beet molasses by S. cerevisiae entrapped in an alginate–maize stem ground tissue matrix. *Enzym. Microb. Technol.* **2011**, *48*, 378–385. [CrossRef]
54. El-Gendy, N.S.; Madian, H.R.; Abu Amr, S.S. Design and Optimization of a Process for Sugarcane Molasses Fermentation by Saccharomyces cerevisiae Using Response Surface Methodology. *Int. J. Microbiol.* **2013**, *2013*, 1–9. [CrossRef]
55. Hamouda, H.I.; Nassar, H.; Madian, H.R.; El-Sayed, M.H.; El-Ghamry, A.A.; El-Gendy, N.S. Isolation of fermentative microbial isolates from sugar cane and beet molasses and evaluation for enhanced production of bioethanol. *Energy Sour.* **2016**, *38*, 2170–2180. [CrossRef]
56. Pholsen, S.; Khota, W.; Pang, H.; Higgs, D.; Cai, Y. Characterization and application of lactic acid bacteria for tropical silage preparation. *Anim. Sci. J.* **2016**, *87*, 1202–1211. [CrossRef]
57. Jin, S.; Chen, H. Structural properties and enzymatic hydrolysis of rice straw. *Process. Biochem.* **2006**, *41*, 1261–1264. [CrossRef]
58. Zajšek, K.; Goršek, A.; Kolar, M. Cultivating conditions effects on kefiran production by the mixed culture of lactic acid bacteria imbedded within kefir grains. *Food Chem.* **2013**, *139*, 970–977. [CrossRef] [PubMed]
59. Deutscher, J. The mechanisms of carbon catabolite repression in bacteria. *Curr. Opin. Microbiol.* **2008**, *11*, 87–93. [CrossRef] [PubMed]
60. Da Silva-Maia, J.K.; Cazarin, C.B.B.; Júnior, S.B.; Augusto, F.; Junior, M.R.M. Passion fruit (Passiflora edulis) peel increases colonic production of short-chain fatty acids in Wistar rats. *LWT* **2014**, *59*, 1252–1257. [CrossRef]
61. Ravindran, R.; Hassan, S.; Williams, G.; Jaiswal, A.K. A Review on Bioconversion of Agro-Industrial Wastes to Industrially Important Enzymes. *Bioengineering* **2018**, *5*, 93. [CrossRef]
62. Horn, S.J.; Vaaje-Kolstad, G.; Westereng, B.; Eijsink, V.G.H. Novel enzymes for the degradation of cellulose. *Biotechnol. Biofuels* **2012**, *5*, 45. [CrossRef]
63. Horn, S.J.; Sikorski, P.; Cederkvist, J.B.; Vaaje-Kolstad, G.; Sørlie, M.; Synstad, B.; Vriend, G.; Vårum, K.M.; Eijsink, V.G.H. Costs and benefits of processivity in enzymatic degradation of recalcitrant polysaccharides. *Proc. Natl. Acad. Sci. USA* **2006**, *103*, 18089–18094. [CrossRef]
64. Payne, C.; Bomble, Y.J.; Taylor, C.B.; McCabe, C.; Himmel, M.E.; Crowley, M.F.; Beckham, G.T. Multiple Functions of Aromatic-Carbohydrate Interactions in a Processive Cellulase Examined with Molecular Simulation. *J. Boil. Chem.* **2011**, *286*, 41028–41035. [CrossRef]
65. Mahajan, P.M.; Desai, K.M.; Lele, S.S. Production of Cell Membrane-Bound α- and β-Glucosidase by Lactobacillus acidophilus. *Food Bioprocess Technol.* **2010**, *5*, 706–718. [CrossRef]
66. Ghose, T.K. Measurement of cellulase activities. *Pure Appl. Chem.* **1987**, *59*, 257–268. [CrossRef]
67. Miller, G.L. Use of Dinitrosalicylic Acid Reagent for Determination of Reducing Sugar. *Anal. Chem.* **1959**, *31*, 426–428. [CrossRef]
68. Nidetzky, B.; Steiner, W.; Hayn, M.; Claeyssens, M. Cellulose hydrolysis by the cellulases from Trichoderma reesei: a new model for synergistic interaction. *Biochem. J.* **1994**, *298*, 705–710. [CrossRef] [PubMed]
69. Eriksson, K.-E. Enzyme mechanisms involved in cellulose hydrolysis by the rot fungusSporotrichum pulverulentum. *Biotechnol. Bioeng.* **1978**, *20*, 317–332. [CrossRef]
70. Lin, L.; Kan, X.; Yan, H.; Wang, D. Characterization of extracellular cellulose-degrading enzymes from Bacillus thuringiensis strains. *Electron. J. Biotechnol.* **2012**, *15*, 2. [CrossRef]
71. Spano, G.; Rinaldi, A.; Ugliano, M.; Moio, L.; Beneduce, L.; Massa, S. A beta-glucosidase gene isolated from wine Lactobacillus plantarum is regulated by abiotic stresses. *J. Appl. Microbiol.* **2005**, *98*, 855–861. [CrossRef] [PubMed]
72. Zanoelo, F.; Polizeli, M.D.; Terenzi, H.F.; Jorge, J.A. β-Glucosidase activity from the thermophilic fungus Scytalidium thermophilum is stimulated by glucose and xylose. *FEMS Microbiol. Lett.* **2004**, *240*, 137–143. [CrossRef]
73. Murashima, K.; Kosugi, A.; Doi, R.H. Synergistic Effects on Crystalline Cellulose Degradation between Cellulosomal Cellulases from Clostridium cellulovorans. *J. Bacteriol.* **2002**, *184*, 5088–5095. [CrossRef]
74. Kim, S.; Lee, M.-H.; Lee, E.-S.; Nam, Y.-D.; Seo, D.-H. Characterization of mannanase from Bacillus sp., a novel Codium fragile cell wall-degrading bacterium. *Food Sci. Biotechnol.* **2017**, *27*, 115–122. [CrossRef]

75. El-Sharouny, E.E.; El-Toukhy, N.M.; El-Sersy, N.A.; El-Gayar, A.A.E.-A. Optimization and purification of mannanase produced by an alkaliphilic-thermotolerant Bacillus cereus N1 isolated from Bani Salama Lake in Wadi El-Natron. *Biotechnol. Biotechnol. Equip.* **2015**, *29*, 315–323. [CrossRef]
76. Rasul, F.; Afroz, A.; Rashid, U.; Mehmood, S.; Sughra, K.; Zeeshan, N. Screening and characterization of cellulase producing bacteria from soil and waste (molasses) of sugar industry. *Int. J. Biosci.* **2015**, *6*, 230–236.
77. Sachslehner, A.; Nidetzky, B.; Kulbe, K.D.; Haltrich, D. Induction of Mannanase, Xylanase, and Endoglucanase Activities in Sclerotium rolfsii. *Appl. Environ. Microbiol.* **1998**, *64*, 594–600. [CrossRef] [PubMed]
78. Allcock, E.R.; Woods, D.R. Carboxymethyl cellulase and cellobiase production by Clostridium acetobutylicum in an industrial fermentation medium. *Appl. Environ. Microbiol.* **1981**, *41*, 539–541. [CrossRef] [PubMed]
79. Cho, S.J. Isolation and characterization of mannanase producing Bacillus amyloliquefaciens CS47 from horse feces. *J. Life Sci.* **2009**, *19*, 1724–1730.
80. Gunst, R.F.; Mason, R.L. Fractional factorial design. *Wiley Interdiscip. Rev. Comput. Stat.* **2009**, *1*, 234–244. [CrossRef]
81. Das, A.K.; Dewanjee, S. Optimization of Extraction Using Mathematical Models and Computation. In *Computational Phytochemistry*; Elsevier BV: Amsterdam, The Netherlands, 2018; pp. 75–106.
82. Rahman, A.S. Effects of nanofibers on properties of geopolymer composites. In *Nanotechnology in Eco-Efficient Construction*; Elsevier BV: Amsterdam, The Netherlands, 2019; pp. 123–140.
83. Watson, S.P.; Clements, M.O.; Foster, S.J. Characterization of the Starvation-Survival Response of Staphylococcus aureus. *J. Bacteriol.* **1998**, *180*, 1750–1758. [CrossRef] [PubMed]
84. Wang, X.; Xia, K.; Yang, X.; Tang, C. Growth strategy of microbes on mixed carbon sources. *Nat. Commun.* **2019**, *10*, 1279. [CrossRef] [PubMed]
85. Stuelke, J.; Hillen, W. Carbon catabolite repression in bacteria. *Curr. Opin. Microbiol.* **1999**, *2*, 195–201. [CrossRef]
86. Jia, X.; Chen, M.; Wan, J.-B.; Su, H.; He, C. Review on the extraction, characterization and application of soybean polysaccharide. *RSC Adv.* **2015**, *5*, 73525–73534. [CrossRef]
87. El Sharkawi, H.M. Effect of Nitrogen Sources on Microbial Biomass Nitrogen under Different Soil Types. *ISRN Soil Sci.* **2012**, *2012*, 1–7. [CrossRef]
88. Li, B.; Qiao, M.; Lu, F. Composition, Nutrition, and Utilization of Okara (Soybean Residue). *Food Rev. Int.* **2012**, *28*, 231–252. [CrossRef]
89. Duc, C.; Pradal, M.; Sanchez, I.; Noble, J.; Tesnière, C.; Blondin, B. A set of nutrient limitations trigger yeast cell death in a nitrogen-dependent manner during wine alcoholic fermentation. *PLoS ONE* **2017**, *12*, e0184838. [CrossRef] [PubMed]
90. Costa, E.; Teixidó, N.; Usall, J.; Atares, E.; Viñas, I. The effect of nitrogen and carbon sources on growth of the biocontrol agent Pantoea agglomerans strain CPA-2. *Lett. Appl. Microbiol.* **2002**, *35*, 117–120. [CrossRef] [PubMed]
91. Brückner, R.; Titgemeyer, F. Carbon catabolite repression in bacteria: choice of the carbon source and autoregulatory limitation of sugar utilization. *FEMS Microbiol. Lett.* **2002**, *209*, 141–148. [CrossRef]
92. Dashtban, M.; Buchkowski, R.; Qin, W. Effect of different carbon sources on cellulase production by Hypocrea jecorina (Trichoderma reesei) strains. *Int. J. Biochem. Mol. Boil.* **2011**, *2*, 274–286.
93. Stanbury, P.F.; Whitaker, A.; Hall, S.J. *Principles of Fermentation Technology*; Elsevier: Amsterdam, The Netherlands, 2013.
94. Egli, T. Microbial growth and physiology: a call for better craftsmanship. *Front. Microbiol.* **2015**, *6*, 287. [CrossRef] [PubMed]
95. Dou, T.-Y.; Chen, J.; Hao, Y.-F.; Qi, X. Effects of Different Carbon Sources on Enzyme Production and Ultrastructure of Cellulosimicrobium cellulans. *Curr. Microbiol.* **2019**, *76*, 355–360. [CrossRef] [PubMed]
96. Ahirwar, S.; Soni, H.; Rawat, H.K.; Prajapati, B.P.; Kango, N. Experimental design of response surface methodology used for utilisation of palm kernel cake as solid substrate for optimised production of fungal mannanase. *Mycology* **2016**, *7*, 143–153. [CrossRef]
97. Lloyd, D.K.; Bergum, J. Application of quality by design (QbD) to the development and validation of analytical methods. In *Specification of Drug Substances and Products*; Elsevier: Amsterdam, The Netherlands, 2014; pp. 29–72.
98. A Mooney, C.; Mansfield, S.D.; Beatson, R.P.; Saddler, J.N. The effect of fiber characteristics on hydrolysis and cellulase accessibility to softwood substrates. *Enzym. Microb. Technol.* **1999**, *25*, 644–650. [CrossRef]

99. Gao, D.; Chundawat, S.P.S.; Krishnan, C.; Balan, V.; Dale, B.E. Mixture optimization of six core glycosyl hydrolases for maximizing saccharification of ammonia fiber expansion (AFEX) pretreated corn stover. *Bioresour. Technol.* **2010**, *101*, 2770–2781. [CrossRef]
100. Kuhad, R.C.; Gupta, R.; Singh, A. Microbial Cellulases and Their Industrial Applications. *Enzym. Res.* **2011**, *2011*, 1–10. [CrossRef]
101. Gao, D.; Uppugundla, N.; Chundawat, S.P.S.; Yu, X.; Hermanson, S.; Gowda, K.; Brumm, P.; Mead, D.; Balan, V.; Dale, B.E. Hemicellulases and auxiliary enzymes for improved conversion of lignocellulosic biomass to monosaccharides. *Biotechnol. Biofuels* **2011**, *4*, 5. [CrossRef] [PubMed]
102. Arotupin, D.; Olaniyi, O.O. Screening and Identification of Mannanase-Producing Fungi Isolated from Selected Agricultural Wastes. *Br. Microbiol. Res. J.* **2013**, *3*, 635–644. [CrossRef]
103. Zakaria, M.M.; Ashiuchi, M.; Yamamoto, S.; Yagi, T. Optimization for β-Mannanase Production of a Psychrophilic Bacterium, Flavobacteriumsp. *Biosci. Biotechnol. Biochem.* **1998**, *62*, 655–660. [CrossRef] [PubMed]
104. Aziz, N.F.H.A.; Abbasiliasi, S.; Ng, H.S.; Phapugrangkul, P.; Abu Bakar, M.H.; Tam, Y.J.; Tan, J.S. Purification of β -mannanase derived from Bacillus subtilis ATCC 11774 using ionic liquid as adjuvant in aqueous two-phase system. *J. Chromatogr. B* **2017**, *1055*, 104–112. [CrossRef] [PubMed]
105. Anderson, R.K.I.; Jayaraman, K. Influence of carbon and nitrogen sources on the growth and sporulation of Bacillus thuringiensis var Galleriae for biopesticide production. *Chem. Biochem. Eng. Q.* **2003**, *17*, 225–232.
106. Bradford, M.M. A rapid and sensitive method for the quantitation of microgram quantities of protein utilizing the principle of protein-dye binding. *Anal. Biochem.* **1976**, *72*, 248–254. [CrossRef]
107. Yin, L.J.; Lin, H.H.; Xiao, Z.R. Purification and characterization of a cellulase from Bacillus subtilis YJ1. *J. Mar. Sci. Tech.* **2010**, *18*, 466–471.

Sample Availability: Samples of the compounds are not available from the authors.

MDPI
St. Alban-Anlage 66
4052 Basel
Switzerland
Tel. +41 61 683 77 34
Fax +41 61 302 89 18
www.mdpi.com

Molecules Editorial Office
E-mail: molecules@mdpi.com
www.mdpi.com/journal/molecules

www.ingramcontent.com/pod-product-compliance
Lightning Source LLC
Chambersburg PA
CBHW061518210926
51850CB00008B/1963
9783039366231